North Sea Oil and Environmental Planning

North Sea Oil and Environmental Planning

THE UNITED KINGDOM EXPERIENCE

by Ian R. Manners

University of Texas Press, Austin

Printed in the United States of America
First edition, 1982

Requests for permission to reproduce material
from this work should be sent to:
Permissions
University of Texas Press
Box 7819
Austin, Texas 78712

LIBRARY OF CONGRESS CATALOGING IN PUBLICATION DATA

Manners, Ian R.
North Sea oil and environmental planning.

Bibliography: p.
Includes index.
1. Oil well drilling—Environmental
aspects—North Sea. I. Title.
TD195.P4M36 333.8'2321'0941 81-16170
ISBN 0-292-76475-8 AACR2

Contents

Figures and Tables

Figure

Table

Conversion Rates and System of Numeration

Approximate conversion rates (those for crude oil based on world average gravity excluding Natural Gas Liquids):

1 tonne	= 7.33 barrels
1 barrel	= 42 U.S. gallons or 35 Imperial gallons
1 U.S. gallon	= 0.833 Imperial gallons
1 million tonnes/year	= 20,000 barrels/day
1 barrel/day	= 50 tonnes/year
1 cubic meter	= 35.31 cubic feet

The American system of numeration has been used throughout the text. One thousand million (10^9) is therefore referred to as one billion and one million million (10^{12}) is referred to as one trillion.

Preface

This study represents an attempt to evaluate British environmental policies in the context of North Sea oil development. How have planners in the United Kingdom dealt with those major social and environmental impacts that inevitably accompany the rapid development of energy resources? How successful have they been in identifying and mitigating the more adverse impacts of offshore activity and in reconciling conflicting demands for energy development and environmental quality? While environmental planning is the central theme, I have attempted to place the study in a wider context. For example, the evolution of licensing and leasing policies is examined in some detail since these critically affect the pace of North Sea exploration and development and, in turn, the character, location, intensity, and timing of onshore impacts. In this way I hope I have avoided an unduly narrow approach to the subject and have treated the wide range of variables influencing the formulation and implementation of environmental policies in what geographers would call a "holistic" manner.

The research for this study was begun in 1978 as part of a broader effort by the Center for Energy Studies at the University of Texas at Austin to analyze the policy implications for the United States of a wide range of energy options. In view of the continuing debate in the United States over the desirability of an accelerated program of outer continental shelf (OCS) exploration and development, it was felt that a great deal could be learned from the North Sea experience of the United Kingdom. As elaborated in the text, an initial assumption (and one frequently advanced in the American literature) was that the United Kingdom, with its more comprehensive framework for land use planning, was in a better position to deal with the impacts of offshore activity. In reality there appears to be little evidence to support this premise, though I am happy to leave the final judgment to the reader.

I would like to acknowledge the assistance received from the Center for Energy Studies and the Department of Geography at the University of Texas in the preparation of this book. Without their generous support it would not have been possible for me to visit the United Kingdom during 1978 and familiarize myself with North Sea oil developments and planning initiatives. I am also grateful to the University Research Institute for a grant to help defray the cost of drafting maps.

I am further indebted to all those people I met during my visit to the United Kingdom who gave so generously of their time and from whom I gained not only information but understanding. A full list of these individuals is given below, but I would like to acknowledge in particular the help I received from Keith Chapman, Brian Clark, and members of the Project Appraisal for Development Control Research Team, Department of Geography, University of Aberdeen; from Allan Campbell and Trevor Sprott of the Department of Physical Planning, Grampian Regional Council; from R. G. H. Turnbull and Richard Hickman of the Scottish Development Department; and from Andrew Graham, formerly Economic Advisor to the Cabinet Office and now Fellow of Balliol College, Oxford.

I should also like to express my deep appreciation to all those individuals who contributed in so many different ways to the completion of this book. My particular thanks go to my own colleagues, Dr. Robin Doughty, Dr. George Hoffman, Dr. James Bill, Dr. Sally Lopreato, and Dr. David Gibson for their encouragement and helpful comments on an earlier version of the manuscript; to Beverly Beaty-Benadom, Wanda Franklin, and Juanita Olivarez for their patience in typing numerous drafts; and to the editorial and production staff of the University of Texas Press. Special thanks are due to my research assistants on this study, Larry Smith and Dennis Moss, and to Mary Fisher who drafted the maps.

A great many people have provided me with ideas and information for this book. The following have been especially generous in sharing their knowledge with me, often taking time out from busy schedules to respond to my requests for help and information. But to all who have helped, I should like to express my gratitude.

Dr. Roger Bailey, Senior Research Officer, Ministry of Agriculture and Fisheries, Aberdeen

Mr. Ron Bissett, Research Officer, Project Appraisal for Development Control, University of Aberdeen

Dr. William A. Bourne, Department of Zoology, University of Aberdeen

Mr. Allan Campbell, Department of Physical Planning, Grampian Regional Council, Aberdeen
Dr. Keith Chapman, Department of Geography, University of Aberdeen
Dr. Brian D. Clark, Department of Geography, University of Aberdeen
Dr. James A. Fox, Department of Mechanical Engineering, Massachusetts Institute of Technology
Mr. Crawford Gordon, North East Scotland Development Association, Aberdeen
Mr. Andrew Graham, Fellow and Tutor in Economics, Balliol College, Oxford
Mr. Richard Hickman, Scottish Development Department, New St. Andrew's House, Edinburgh
Professor John House, Halford Mackinder Professor of Geography, Oxford University
Mr. S. C. Kyle, Scottish Economic Planning Department, New St. Andrew's House, Edinburgh
Mr. Gordon P. Lindlom, Exxon Chemical Company, Houston, Texas
Mr. Maitland Mackie, Cramond House, Aberdeen
Dr. Iain H. McNicoll, Department of Business Studies, University of Edinburgh
Dr. Robert Moore, Department of Sociology, University of Aberdeen
Mr. Keith Pearce, Research Officer, Project Appraisal for Development Control, University of Aberdeen
Dr. Sonia Z. Pritchard, Park Hill, London
Professor Colin Robinson, Department of Economics, University of Surrey, Guildford
Dr. David H. Rosen, M.D., Langley Porter Neuropsychiatric Institute, University of California, San Francisco
Mr. David N. Skinner, Architect and Landscape Architect, Edinburgh
Mr. Trevor Sprott, Director, Department of Physical Planning, Grampian Regional Council, Aberdeen
Mr. Kevin T. Standring, The Royal Society for the Protection of Birds, Sandy, Bedfordshire
Ms. Lorna Thompson, Scottish Economic Planning Department, New St. Andrew's House, Edinburgh
Mr. R. G. H. Turnbull, Scottish Development Department, New St. Andrew's House, Edinburgh
Mr. G. A. Walker, Division of Planning, An Foras Forbartha Teoranta, Dublin
Dr. Peter Wathern, Research Officer, Project Appraisal for Development Control, University of Aberdeen

North Sea Oil and Environmental Planning

Introduction

Oil production from under the sea is not entirely a recent phenomenon. Exploration and production wells were drilled from permanent wharves extending into the Santa Barbara Channel as early as the last decade of the nineteenth century. Drilling from floating platforms was successfully pioneered in the marshlands, bayous, and lakes bordering the Gulf of Mexico shortly after the turn of the century. Development of the oil reserves underlying Lake Maracaibo, Venezuela, began in the early 1920s. Perhaps the offshore age really began, however, in November 1947 when a wildcat struck oil in the Ship Shoal area of the Creole field some twelve miles off the Louisiana coastline. This successful well, Ship Shoal 32, was the first drilled out of the sight of land and initiated major changes in the structure and orientation of the petroleum industry.[1]

At first the technical difficulties and high costs of offshore production restricted activity to shallower waters bordering areas of proven potential—notably the Gulf of Mexico and the Arab/Persian Gulf. According to one estimate just over 50 percent of the world's proven offshore reserves are associated with marine extensions of onshore fields identified in this initial phase of offshore exploration (Klemme, 1977, p. 108). In the 1970s however the picture changed quite dramatically. Spurred by rapid and continuing increases in oil prices and by a tightening supply/demand situation, offshore exploration is now taking place in areas far removed from the traditional centers of production and in much deeper waters, where both environmental conditions and the logistics of supply and transportation pose a formidable challenge (Manners, 1980a). By the end of the 1970s exploration drilling was underway on the continental shelf of some fifty-nine nations and in water depths approaching five thousand feet.[2] Although much of this exploration activity has yet to be translated into development and production, output from offshore fields in 1980 was averaging around 13.687 million barrels per day (bpd), equivalent to 23 percent of total world oil production. Moreover, for all the uncertainty and risk that surrounds oil exploration, the likelihood of discovering further large fields capable of influencing global energy flows appears much greater for the less intensively explored continental shelf than for the well-drilled onshore producing regions.

These trends in offshore exploration and development raise major policy questions for the United States. At present approximately 13 percent of U.S. oil production comes from offshore fields, primarily the maturing fields off the Gulf Coast. However U.S. offshore production peaked in the early 1970s. Since that time exploration drilling and the identification and development of new fields have not kept pace with the rate of pumping from proven discoveries. In part at least this decline reflects a political decision, arrived at in the aftermath of the Santa Barbara blow-out, to curtail further exploration and development off the California coastline and to pursue a more restrictive leasing policy elsewhere. Yet, to the extent that recoverable reserves can be identified, the outer continental shelf (OCS) appears to offer the greatest potential (outside perhaps mainland Alaska) for reducing the nation's dependency upon imported oil.

Those who advocate an accelerated program of OCS leasing and exploration argue that "legislation implemented during a period of low-cost, abundant energy should not be rigidly adhered to a quarter of a century later for frontier areas when prices, imports, and technology are completely different" (Wassall, 1979, p. 43). Much potentially productive acreage on the continental shelf still remains to be drilled:

> . . . at the 1978 snail's pace of awarding some 4,000 square miles of OCS leases per year, it will take four and a half centuries to explore the total 1.8 million square miles of prospective offshore areas that may hold substantial solutions for our energy problems.
>
> The fact that the small area already leased in the shallow part of the OCS produced some 4 billion bbl [4×10^9 barrels] of oil and 31 Tcf [31×10^{12} cubic feet] of gas as of 1977 emphasizes why we should speed exploration of the unleased areas in shallow and deep water. (Wassall, 1979, pp. 43–44)

Such an assessment reflects the prevailing sense of optimism about the hydrocarbon potential of the continental shelf. Yet the search for oil involves tremendous uncertainties—not the least of which are the actual existence of oil and gas, the development of appropriate risk-free production and transportation technologies, and the environmental and social impact of offshore activity. And in the United States in particular there is now a far greater awareness of quality-of-life issues than in the 1950s when the oil industry first moved out into the Gulf of Mexico. The environmental movement of the late 1960s and early 1970s both reflected and articulated a feeling that far greater attention should be paid to the preservation of such intangible attributes of the environment as clean air, clean water, unspoiled coastlines, and unique habitats. Implicit in public expressions of concern over environmental degradation was the demand for a more realistic appraisal of the social and environmental costs of resource development. The level of demand for environmental quality is of course extremely difficult to assess, being intimately related to individual and community values and preferences and subject to constant reappraisal as a result of changing circumstances and actual experiences. Yet, for all the talk of an "energy crisis," of the regulatory excesses of the federal government that shackle U.S. industry, of a deepening economic recession, it is clear that there remains a strong public commitment to maintaining the quality of the environment.[3]

In these circumstances it is perhaps not surprising that proposals for an accelerated OCS leasing program have encountered such opposition. The controversy is multifaceted. It arises from the difficulty of predicting the social and environmental impacts of new exploration and production technologies; it is complicated by the uneven distribution of both the costs and the benefits of resource development over space and through time; it is inflamed by "high-stress" incidents such as oil spills and blow-outs; and it is obscured by inevitable disagreements over the relative weighting of social, ecological, and economic variables. In short the development of offshore energy resources typifies the difficult decisions and trade-offs that are involved when one kind of development or use of the environment appears to threaten other desired uses. Undoubtedly the greatest apprehension has been expressed in small coastal communities facing the possibility of rapid OCS development and all the problems that are associated with energy-based growth booms. Such communities are likely to bear the brunt of the social and environmental costs of offshore development, yet all too frequently they lack the full-time public officials, planners, and administrators or the budgeting and land use programs that their situation may require.

Resolving these issues is of critical importance to the United States. An obvious question in this situation is whether there are not some useful lessons to be derived from the way in which the United Kingdom has approached North Sea oil development. More specifically, in which ways and with what success has the United Kingdom been able to mitigate the more adverse impacts of offshore activity and reconcile the conflicting demands for energy development and environmental quality?

Several aspects of North Sea oil development appear to be of particular significance in the context of U.S. efforts to appraise the potential of the outer continental shelf. In the first place, the North Sea represents the "technological frontier" of offshore exploration and production. "In the North Sea equipment has to be capable of coping with the legendary storm that occurs once every hundred years—in other words the one-hundred-foot wave—and there are additional problems arising from the variable nature of the seabed which can consist of mud, quicksand, sand waves, [or] exposed rock" (MacKay and Mackay, 1975, p. 54). Virtually every account refers to North Sea development as being close to the margins of existing offshore technology or "stretching the ingenuity of the oil industry to the limit" (Chapman, 1976, p. 75). In these circumstances, many of the potential impacts associated with deepwater exploration and production (whether occurring offshore in the form of marine oil pollution or onshore as a result of the infrastructure required to sustain offshore activity) have already been identified and confronted in the North Sea.

Second, the pace and circumstances of North Sea oil development contrast sharply with the United States' experience in the Gulf of Mexico. Oil production from the Gulf of Mexico built up extremely slowly (at least when compared with the North Sea); moreover, it occurred within a pre-existing infrastructure (services and institutions) developed to meet the needs of proven onshore fields and at a time when environmental concerns were less well articulated. For all of these reasons it seems likely that the North Sea will provide a better guide to the current situation in frontier areas than the earlier U.S. experience in the Gulf of Mexico.

Third, when viewed from this side of the Atlantic, U.K. offshore policies appear to have been remarkably successful, at least as measured in terms of facilitating North Sea oil development and moving relatively swiftly from the initial discovery of oil in 1969, to the start-up of production from the Argyll and Forties fields in 1975, to a position of self-sufficiency in hydrocarbons in the early 1980s. There is the impression that this build-up has occurred without serious opposition or controversy over the social and environmental costs of offshore development. A number of American commentators have even suggested that the more comprehensive British approach to planning served to expedite development while simultaneously anticipating and absorbing major onshore impact without significant social dislocation or environmental degradation. That energy development proceeded rapidly and smoothly despite the tensions associated with the growth of Scottish nationalist sentiments and demands for greater regional autonomy in decision-making would seem to reinforce the value of the U.K. experience since offshore oil development in the United States inevitably interfaces with the controversial issue of state and local rights and responsibilities. In these circumstances it is perhaps not surprising to encounter suggestions that the North Sea provides a "model" for U.S. offshore development.

It was to examine this premise that the present study was begun in the summer of 1978. The intention was to look more closely at environmental and land use policies in the United Kingdom as these related to the development of North Sea oil. What role, for example, had impact assessment methods and procedures played in the design and implementation of land use policies in the coastal zone? How responsive had planners been to regional and local interests and priorities? Had the "more comprehensive" planning framework helped in resolving potential conflicts over the siting of energy-related facilities and in reducing environmental losses? Had effective contingency planning and the adoption of available pollution control measures reduced the risk of marine oil pollution to an acceptable level? The level and timing of public involvement and the adequacy of the information bases available to all those involved in the planning process were thought to be key issues in achieving a balanced and equitable distribution of the costs and benefits of North Sea oil development. But the underlying concern was to see whether the U.K. experience provided a framework for dealing with the problems of offshore oil development and avoiding the sort of adversary relationship between developers, planners, and affected communities that had delayed exploration of the Atlantic outer continental shelf.

As the background research for this book proceeded, however, it quickly became apparent that many of the initial assumptions needed modification. One correspondent in Aberdeen, for example, wrote, "I will be very curious to hear what has led you to believe that the United Kingdom has been successful in planning for and accommodating the impacts of offshore oil development. It doesn't look at all like that from here." And a visit to Scotland during 1978 soon confirmed that U.K. offshore policies had not been as successful in reducing conflict or mitigating impacts as suggested by some American observers. In the early phases of offshore oil development in particular, planners had evidently been overwhelmed by the number, size, and complexity of oil-related planning applications. In contrast to the situation in the United States, environmental impact procedures have still to be formally incorporated into the U.K. planning framework. The relative lack of opposition to North Sea oil development (at least in the early exploration and development phases) appears to have been as much a function of the economic environment and political culture and a reflection of the lack of opportunity for public involvement and comment, as it was a result of effective planning mechanisms. As the less desirable consequences of North Sea oil development have become more visible—notably the siting of platform fabrication yards, storage terminals, and processing facilities in small, isolated communities that must then deal with the influx of a large, often temporary, construction force, and the increasing frequency of oil pollution incidents in coastal waters—so dissatisfaction with the way in which the costs and benefits of offshore energy development are being distributed and the manner in which decisions are being made has become more widespread. Yet local concerns and apprehensions, even when legitimized by the findings of a public inquiry, have all too frequently been overridden by the U.K. government.

That is not to say that no lessons are to be derived from the North Sea experience, but they may be primarily lessons of avoidance. The following chapters therefore outline some of the major policy questions that have arisen in the course of North Sea oil exploration and development. Since there is already a considerable body of literature on the economics of North Sea oil, the emphasis is on environmental planning. In one sense of course planning has always involved environmental issues, but in a larger sense there is today a far greater awareness of the complex and varied systems of interaction between a society and its environment. In this context it is important for all those involved in the planning process to recognize the ecological and functional linkages between the marine environment and the coastal zone. This book attempts to avoid the sort of fragmented, discipline-oriented approach that has all too frequently characterized environmental policy-making in the past. Identifying the key environmental issues in the offshore development process should set the stage for a careful review of U.S. offshore planning initiatives and for an evaluation of whether the United States, through its federal and state programs, will be able to avoid the problems encountered in the United Kingdom. Such an evaluation should provide constructive input into the ongoing process of reconciling social and environmental objectives with the development of this nation's energy resources.

1. The Development of North Sea Oil and Gas

Introduction

By the end of 1980 oil production from the United Kingdom sector of the North Sea was averaging 1.8 million barrels per day, a level of production sufficient to guarantee the United Kingdom a place among the world's leading oil producers (*Offshore*, vol. 41, no. 7, 1981, p. 111). British commentators were quick to point out that the United Kingdom stood ahead of several OPEC nations, including Abu Dhabi, Algeria, Ecuador, Qatar, and Dubai, in the ranks of oil-producing nations (Table 1). With production from the Norwegian and Danish sectors added to that from the U.K. sector, North Sea fields in 1980 accounted for 16.7 percent of the world's offshore output of crude petroleum (Table 2), an achievement that is particularly impressive when it is recalled that the first major oil strike, at Ekofisk, had occurred only a decade earlier.

The emergence of the North Sea as a major oil province epitomizes the changes that are occurring in the structure and orientation of the petroleum industry. As a leading oil journal observed in a recent editorial, "Offshore is as magic a word today as Westward ho was to the men of those violent days after America's Civil War in the 1860s" (*Oil and Gas Journal*, vol. 76, no. 19, 1978, p. 1). The same editorial estimated that "about 23% of the world's crude oil reserves and 14% of the world's gas reserves lie offshore. But only about 15% of the world's production . . . came from offshore fields in 1976. As of 1977, it is estimated that 43% of all conventionally recoverable oil and gas found on land has been produced, but only 18% of the total reserves found offshore has been produced" (ibid., p. 113). Moreover, for all the uncertainty and risk that surrounds oil exploration, the likelihood of discovering further large fields capable of influencing global energy flows appears much greater for the less intensively ex-

Table 1. World Oil Production, 1980

	Estimated Production (1,000 bpd)	*Percentage of World Production*
Middle East/North Africa		
Saudi Arabia	9,620	16.1
Iraq	2,600	4.4
Libya	1,780	3.0
Kuwait	1,400	2.3
Abu Dhabi	1,380	2.3
Iran	1,280	2.1
Algeria	1,000	1.7
Egypt	585	1.0
Others (including Qatar, Dubai, Oman)	2,017	3.4
Subtotal	21,662	36.3
Asia/Australia		
Indonesia	1,570	2.6
Others (including Australia, Brunei)	1,149	1.9
Subtotal	2,719	4.5
Sub-Sahara Africa		
Nigeria	2,100	3.5
Others (including Angola, Gabon)	432	0.7
Subtotal	2,532	4.2
Europe		
United Kingdom	1,600	2.7
Norway	530	0.9
Others	317	0.5
Subtotal	2,447	4.1
America		
United States	8,650	14.5
Venezuela	2,150	3.6
Mexico	1,960	3.3
Canada	1,470	2.5
Others	1,504	2.5
Subtotal	15,734	26.4
Communist World		
U.S.S.R.	12,050	20.2
China	2,170	3.6
Others	360	0.6
Subtotal	14,580	24.4
Total world production	59,674	

Note: Only nations with an estimated production in excess of 500,000 barrels per day are listed individually.

Source: Compiled from data in *Oil and Gas Journal*, vol. 78, no. 52, December 29, 1980.

Table 2. Offshore Oil Production, 1970–1980

	1970 (1,000 bpd)	*% Offshore Total*	*1975 (1,000 bpd)*	*% Offshore Total*	*1980 (1,000 bpd)*	*% Offshore Total*
Saudi Arabia	1,251	16.6	1,386	16.7	2,958	21.6
United Kingdom	—	—	83	1.0	1,650	12.1
Abu Dhabi	269	3.6	463	5.6	1,322	9.7
Venezuela	2,460	32.7	1,737	21.0	1,096	8.0
U.S.A.	1,577	20.9	910	11.0	1,038	7.6
Norway	—	—	190	2.3	629	4.6
Nigeria	275	3.7	431	5.2	579	4.2
Indonesia	—	—	246	3.3	533	3.9
Mexico	35	0.5	45	0.5	500	3.7
Divided Zone	—	—	315	3.8	403	2.9
Egypt	257	3.4	165	2.0	390	2.8
Dubai	70	0.9	249	3.0	345	2.5
Australia	216	2.9	413	5.0	323	2.4
Malaysia	—	—	84	1.0	280	2.0
Qatar	172	2.3	n.a.	—	248	1.8
U.S.S.R.	258[a]	3.4	228[a]	2.8	200[a]	1.5
Brunei	146[b]	1.9	141	1.7	192	1.4
Gabon	29	0.4	180	2.2	178	1.3
Trinidad/Tobago	76	1.0	174	2.1	167	1.2
Iran	322	4.3	481	5.8	150[a]	1.1
India	—	—	—	—	142	1.0
Cabinda	96	1.3	143	1.7	97	0.7
Brazil	8	0.1	19	0.2	73	0.5
Tunisia	—	—	43	0.5	44	0.3
Spain	—	—	33	0.4	31	0.2
Peru	—	—	29	0.4	30	0.2
Congo	—	—	37	0.4	27	0.2
Zaire	—	—	—	—	22	0.2
Sharjah	—	—	38	0.5	10	<0.1
Denmark	—	—	3	<0.1	7	<0.1
Italy	12	0.2	10	0.1	6	<0.1
Ivory Coast	—	—	—	—	6	<0.1
Philippines	—	—	—	—	4	<0.1
New Zealand	—	—	—	—	3	<0.1
China	—	—	—	—	2	<0.1
Ghana	—	—	—	—	1	<0.1
Japan	3	<0.1	1	<0.1	1	<0.1
Total offshore	7,532		8,277		13,687	
Total world	45,021		53,850		59,812	
% offshore	16.7		15.4		22.9	

[a]Estimate.
[b]Includes Malaysia.
—Indicates no production or less than 1,000 bpd.
Sources: Compiled from *Offshore*, vol. 35, no. 7 (1975); vol. 38, no. 7 (1978); and vol. 41, no. 7 (1981).

plored outer continental shelf than for the well-drilled onshore producing regions.

As the search for oil and gas has moved into deeper and more turbulent offshore waters, the North Sea has been the testing ground for new techniques in exploration and development drilling. Innovative concepts, ranging from semisubmersible drilling rigs to concrete production platforms and subsea completions, have been required to deal with the challenge of operating in deep-water environments.[1] Whereas in the 1950s and early 1960s exploration activity was confined to the shallower waters bordering areas of proven potential, by the beginning of the 1980s, largely as a result of the North Sea experience, offshore operators had gained the expertise and the confidence to explore for oil in environments as hostile (and as contrasted) as the Davis Strait between Greenland and Labrador and the Exmouth Plateau off Western Australia.

Much of this exploration activity has still to be translated into production. As should be evident from Table 3, however, the emergence of new producing regions led by the North Sea has already had a major impact on the scale and spatial character of offshore production. With Chile's Ostion oil field in the Strait of Magellan coming onstream in January 1979, and with the start of production from the Nido field in the South China Sea off the Philippine island of Palawan in February 1979, the number of nations with commercial offshore fields had nearly doubled in the course of the decade. Many of these producing nations are still comparatively small by global standards, yet their output is significant in terms of domestic needs and dependency upon imported oil or in terms of their potential role as non-OPEC exporters. Moreover, collectively the output from the new offshore fields has more than offset the marked decline in output from such traditional producers as Venezuela and the United States. Both continue to rank as leading offshore nations, but production has fallen as the rate of exploration drilling and the discovery and development of new fields has failed to keep pace with the rate of pumping from established fields (Table 2).

The international implications of the emergence of the United Kingdom as a major oil producer remain unclear and indeed fall outside the scope of this book. Given the uncertainties that exist with respect to both the ultimate size of the recoverable reserves and future government policy on production levels, such an assessment would in any case be highly speculative.[2] Of immediate domestic significance, however, was the progress toward net self-sufficiency in oil.[3] In 1979 production of oil from the U.K. sector of the North Sea (including condensate and heavier associated gases extracted at

Table 3. Offshore Oil Production by Region, 1970–1980 (1,000 bpd/% offshore output)

Region	*1970*	*1971*	*1972*	*1973*	*1974*	*1975*	*1976*	*1977*	*1978*	*1979*	*1980*
Middle East and North Africa	2,341 (31%)	2,830 (34%)	3,209 (35%)	4,022 (39%)	3,607 (39%)	3,140 (38%)	3,542 (38%)	4,960 (43%)	5,195 (45%)	5,299 (42%)	5,870 (43%)
U.S.A.	1,577 (21%)	1,692 (20%)	1,665 (18%)	1,697 (16%)	1,428 (15%)	910 (11%)	1,064 (11%)	1,238 (11%)	1,124 (10%)	1,066 (8%)	1,038 (8%)
Central and South America	2,579 (34%)	2,680 (32%)	2,725 (30%)	2,880 (28%)	2,272 (25%)	2,004 (24%)	1,969 (21%)	1,545 (14%)	1,368 (12%)	1,748 (14%)	1,866 (14%)
Sub-Sahara Africa	400 (5%)	520 (6%)	598 (7%)	755 (7%)	895 (10%)	791 (10%)	783 (8%)	898 (8%)	693 (6%)	975 (7%)	910 (7%)
Europe	12 (—)	33 (—)	43 (—)	68 (1%)	82 (1%)	319 (4%)	740 (8%)	1,085 (9%)	1,458 (13%)	2,017 (16%)	2,323 (17%)
South and Southeast Asia, Oceania	365 (5%)	434 (5%)	592 (7%)	788 (8%)	754 (8%)	886 (11%)	1,114 (12%)	1,506 (13%)	1,443 (13%)	1,347 (11%)	1,480 (11%)
U.S.S.R.	258 (3%)	250 (3%)	236 (3%)	236 (2%)	231 (2%)	228 (3%)	220 (2%)	205 (2%)	200 (2%)	195 (2%)	200 (1%)
Total	7,532	8,439	9,068	10,446	9,269	8,278	9,432	11,437	11,481	12,647	13,687

Source: Compiled from data in *Offshore,* 1970–1980.

Table 4. U.K. Crude Oil Production, 1975–1979 (million tonnes)

Field	*Onstream*	*1975*	*1976*	*1977*	*1978*	*1979*
Argyll	6/75	0.5	1.1	0.8	0.7	0.8
Forties	11/75	0.6	8.6	20.1	24.5	24.5
Auk	2/76	—	1.2	2.3	1.3	0.8
Beryl A	6/76	—	0.4	3.0	2.6	4.7
Montrose	6/76	—	0.1	0.8	1.2	1.3
Brent	11/76	—	0.1	1.3	3.8	8.8
Piper	12/76	—	0.1	8.6	12.2	13.2
Claymore	11/77	—	—	0.3	3.0	4.0
Thistle	2/78	—	—	—	2.6	3.9
Dunlin	8/78	—	—	—	0.7	5.7
Heather	10/78	—	—	—	0.1	0.8
Ninian	12/78	—	—	—	0.04	7.7
U.K. Statfjord	11/79	—	—	—	—	0.04
South Cormorant	12/79	—	—	—	—	0.04
Total offshore production		1.1	11.6	37.3[c]	52.8[c]	76.5[c]
Petroleum gases[a]		—	—	—	0.6	0.8
Condensate[b]		0.3	0.3	0.4	0.4	0.4
Onshore production		0.1	0.1	0.1	0.1	0.1
Total U.K. production		1.5	12.0	37.8	53.9	77.9[c]

[a]Petroleum gases are ethane, propane, and butane produced in the treatment of liquid or gaseous hydrocarbons at pipeline terminals.

[b]Condensate arises mainly from the treatment of gas produced from the Frigg and Southern North Sea Basin fields.

[c]In most instances production figures have been rounded by the Department of Energy to the nearest 100,000 tonnes. This has apparently resulted in some discrepancies between individual field figures and total U.K. production.

Source: Compiled from data in Department of Energy, 1980.

pipeline terminals) amounted to nearly 78 million tonnes (Table 4). This amount corresponded to about 83 percent of the total oil requirements of the United Kingdom and to 35 percent of the nation's total primary fuel demand (Table 5).

The "Brown Book" published annually by the Department of Energy confidently forecast that production would continue to increase in 1980 "so that although the UK will still require net imports for the year as a whole towards the end of the year a rate of production roughly equivalent to net self-sufficiency in oil for the UK should be attained" (Department of Energy, 1980, p. 21). In fact a combination of adverse weather conditions, unanticipated technical difficulties, and labor problems delayed development and curtailed production at several fields, with the result that U.K. output in 1980 was slightly lower than anticipated (*Offshore*, vol. 41, no. 2, 1981, p. 89). Nevertheless, the fact that the current debate is over the most appropriate depletion policies and controls for the North Sea fields (as opposed to achieving self-sufficiency as rapidly as possible) is an indication of the pace of development over the last decade.

The dramatic turn-around in U.K. energy fortunes is illustrated in Figure 1, which depicts the relationship between production and consumption of primary fuels (see also Table 5). As recently as 1974, U.K. oil production was insignificant, amounting to a mere 0.4 mil-

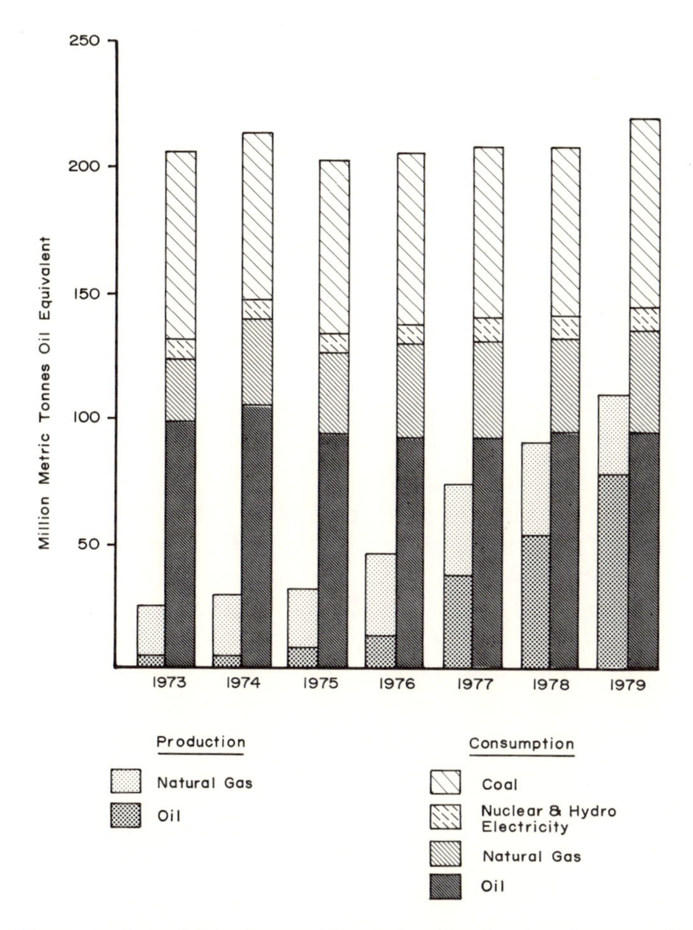

Figure 1. United Kingdom: Oil and Gas Production Compared to Energy Consumption, 1973–1979 (data from Table 5)

lion tonnes from small inland fields and condensate from the gas fields in the southern North Sea. Even before 1973 and the rapid and unprecedented rise in oil prices, the United Kingdom's dependency on foreign suppliers of oil produced a drain on the nation's balance of payments in the amount of £953 million in 1972.[4] In 1974, the U.K. balance of payments deficit (on current account) amounted to a whopping £3,350 million (5 percent of Gross Domestic Production), although this amount was financed without too much difficulty, despite rapid inflation, since North Sea oil was on the horizon (Robinson and Morgan, 1976, p. 1).

In these circumstances, the discovery and development of oil in the North Sea can best be described as timely:

> Although it would have been preferable for the first discoveries to have been made a few years earlier, so that production could have begun in the early 1970s, when oil prices were already rising, rather than in 1975, nevertheless by 1980 the United Kingdom and Norway seemed likely to become the only substantial oil producers in Western Europe. North Sea oil seemed to offer the prospect of escape from the balance-of-payments constraint which had apparently hampered Britain's economic growth for many years. Whereas her overseas competitors would still have to import expensive OPEC crude oil, Britain would be able to substitute indigenous oil output for the imports she would otherwise have had to make and she might well, eventually, become a net exporter of crude oil. (Robinson and Morgan, 1978, p. 147)

Moreover, with so much attention focused on the oil potential of the North Sea, it is all too easy to overlook the fact that since 1967, when the first natural gas was piped ashore from the West Sole field, gas production from North Sea fields has increased to the point where the British gas industry is virtually self-sufficient. In total, 39 billion cubic meters (39×10^9) of natural gas was produced from the U.K. continental shelf in 1979 (Table 6). This amount corresponded to 81 percent of primary gas demand in the United Kingdom, or a further 15 percent of total primary fuel demand.[5] Since 1977, production has also included gas from a field in the northern sector of the North Sea, the Frigg field, which straddles the U.K. and Norwegian sectors and is linked to a gas terminal at St. Fergus in the Grampian region of Scotland. More recently the gas found in association with crude oil—which usually contains substantial quantities of ethane, propane, and butane as well as the methane found in a "dry gas" field[6]—has become the focus of conservation and recovery

Table 5. U.K. Oil and Natural Gas Production Compared to U.K. Energy Consumption (million tonnes of oil or oil equivalent)

	U.K. Energy Consumption[a]						*U.K. Energy Production*	
	Total	*Coal*	*Petroleum*	*Natural Gas*	*Nuclear Electricity*	*Hydro-Electricity*	*Oil*[b]	*Natural Gas*[c]
1973	207.9	78.2	96.6	26.0	5.9	1.2	0.4	24.6
1974	215.3	69.3	106.4	31.2	7.1	1.3	0.4	30.6
1975	204.1	70.6	93.3	32.6	6.4	1.2	1.6	31.8
1976	207.6	71.8	92.5	34.6	7.6	1.1	12.1	33.7
1977	211.6	72.2	92.9	36.9	8.4	1.2	38.2	35.3
1978	211.9	70.5	94.0	38.3	7.9	1.2	54.0	33.9
1979 (provisional)	221.6	76.2	94.0	42.0	8.1	1.3	77.9	33.9

[a]Primary fuel input basis, including oil and gas for non-energy uses.

[b]Includes condensate and petroleum gases extracted at terminal separation plants.

[c]Includes land and colliery methane as well as associated gas produced and used on production platforms, but excludes gas flared or reinjected.

Source: Department of Energy, 1980, p. 21.

Table 6. U.K. Natural Gas Production, 1968–1979 (10 million cubic meters)

Field	*1968*	*1969*	*1970*	*1971*	*1972*	*1973*	*1974*	*1975*	*1976*	*1977*	*1978*	*1979*
West Sole	128	158	118	186	228	189	183	183	201	195	153	137
Leman Bank	74	292	797	1,289	1,313	1,310	1,561	1,501	1,537	1,558	1,472	1,383
Hewett		56	195	338	515	571	706	764	811	785	639	629
Indefatigable				16	451	456	555	625	635	678	645	601
Viking					139	359	476	551	605	633	524	440
Rough								1	51	106	93	101
Frigg[a]										61	291	535
Piper												54
Others[b]									1	14	33	46
Totals[c]	202	506	1,110	1,829	2,646	2,885	3,481	3,625	3,841	4,030	3,850	3,923

[a]U.K. share only.
[b]Associated gas produced and used on Northern Basin oil production platforms.
[c]Rounding of totals to nearest 10 million cubic meters may result in some discrepancy between individual field totals and overal U.K. production.

Source: Department of Energy, 1980, p. 33.

efforts. Associated gas from the Brent oil field, the major source of such gas currently under development, will flow to the St. Fergus terminal through a new trunk pipeline. This 281-mile pipeline, referred to as the Far North Liquid and Associated Gas System, or FLAGS, was commissioned in 1980 with the first deliveries to St. Fergus scheduled for 1982. Ultimately, associated gas from the Ninian and South Cormorant oil fields, where gas deposits are too small to justify individual trunk lines, will flow into the FLAG system. A similar collector network will allow associated gas from Piper, Claymore, and Tartan to be brought ashore through the Frigg pipeline. More recently the government has announced its support for an entirely new gas-gathering network linking a number of smaller fields in the East Shetland Basin as well as undeveloped oil and condensate fields in the central North Sea with the St. Fergus terminal.[7] As the 1979 government "Brown Book" notes, "prior to the development of North Sea fields the only practical petrochemical feedstock for the UK was naptha, a light distillate fraction produced by the refining of crude oil. The UK now has the option of also using associated gases from North Sea oil fields as petrochemical feedstocks and the use of these gases has been encouraged in order to strengthen the [U.K.] petrochemical industry" (Department of Energy, 1979, p. 23).

Given the United Kingdom's energy (and hence balance-of-payments) situation in the early 1970s, it was almost inevitable that the policy of successive British governments would be to facilitate offshore exploration and to promote the rapid development of North Sea oil and gas discoveries. Such a policy has major implications not only for offshore licensing strategies but for onshore planning initiatives intended to ensure the timely provision of such essential support facilities as supply bases and platform construction sites.

Offshore Development

The energy fortunes of the United States and the United Kingdom in the last decade could not have contrasted more sharply. Whereas rapid exploration and development of North Sea hydrocarbons has enabled the United Kingdom to move toward a position of net self-sufficiency in oil and gas, the United States, lacking a coherent energy policy and frustrated in its efforts to assess the hydrocarbon potential of the outer continental shelf areas, has seen its position of hydrocarbon self-sufficiency gradually eroded. Thus in the United States, oil imports have increased from 4.7 million barrels per day in 1972 to 8.2 million barrels per day in 1978, by which time imports of crude and products accounted for about 43 percent of total U.S. oil needs (*International Petroleum Encyclopedia*, 1980, p. 295). The argument for an accelerated rate of leasing on the U.S. outer continental shelf (OCS) is reflected in American Petroleum Institute (API) figures for 1978 indicating that the amount of oil pumped from U.S. fields was nearly two and a half times the amount added to reserves through new discoveries and reappraisal of known fields (*Offshore*, vol. 39, no. 7, 1979, p. 49). Moreover, even assuming sizable discoveries in OCS frontier areas, several years would be required for appraisal and development before the overall decline in U.S. offshore production could be reversed.

Lest this appear too pessimistic an assessment, it is perhaps worth reiterating that the United Kingdom's current position of strength has been achieved within the space of a decade. The existence of oil beneath the North Sea was confirmed only with the Montrose strike in September 1969. Actual production from the U.K. sector began as recently as June 1975, when oil from the Argyll field was delivered by tanker to refineries on the Isle of Grain in the Thames estuary. The same year saw the completion of the first crude oil pipelines: a thirty-four-inch pipeline linking the Ekofisk field in the Norwegian sector with Teesside in northeast England and a thirty-two-inch pipeline from the Forties field to the Scottish coastline at Cruden Bay, whence an overland pipeline carries the crude oil to the British Petroleum refinery at Grangemouth on the Firth of Forth.

Table 7 provides a chronology of key events in the discovery and development of North Sea oil and gas. In the immediate post–World War II period the sheer size and low production cost of Middle Eastern oil fields tended to dominate the interest of both companies and governments, despite suggestions from petroleum geologists that the North Sea represented a potentially interesting oil and gas province. Moreover, those onshore fields that were discovered in the course of rather low-key exploration programs tended to be too small to justify the high costs of exploration in similar geological conditions offshore (Chapman, 1976, p. 40).[8]

Interest in the North Sea therefore really dates back only to 1959, when the discovery of a sizable natural gas field at Slochteren in the Groningen province of the Netherlands led to a reassessment of the hydrocarbon potential of the North Sea. As has subsequently been acknowledged, Dutch exploration efforts had been so unsuccessful that the program would probably have been discontinued had the crucial Slochteren No. 1 well not struck gas. In the early 1960s, following an evaluation of the geological features of the Slochteren field, an intensive program of geophysical investigation and magnetic gravitational and seismic surveying gradually acquired momentum in both coastal and offshore waters.[9] As noted by Keith Chapman, "initial results were disappointing, but a gas strike on the Frisian Island of Ameland in 1964 was significant both in a geological and psychological sense . . . these islands formed intermediate points of great value in indicating a seaward extension of mainland structures" (Chapman, 1976, p. 40).

From 1964 onward, events gathered speed. Initial attention was focused on the relatively shallow waters of the southern North Sea (later to be identified by geologists as the Southern North Sea Basin). Thus in the first two licensing rounds held by the British government during 1964 and 1965 a total of 2,062 blocks were placed on offer, but very little interest was shown in any area outside the southern North Sea. By the end of the second round of exploration licensing this area was almost fully taken up. Discoveries of natural gas came in rapid succession following the initial British Petroleum gas strike in the West Sole field in October 1965 (Figure 2). With the large fields (Leman Bank, Indefatigable, Hewett, Viking, and Rough) identified during 1966 and 1968, the Southern Basin was firmly established as a major gas-bearing province, with proven recoverable reserves in excess of 700 billion cubic meters, sufficient to support output of some 100 million cubic meters per day for a nominal twenty-year depletion period.[10]

Development of these fields was equally rapid. The first deliveries of gas from the West Sole field (via a sixteen-inch pipeline to Easington, Yorkshire) were made in March 1967, representing only an eighteen-month interval between discovery and production start-up. By the end of the decade Leman Bank and Hewett had been brought into production and linked to a new gas-receiving terminal at Bacton in Norfolk. Indefatigable was linked to the Bacton collection system in 1971; Viking was brought into production in 1972, its gas being delivered through a separate pipeline to Theddlethorpe in Lincolnshire (Figure 2). Although there have been gas strikes outside the Southern Basin, including the discovery of the major Frigg field on the British/Norwegian median line, this first phase of exploration and development was essentially gas-oriented and spatially confined to the more southerly and shallower waters of the North Sea.

By the beginning of the1970s, however, the focus of exploration activities was beginning to shift to the deeper waters off the Scottish coastline north of 56°N (Figure 3). In part this shift may have reflected a feeling that the major gas-bearing structures in the southern North Sea had already been identified. Certainly by the end of

Table 7. North Sea Oil and Gas: Development Chronology

Date	*Event*
1959	Discovery of the Slochteren natural gas field in the Groningen province of the Netherlands.
1961	First exploratory well drilled off the Netherlands coast.
1964	First licensing round for the U.K. sector.
	First offshore discovery of gas (off East Frisian), commercially insignificant but a psychological boost to exploration.
1965	Discovery by British Petroleum (BP) of West Sole gas field (October). Development delayed by first serious mishap in North Sea operations, the loss of the drilling rig Sea Gem.
	Second licensing round for the U.K. sector.
	Southern North Sea Basin almost fully licensed but little interest shown in areas north of 56°N.
1966	Successive discoveries of natural gas fields in southern portion of U.K. sector, including the major fields of Leman Bank (April), Indefatigable (June), and Hewett (October).
1967	First onshore deliveries of natural gas from West Sole field via newly commissioned 16-inch pipeline to Easington, Yorkshire (March).
1968	Further major gas discoveries in the Southern North Sea Basin, U.K. sector, including Viking (May) and Rough (May).
	First discovery in Norwegian sector (the Cod gas condensate field) stimulates interest north of 56°N.
1969	First oil strike in the U.K. sector announced in September by an Amoco/British Gas Corporation consortium, the Montrose field.
	Ekofisk, the first major North Sea oilfield discovered by the Phillips consortium in the Norwegian sector (November).
1969–1970	Third licensing round in the U.K. sector: choice of blocks to be offered made prior to Montrose, Ekofisk discoveries, but focus of exploration shifts north.
1970	Forties, the first major oil strike in the U.K. sector, discovered by BP in November.
1971	Announcement of fourth licensing round in the U.K. sector: includes first auctioning of blocks.
	Further significant oil discoveries in U.K. sector including Brent (July), the first strike in the highly productive East Shetland Basin.
1972	Discovery of Frigg gas field astride U.K./Norwegian line; the first significant gas discovery outside the Southern North Sea Basin.
1973–1974	Series of significant discoveries including Thistle (July 1973), Dunlin (July 1973), Heather (December 1973), Ninian

	(January 1974), Magnus (March 1974), Statfjord (April 1974), and North Cormorant (July 1974) confirm the potential of the East Shetland Basin.
1974	Installation of first oil production platforms at Forties and Auk.
1975	First oil landed from U.K. sector: Argyll field came on stream (June) with oil offloaded onto tankers, followed by Forties start-up in November.
	Completion of 110-mile pipeline linking Forties field with Cruden Bay.
	Enactment of Petroleum and Submarine Pipe-lines Act creating British National Oil Corporation (BNOC).
1976	Auk (February), Beryl A (June), Montrose (June), Brent (November), and Piper (December) come on stream.
1976–1977	Fifth round of licensing in the U.K. sector.
1977	Blow-out on the Ekofisk production platform *Bravo* in the Norwegian sector (April).
	Claymore comes on stream (November).
	Commissioning of first gas pipeline in northern North Sea linking Frigg with a terminal at St. Fergus.
	First reported strike to the west of the Shetlands.
1978	Production for U.K. sector reaches 1 million bpd (May).
	Thistle (February), Dunlin (August), Heather (October), and Ninian (December) come on stream.
	BNOC announces its first oil strike in a block where it was granted the original license as operator (June).
	Sullom Voe, the massive oil terminal in the Shetland Islands, receives first oil by pipeline from the Dunlin field (November).
	First production well in BP's West Sole gas field shut down and steel platform removed.
1978–1979	Sixth licensing round for the U.K. sector.
1979	First oil spill at Sullom Voe terminal (January).
	Production from U.K. sector reaches 1.5 million bpd for the first time (February).
	Statfjord (November) and South Cormorant (December) fields enter production.
	Announcement of seventh licensing round (December).
1980	Announcement of criteria and areas for seventh licensing round.
	Collapse of *Alexander L. Kielland* with loss of 123 lives (March); a semisubmersible drilling rig, the *Kielland* had been converted to serve as an accommodation vessel for the Edda field in the Ekofisk area of the Norwegian sector.
	Murchison (October) enters production.

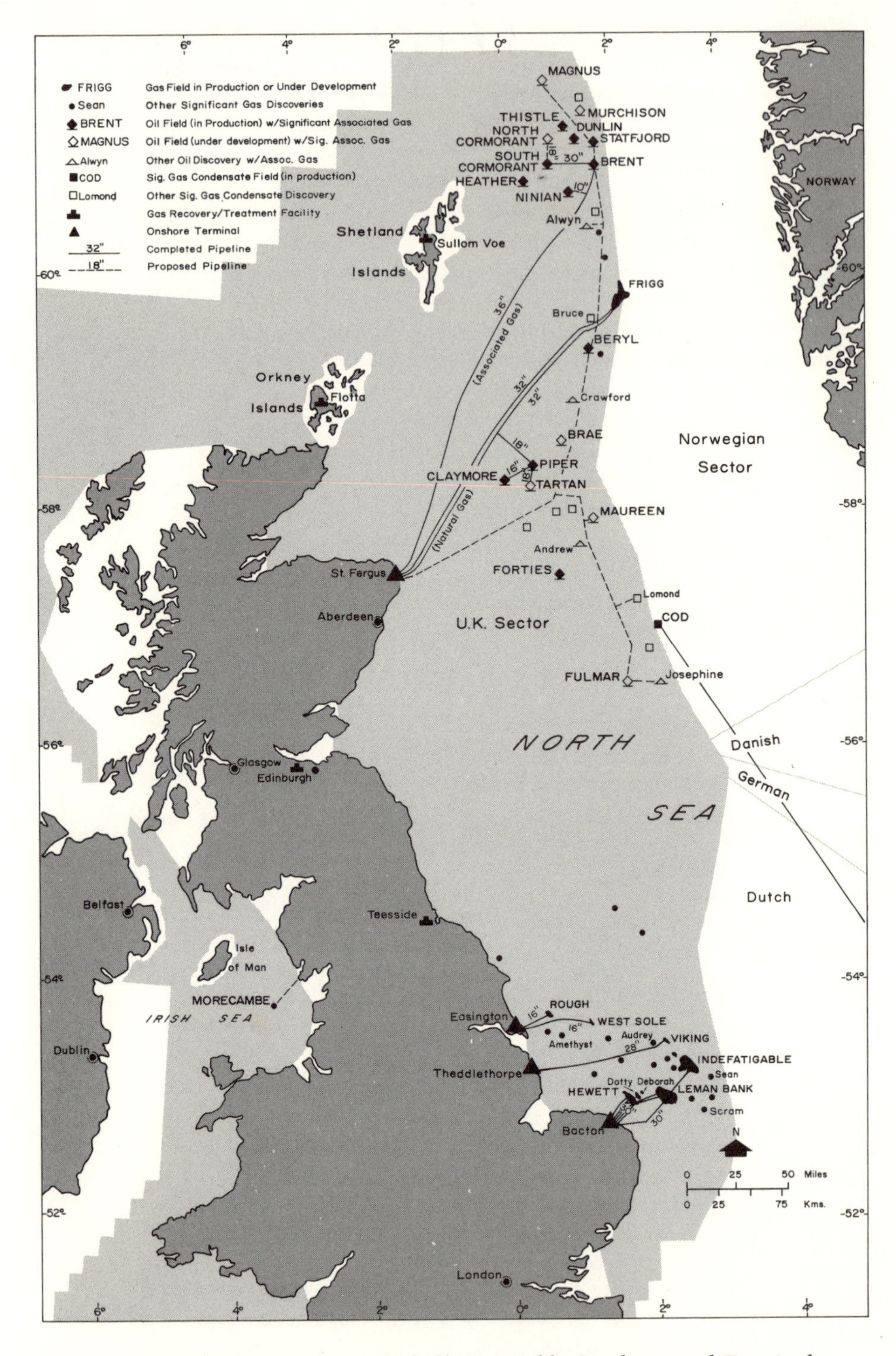

Figure 2. U.K. Continental Shelf: Gas Fields, Pipelines, and Terminals, 1980 (data from Department of Energy, 1980)

the decade it had become apparent that all the major discoveries lay within a relatively narrow northwest/southeast quadrant approximately fifteen to forty miles off the Norfolk and Lincolnshire coastline.

It has been suggested that there was a sense of frustration on the part of the companies involved in exploration work on account of the price paid for natural gas by the Gas Council, which the government had decided should be a monopoly buyer. However it seems equally true that activities in the southern North Sea had given the companies a chance to improve offshore exploration and production technologies and the confidence to move into the deeper northern waters where the prospects seemed better for oil than for gas. This shift in interest and area of activity was further encouraged by the discovery of the Cod gas condensate field in the Norwegian sector of the North Sea in May 1968. The first major North Sea oil strike, the Ekofisk field, was discovered by the Phillips consortium in November 1969 in the Norwegian sector in the same general area as the Cod field. Both these strikes were close to the median line, and any doubt that similar geological conditions extended to the U.K. sector was dispelled by the discovery of the Montrose field in one of the most northerly blocks to be awarded in the first U.K. licensing round.

Although exploration and development activity has moved steadily northward since 1970, the potential of the Southern North Sea Basin has by no means been fully exhausted. As noted by Chapman, "Trends in exploration must be seen in the context of developments elsewhere since the available financial and technical resources are limited. The opening of the Dutch sector in 1968 had an immediate impact upon activity in British waters while interest in oil has recently tended to overshadow gas developments" (Chapman, 1976, p. 60). Many smaller discoveries remain to be developed. A number of these, such as Deborah and Dotty, are closely related geologically to major fields already in production; others, such as Amethyst, Sean, and Scram represent independent fields in terms of location and geologic structure. "The possibility of further discoveries was clearly reflected in the tendency of most companies with blocks in the main producing zone to retain the full 50 percent of the areas originally licensed in 1964 and 1965—a sharp contrast with the wholesale release of entire blocks outside this zone. Furthermore, these relinquished areas were virtually all relicensed as soon as they came on offer again in 1972" (Chapman, 1976, p. 61). (See Chapter 2 of this book for explanation of licensing procedures.) This continuing interest has been generally justified. Despite a greatly reduced level of exploration drilling, gas strikes continue to be reported in the southern North Sea, including three "significant discoveries" since 1975.[11] In the near future, however, development of these smaller fields appears unlikely given the rapid build-up of gas supplies from Frigg.

All oil discoveries in the U.K. sector of the North Sea have been made north of the fifty-sixth parallel. Initially, exploration in this area was restricted to the limited number of blocks taken up by companies in the first three licensing rounds. Even the third round of exploration licensing in 1970 did not fully reflect the growing interest in the northern waters arising from the Montrose and Ekofisk discoveries, as blocks to be offered had already been selected prior to the announcement of these finds. The fourth round, however, which was held in 1971–1972 and which for the first time included a number of blocks to be leased through an auction bidding system, revealed the extent of the interest in the northern North Sea in terms of both numbers of applications and the size of the bids. One bid of £21 million by Shell for a block one hundred miles east of the Shetlands was more than three thousand times the standard license fee.

The initial result of the northward extension of exploration drilling was a series of oil strikes in the early 1970s off the east coast of Scotland in what geologists now refer to as the Central North Sea Basin (Figure 4). Within this broad province, a number of oil and gas condensate strikes were grouped close to the median line in the same general area as the earlier Ekofisk and Montrose discoveries. This area is also referred to as the Central Graben (a graben is a long, narrow area of the earth's crust that has subsided between two bordering faults). These strikes included the Forties field, discovered by British Petroleum in November 1970 and still the largest field in the U.K. sector in terms of proven recoverable reserves. Other discoveries that have proved to be commercially viable include Auk (discovered in February 1971), Argyll (October 1971), and Fulmar (November 1975). More or less simultaneously, a second cluster of fields was identified further to the northwest on the eastern margins of the Moray Firth (Figure 4). These included Piper (January 1973), Claymore (July 1974), Buchan (August 1974), and Tartan (December 1974). A more recent discovery to the west of this group is the small but significant Beatrice field, which poses special development problems because of its location only twelve miles off the Scottish coastline.

As is evident from Figure 3, this area remains under active exploration and development, and new finds are still being announced. By far the bulk of current U.K. production comes from fields in this general province. Experience with these fields suggests that it is realistic to think in terms of a five-year interval between the discovery

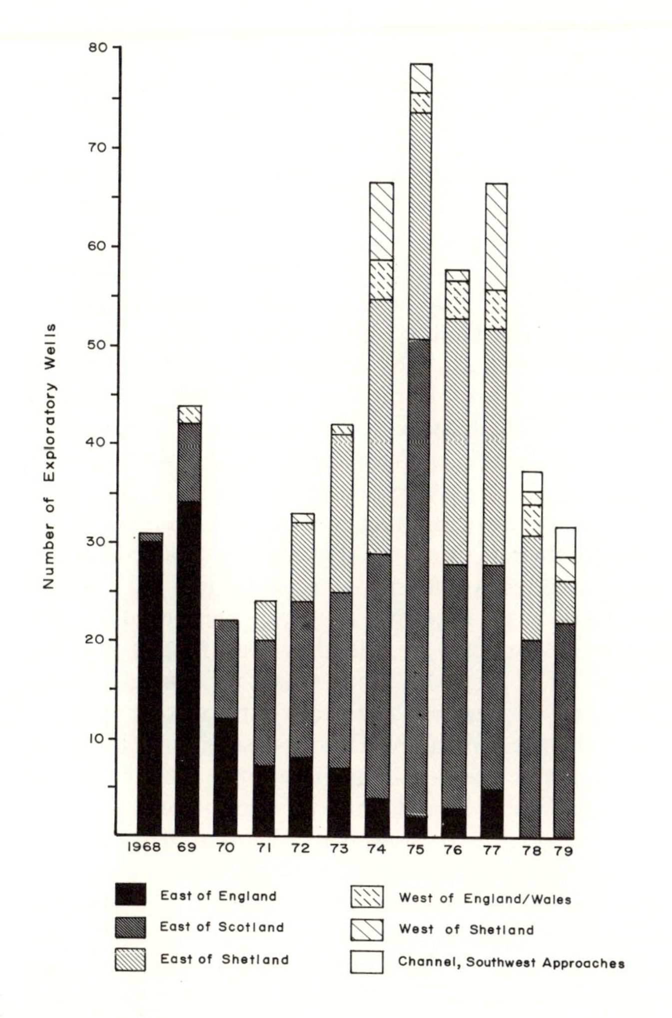

Figure 3. U.K. Continental Shelf: Exploratory Drilling Activity, 1968–1979 (data from Table 8)

of a commercially viable field and the start-up of production. Both Forties and Auk were originally scheduled to come on stream in 1973, but delays resulting from both technical and commercial problems prevented their entering production until 1975 and 1976 respectively. Where fewer problems were encountered (as at Piper), fields have been brought into production within four years of discovery.

Although the discoveries in the Central Basin currently contribute the bulk of U.K. offshore production, the most significant strikes since 1972 have been in the vicinity of the Viking Graben, a structure trending north and south, approximately four hundred kilometers in length, that underlies the median line to the east of the Shetlands and Orkneys. The first "wildcat" to penetrate the graben on the U.K. side established an encouraging precedent with the discovery of the large Brent oil field in September 1971. Strikes have been made along the length of the graben from Maureen (February 1973) in the south to Magnus (June 1974) in the north, although the most significant discoveries (including Brent, September 1971; South Cormorant, September 1972; Dunlin, July 1973; Thistle, July 1973; Northwest Hutton, September 1973; Heather, December 1973; Ninian, January 1974; Statfjord, May 1974; North Cormorant, July 1974; and Murchison, September 1975) cluster together approximately one hundred miles to the north and east of the Shetlands (Figure 4). These in a very real sense constitute the second generation of U.K. fields and will be brought into production over the next five years. It is still impossible to assess the full potential of this area as active exploration continues.

Success in the North Sea has stimulated interest in other portions of the continental shelf. In 1977 a strike by British Petroleum in the area lying to the west of the Shetlands led to speculation in the national media about a "major discovery."[12] Some early estimates suggested that the oil in place exceeded 13 billion barrels (more than six times the recoverable reserves of the Forties field). Additional strikes by ESSO and Elf, the licensees of neighboring blocks, focused attention on a small area about thirty miles west of the islands. However, subsequent appraisal drilling by British Petroleum during 1978 confirmed that the crude was much heavier and more viscous than that discovered in the North Sea. The current assessment appears to be that, despite the existence of substantial amounts of oil, recovering a reasonable proportion poses major technical problems that are unlikely to be resolved until the late 1980s at the earliest.[13]

Elsewhere the results have been less spectacular. In the Irish

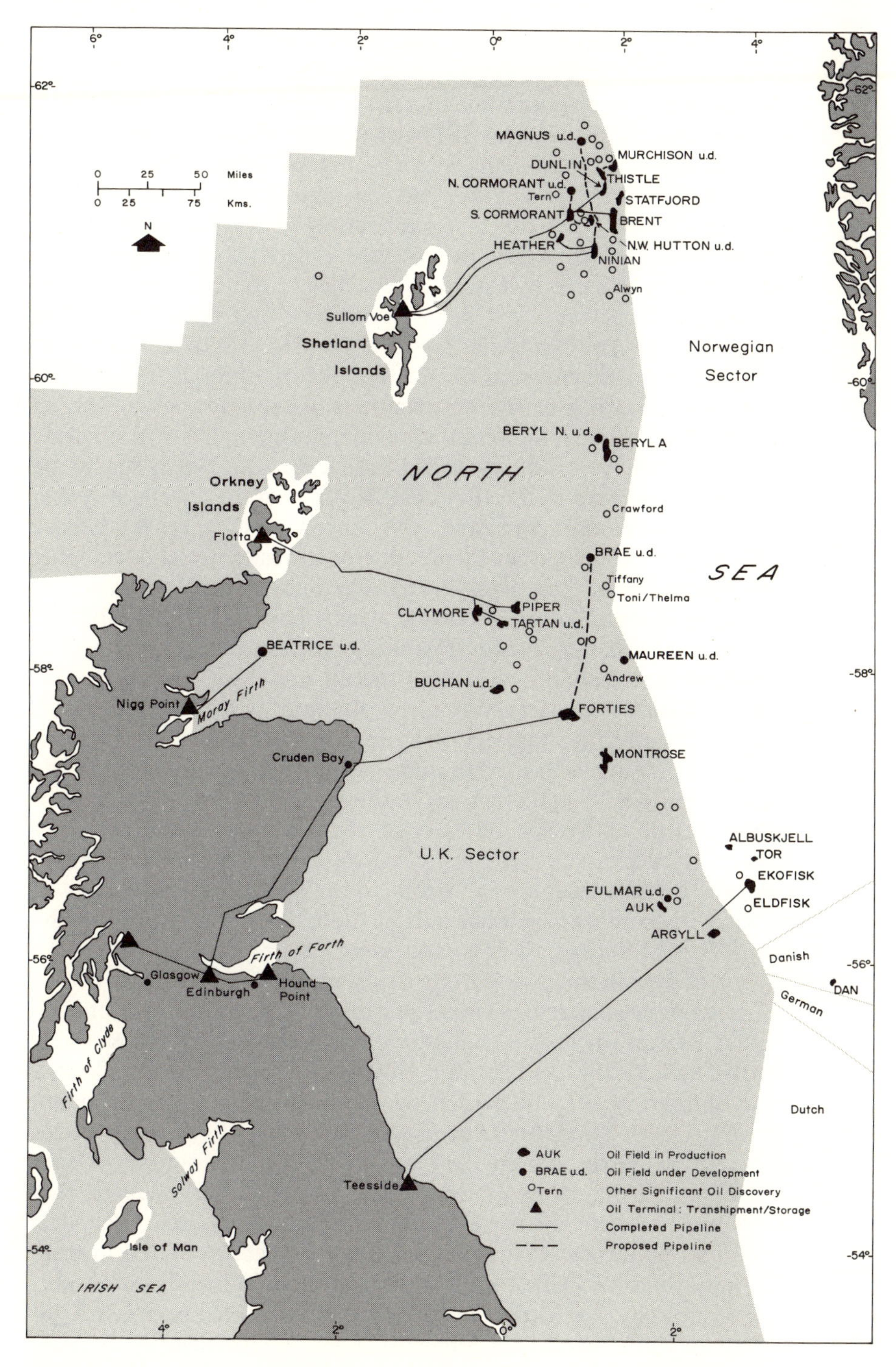

Figure 4. U.K. Continental Shelf: Oil Fields and Pipelines, 1980 (data from Department of Energy, 1980)

Sea, for example, the outcome of a fairly intensive search has been a single commercial gas find in Liverpool Bay, the Morecambe field, for which development plans were announced in 1978. By the end of the decade exploration activity had completely "encircled" the British Isles. In late 1978 the first exploration well was drilled by the British National Oil Corporation (BNOC) about 150 miles off the Scilly Isles in the southwestern approaches to the English Channel, followed by a second well begun by the British Gas Corporation (BGC) about twenty-five miles south of the Isle of Wight.

With the discovery and verification of commercially exploitable hydrocarbon resources, the uncertainties of exploration give way to the challenge of development. This changing emphasis in offshore activity is best illustrated for the U.K. sector of the North Sea by the significant increase in development drilling from 21 wells in 1975, to 54 in 1976, to 96 in 1977 and 1978, to 107 in 1979 (Table 8, Figure 5). Transporting oil, either by pipeline or by tanker, and developing facilities for storage, processing, and distribution represent major logistical problems that must now be resolved by the oil companies in concert with onshore authorities and communities. At the end of 1979 approximately 80 percent of the oil produced and all the gas brought ashore were transported by pipeline, with only five fields loading directly into tankers offshore. The proportion of oil transported by pipeline will further increase following completion of the massive Sullom Voe oil terminal in the Shetlands. Although the terminal was officially opened early in 1979, it appears unlikely that the facility, with its current design capacity of 1.4 million barrels per day, will be fully operational before 1981. When completed the terminal will handle more than half of the U.K. production from the North Sea. Moreover, the gas processing facilities will allow the recovery of petroleum gases from fields in the Northern North Sea Basin where associated gas must presently be flared or reinjected. A further indication of the changing circumstances and character of North Sea activity has been the growing reluctance of the Department of Energy to extend gas flaring permits (except where the quantity of gas is too small for recovery to be practical). In the early stages of North Sea development the Department of Energy gave its consent to gas flaring on an almost routine basis in order to ensure that fields were brought into production as rapidly as possible. The shift in official policy became apparent in 1977, when production from the Brent field was halted until the operator could install gas reinjection equipment.[14] Clearly the quantity of associated petroleum gas landed in the United Kingdom will increase significantly as more fields pipe their oil ashore, as the Brent gas pipeline to St. Fergus comes into operation, and as other, more ambitious, gas gathering schemes are implemented. These developments have major implications for land use policies and programs in the coastal zone. The siting of gas separation facilities and downstream petrochemical plants, such as the ethane cracker at Mossmorran, Fife, has already generated considerable controversy.

Even as development work has proceeded, new arguments have emerged with regard to what constitutes an appropriate rate of exploration drilling for the U.K. sector of the North Sea. The unexpectedly sharp decline in exploration activity during 1978 stimulated considerable discussion about the United Kingdom's prospects for maintaining self-sufficiency into the 1990s as the first- and second-generation oil fields are depleted. Whether the decline should be interpreted as a short-term phenomenon, resulting from the demands of development work on the time and resources of offshore operators, or as a more permanent withdrawal of rigs from U.K. waters was sharply debated. The oil journal *Offshore* spoke in terms of "the producers' rebellion," echoing the argument of the United Kingdom Offshore Operators Association (UKOOA) that the decline in exploration drilling was attributable to the increasingly stringent conditions governing licensing and participation agreements, to proposed increases in the petroleum revenue tax (PRT) rate, and to the restricted amount of "prime acreage" being offered for licensing (*Offshore*, vol. 38, no. 8, 1978, p. 41). Others noted that although the annual rate of drilling had fallen substantially below the sixty to ninety-five exploration wells suggested by the UKOOA as being necessary to ensure continued self-sufficiency, it was still comparable to the rate achieved during the period 1969–1973, when eleven out of the fourteen currently producing fields had been discovered. This argument will be returned to in a later chapter, since it demonstrates the difficulty of formulating an offshore policy (as expressed in the rate and regulatory terms of exploration and production licensing) that will stimulate the desired level of exploration while ensuring a satisfactory and equitable return to both the nation and the private investor.

Future Prospects

The short-term oil outlook (i.e., for the 1980s) appears bullish. By mid-1981 seventeen fields had been brought into production in the U.K. sector, with a further nine fields due to come on stream before 1985 (Table 9). In 1980 proven reserves (i.e., oil that can be recovered

Table 8. Drilling Activity: U.K. Sector, 1968–1979 (number of wells)

Activity/Region	*1968*	*1969*	*1970*	*1971*	*1972*	*1973*	*1974*	*1975*	*1976*	*1977*	*1978*	*1979*
Exploration drilling:												
East of England	30	34	12	7	8	7	4	2	3	5	—	—
East of Scotland	1	8	10	13	16	18	25	49	25	23	20	22
East of Shetland	—	—	—	4	8	16	26	23	25	24	11	4
West of Shetland	—	—	—	—	1	—	8	3	1	11	1	3
West of England/Wales	—	2	—	—	—	1	4	2	4	4	3	—
Channel, southwest approaches	—	—	—	—	—	—	—	—	—	—	2	4
Total all areas	31	44	22	24	33	42	67	79	58	67	37	33
Development drilling:												
East of England	36	27	28	34	36	21	20	13	7	7	7	2
East of Scotland	—	—	—	—	—	—	—	7	37	60	35	39
East of Shetland	—	—	—	—	—	—	—	1	10	29	54	66
Total all areas	36	27	28	34	36	21	20	21	54	96	96	107

Source: Department of Energy, 1980, pp. 36–38.

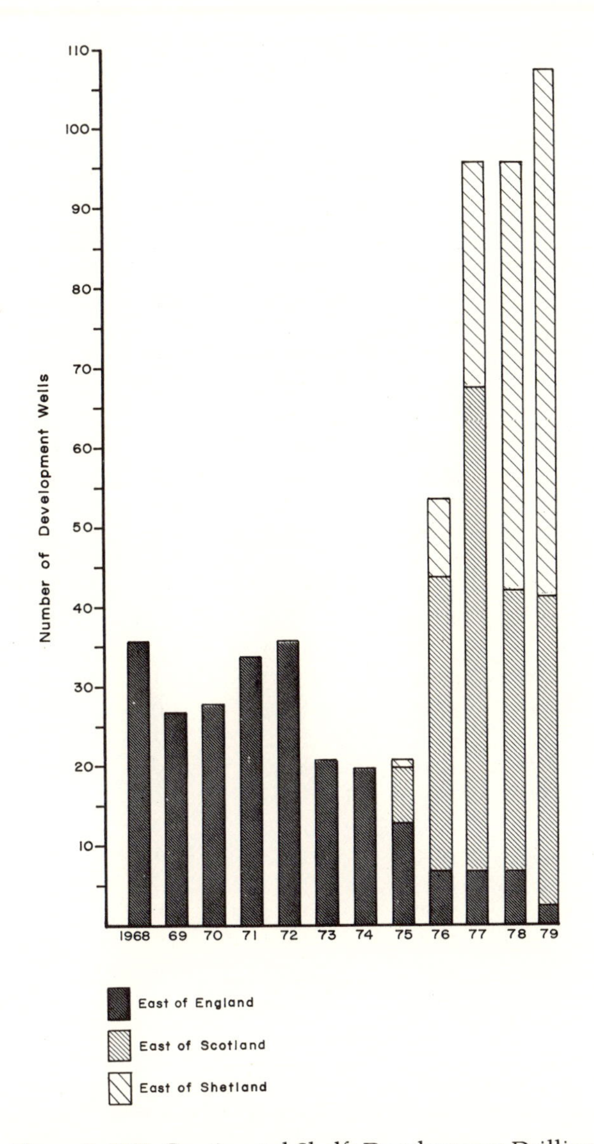

Figure 5. U.K. Continental Shelf: Development Drilling Activity, 1968–1979 (data from Table 8)

Table 9. Offshore Oilfields in Production or under Development, 1980

Field	*Date of Discovery*	*Production Start-up*	*First Year of Peak Production*	*Operator's Estimate: Peak Production (million tonnes/year)*	*Operator's Estimate: Proven Recoverable Reserves (million tonnes)*[a]
Argyll	10/71	6/75	1977	1.1	—[b]
Forties	11/70	11/75	1978	24.0	240.0
Auk	2/71	2/76	1977	2.3	8.0[c]
Beryl A	9/72	6/76	1980	5.0	66.0
Montrose	9/69	6/76	1979	1.4	12.1
Brent	7/71	11/76	1984	23.0	229.0[c]
Piper	1/73	12/76	1979	12.6	88.0
Claymore	5/74	11/77	1980	4.5	55.0
Thistle	7/73	2/78	1982	8.7	69.0
Dunlin	7/73	8/78	1980	5.9	41.0[bc]
Heather	12/73	10/78	1982	1.7	14.0
Ninian	1/74	12/78	1981	17.7	155.0
Statfjord (U.K.)	4/74	11/79	1988	4.3	65.5[d]
South Cormorant	9/72	12/79	1982	3.0	12.0[c]
Murchison (U.K.)	9/75	10/80	1982	7.2	51.0[e]
Tartan	12/74	1/81	1982	4.0	27.0
Buchan	8/74	5/81	1982	2.2	6.8[c]
Beatrice	9/76	1981	1982	3.9	21.0
Fulmar	11/75	1982	1984	8.6	70.0[c]
North Cormorant	7/74	1982	1986	7.3	55.0[c]
Northwest Hutton	4/75	1982	1983	5.1	37.5
Brae	4/75	1983	1983	4.9	36.0
Magnus	3/74	1983	1984	5.9	60.0
Maureen	2/73	1983/1984	1984	4.0	21.0
Beryl (North)	5/75	1984	1984	4.0	—[b]
Hutton	11/73	1984	1985	5.0	—[b]
					1,439.9

Note: The expression "under development" indicates fields where significant development work has occurred, including the placement of contracts for offshore equipment.

[a]Figures may not be precisely comparable because of different methods of estimation.

[b]Under assessment or reassessment.

[c]Total discounted reserves, i.e., proven plus suitable discounted figures for probable and possible reserves.

[d]Estimated U.K. sector reserves: peak production and reserves for entire field including Norwegian sector are estimated to be 27 million tonnes/year and 412 million tonnes.

[e]Including Norwegian sector.

Source: Department of Energy, 1980, pp. 50–54, updated.

Table 10. U.K. Continental Shelf: Official Estimates of Oil Reserves, 1980 (million tonnes)

	Proven[a]	*Probable*[a]	*Possible*[a]	*Possible Total*
Fields in production or under development	1,075	225	150	1,450
Other significant discoveries not yet fully appraised	125	400	425	950
Total present discoveries	1,200	625	575	2,400
Expected discoveries from present licenses (including sixth round)	—	—	—	300–750
Expected discoveries from remainder of U.K. continental shelf		—	—	500–1,050[b]

[a]The terms "proven," "probable," and "possible" are given the internationally accepted meanings in this context of:

Proven—those which on the available evidence are virtually certain to be technically and economically producible.

Probable—those which are estimated to have a better than 50 percent chance of being technically and economically producible.

Possible—those which at present are estimated to have a significant but less than 50 percent chance of being technically and economically producible.

[b]An estimated half of these anticipated discoveries are expected to be in water depths in excess of 1,000 feet.

Source: Department of Energy, 1980, p. 4.

using available techniques at present costs and prices) were officially placed at 1,200 million tonnes (Table 10). Despite unanticipated delays in bringing some fields into production, the British government anticipates that by 1984 production from the U.K. continental shelf will be in the order of 90–120 million tonnes (*Times*, March 6, 1981, p. 15).

For all the optimism about the availability of North Sea oil, there can be no guarantees with respect to future production levels. Experience suggests that even for the established fields there will continue to be unanticipated problems. Technical difficulties may delay development schedules; or production may be disrupted by accidents such as the blow-out that occurred on the Ekofisk platform *Bravo* in 1977 or the collapse of the converted semisubmersible accommodate vessel *Alexander L. Kielland* in 1980. Only as a field is developed is it possible to get a clear picture of reservoir characteristics. Some fields may exceed expectations; others, such as the Argyll and Buchan fields, may yield far less oil than had been anticipated.[15]

Table 11. U.K. Continental Shelf: Official Production Forecasts for Oil (million tonnes)

	1976	*1977*	*1978*	*1979*	*1980*	*1981*	*1982*	*1983*	*1984*
1976 forecast	15–20	35–45	55–70	75–95	95–115	—	—	—	—
1977 forecast	—	40–45	60–70	80–95	90–110	100–120	105–125	—	—
1978 forecast	—	—	55–65	80–95	90–110	100–120	105–125	—	—
1979 forecast	—	—	—	70–80	85–105	95–115	115–140	—	—
1980 forecast	—	—	—	—	80–85	85–105	90–120	95–130	95–135
1981 forecast	—	—	—	—	—	80–95	85–100	85–115	90–120
Actual production	11.6	37.3	52.8	76.5	80.5	—	—	—	—

Source: Department of Energy, 1977, 1978, 1979, 1980, updated.

Even the weather may play a critical role, with adverse conditions disrupting production on those fields where crude must be offloaded onto tankers. Moreover, the rate of production build-up is not independent of other considerations. Operators must weigh the short-term advantages of a rapid build-up to peak output against the need to protect the production characteristics of a reservoir over a longer period of time. A similar compromise must be sought between bringing a field quickly into production and the conservation of associated petroleum gases. Onshore delays, as in the case of construction of the Sullom Voe oil terminal complex, result in further delays in the production build-up. For a variety of reasons therefore, any forecast of future production levels, even for established fields, will necessarily be subject to considerable error. In general, recent official forecasts have been overoptimistic with respect to the rate at which oil development will occur. In 1977 and again in 1978 actual production levels were lower than forecast (Table 11). In consequence the 1979 Brown Book revised downward its forecast for 1980 and 1981 to "take more account of possible delays to offshore developments, particularly during the early years when oil production is being built up rapidly" (Department of Energy, 1979, p. 3). The official forecast was again reduced in 1980 "because experience has shown that with technical and weather difficulties and some problems of reservoir performance at some existing fields, greater account should be taken of possible delays which would defer some of our production beyond the first half of the 1980s" (Department of Energy, 1980, p. 9).

The short-term outlook for gas from North Sea fields can perhaps be assessed with rather more confidence in view of the longer operating experience. The fields currently under development should support an average production rate of 160 to 170 million cubic meters per day (6 to 7 billion cubic feet per day) by the early 1980s. Most estimates suggest that further contracts and the development of fields in the Central North Sea Basin (including those in the Norwegian sector) will enable production to be maintained at or about this level throughout the decade. Proven gas reserves are presently estimated to be 754 billion cubic meters, equal to about nineteen years' supply at current rates of consumption (Table 12). Moreover, despite the gradual depletion of fields in the southern North Sea (where proven reserves declined by 65 billion cubic meters in 1977, by 38 billion cubic meters in 1978, and by a further 33 billion cubic meters in 1979), there was actually an increase in the 1980 estimate of proven reserves for the U.K. continental shelf as a whole. This was due to a reassessment of the Morecambe field in the Irish Sea and of several condensate fields in the North Sea.

Table 12. U.K. Continental Shelf: Official Estimates of Gas Reserves (billion cubic meters)

	Proven	*Probable*	*Possible*	*Possible Total*
1. *Southern Basin:*				
Fields under contract to British Gas Corporation	354	14	25	393
Other discoveries	51	96	40	187
Subtotal	405	110	65	580
2. *Other basins:*				
Brent, Frigg (U.K. share)	169	—	3	172
Morecambe and other gas and gas condensate fields	105	159	323	587
Gas associated with oil	75	54	45	174
Subtotal	349	213	371	933
Total reserves for present discoveries	754	323	436	1,513
Expected discoveries				575
Cumulative production through 1979				335
Ultimately recoverable reserves				2,423

Source: Department of Energy, 1980, p. 5.

However, the decline in gas reserves for the southern North Sea, together with the shut-down in 1978 of the first offshore gas producing well on the West Sole field, draws attention to the even greater uncertainties surrounding North Sea oil and gas supplies over the long term. There appears to be a general consensus that crude oil production from established fields (i.e., in production or under development) will probably have reached a peak of around 110–120 million tonnes by 1982–1983. By that time production from several of the smaller, first-generation fields will already be declining.[16] Maintaining production levels into the latter half of the 1980s, or achieving any further increase, will therefore be dependent upon more efficient recovery methods for established fields, new discoveries, or the development of fields currently considered marginal due to their size, location, or structural characteristics. Remaining recoverable oil reserves on the U.K. continental shelf are officially estimated to be in the 2,400 to 4,200 million tonnes range (Table 10).[17] Other estimates suggest that the amount of crude oil ultimately recovered from the North Sea may be as much as 7,000 million tonnes (Moody, 1975). However, these figures should be interpreted with considerable caution. The higher government figures, for example, include both "possible" and "probable" reserves for all significant discoveries. Yet there must remain considerable uncertainty, on both technical and economic grounds, as to what will actually be recovered. The complete recovery of all possible reserves is an optimistic (or more accurately an unrealistic) assumption. Even more speculative are the estimates of oil resources in place on the unexplored portions of the continental shelf. The existence of hydrocarbons can be determined only through exploration drilling. Colin Robinson and Jon R. Morgan summarize the situation well: "No one really knows what may eventually be found in the North Sea, still less what finds might occur in the offshore area to the west of the United Kingdom. The general view is that the westerly areas are not very promising and that the North Sea proper will yield only relatively small discoveries but the opinions of even the best-informed observer contain large amounts of speculation" (1978, p. 187). All too often even promising strikes, such as the discovery of the Brae field, may turn out to be disappointing.[18] Robinson and Morgan suggest that North Sea oil output from the U.K. sector could reach 150 million tonnes by the late 1980s primarily as a result of further exploration efforts and the development of smaller discoveries.[19] As the authors themselves emphasize, however, quite apart from geological uncertainties any forecast will be extremely sensitive to assumptions about (1) the nature and rate of technological change in exploration and production methods; (2) the relationship of exploitation costs (including tax and royalty rates) to oil prices; and (3) future government depletion policies intended to extend production over a longer period of time.

While the short-term outlook in terms of available hydrocarbon reserves is excellent, it is quite impossible to predict how or to what extent these natural riches will be translated into tangible economic benefits. The major value of North Sea reserves to the United Kingdom lies in the high price of hydrocarbons, and particularly oil, in the world market. Production costs, though high when compared with Middle Eastern fields, are still low relative to world prices. According to the Department of Energy, oil production from the U.K. shelf was valued at £2.8 billion during 1978, while natural gas production (estimated in terms of the cost of imported oil equivalent) was valued at around £2.3 billion. Even allowing for the substantial costs of production there was still "a very significant excess of revenues over costs, a large part of which accrues to the nation in the form of royalties and taxes on profit" (Department of Energy, 1979, p. 18).[20]

Quite apart from the direct revenues accruing to the government, it is usually assumed that North Sea oil and gas will automatically result in a significant improvement in the balance of payments. As Robinson and Morgan emphasize, "there are serious difficulties in quantifying balance-of-payments and other economic effects which stem largely from uncertainty about what the state of the United Kingdom's economy might have been in the absence of offshore oil" (1978, p. 148). Moreover estimates of potential benefits will be extremely sensitive to varying assumptions about both oil output and oil prices. Nevertheless, there has been much speculation, often fueled by optimistic ministerial speeches, about increasing tax revenues and sizable balance-of-payments benefits. The general sense of euphoria—that the only real problem was to decide how to allocate this unexpected windfall—was perhaps encouraged by the immediate effect of North Sea oil production in reducing oil imports (Robinson and Morgan, 1978, p. 203). In 1977, for example, the value of North Sea oil production in terms of import saving was estimated to be in the order of £2 billion, and for the first time exceeded expenditures upon imported goods and services necessary to sustain the exploration and development program (Department of Energy, 1978, p. 28). With respect to future balance-of-payments effects, perhaps the most detailed analysis (under varying assumptions about oil output, prices, and other variables) is that undertaken by Robinson and Morgan. They estimate that potential benefits (by comparison with a non–North Sea oil state) may be in the order of £6 to £15 billion by 1985 (roughly equivalent to 2–6 percent of current GNP) and £6 to £22 billion by 1990 (Robinson and Morgan, 1978, p. 181).

These potential benefits are not insignificant, although as the authors themselves acknowledge the range of estimates is hardly likely to satisfy those "who like to see apparently precise point estimates of future events" (Robinson and Morgan, 1978, p. 181). Moreover they are potential benefits that will not be available to other nations that must continue to import expensive oil. However, it is perhaps unwise to anticipate such benefits, "since the advent of North Sea oil in no way guarantees any *actual* balance of payments benefits" (Robinson and Morgan, 1977b, p. 8). And indeed, as the decade ended events tended to suggest a more pessimistic scenario, namely that the appreciation of sterling (in itself a result of North Sea oil although reinforced by extremely high interest rates) was contributing at least in the short term to a worsening of the non-oil balance of payments. "It is even conceivable that the overall balance of payments will, in a few years' time, be worse with North Sea oil than it would otherwise have been" (Robinson and Morgan, 1978, p. 206). As these authors acknowledge this worst-case outcome is highly unlikely, but it clearly "is important to beware of simple conclusions about the impact of North Sea oil on the economy. A North Sea oil state will be different from a non–North Sea oil state in many ways which we can hardly begin to guess at present: the crucial factor in determining how much of the potential gain is realised, in terms of balance of payments improvement or otherwise, will most probably be the quality of the British government's management of the economy" (Robinson and Morgan, 1977a, p. 392).

Realization of the large potential balance-of-payments gains, however, may have important ramifications for future depletion policy. It has been widely assumed in the United Kingdom that once self-sufficiency has been achieved in the early 1980s, the government of the day will be anxious to restrict oil output to a level below that demanded by the producing companies. This was one consideration in the establishment of the British National Oil Corporation and in the system of depletion control set out in the Petroleum and Submarine Pipe-lines Act of 1975. As Robinson and Morgan point out, however,

> one must question whether it is realistic to assume that governments will actually be willing to hold down North Sea oil production, given that they will thereby postpone the balance of payments benefits they could have had. . . . A government which exercises downward depletion control will, in effect, be relinquishing short term balance of payments gains for longer term gains. It is possible to think of circumstances in which such action might be taken by a wise central authority—if, for example, oil prices were expected to increase at a percentage rate more rapid than the government's discount rate.
>
> Nevertheless, we doubt whether it is likely policy for a government which most probably has a short time horizon (in economist's language, a high discount rate) and is anxious to claim benefits for itself rather than pass them on to its successors, especially since the latter may not be of the same political party. . . .
>
> Looking at the prospects for the British economy, however, one wonders whether even in the 1980s governments will be holding down the production of North Sea oil. Whatever one's views on the desirability of depletion control (upward or downward), there appears to us to be a high probability that for

> many years to come, far from trying to reduce North Sea oil output, governments will be working hard behind the scenes in an endeavour to make the companies maximise production. (Robinson and Morgan, 1977b, p. 9)

Thus, the "economic imperative" that has shaped existing offshore and onshore development policies may well continue to provide the incentive for rapid exploitation of North Sea reserves in the future.

The only certain thing about the future is that it is unlikely to conform to predictions. Even in the period since publication of Robinson and Morgan's analysis, further sharp increases in oil prices have resulted in the reappraisal of several discoveries previously considered to be too marginal for production, while innovative production techniques, such as the prototype tension leg platform designed for Conoco's Hutton field, are being introduced and tested for deeper water conditions (*Offshore*, vol. 39, no. 14, 1979, p. 5).[21] The dilemma confronting those who must formulate a long-term policy for North Sea oil is that, in a world of rapid scientific and technological advance, of changing economic circumstances, and of constant social reappraisal of resources, they must select the most probable elements of change and determine the relevant time and space dimensions within which such changes will occur. As in so many other areas, we are forced to recognize the truth of James Clerk Maxwell's dictum that "the true logic of this world is in the calculus of probabilities."

Implications for Planning

The preceding description has focused on the spatial character and temporal progress of offshore activities in the U.K. sector of the North Sea, together with the economic environment within which those activities have occurred. At this level of analysis the impression is that the development of North Sea oil and gas—from the initial exploratory activities in the early 1960s to the deliveries of crude oil in the mid-1970s—has proceeded smoothly, even rapidly, and certainly without the disruptions and controversies that have recently characterized offshore oil exploration and development in the United States.

There have been the almost inevitable disputes over such matters as the size of North Sea reserves, which are notoriously difficult to estimate but are significant in the context of any attempt to develop a depletion policy, as well as over product pricing and royalty payments (essentially a government/company dialogue and again almost a sine qua non of oil development in the second half of the twentieth century). What seems to be lacking when viewed from the U.S. perspective is the absence of any significant or organized opposition to offshore oil development. For those who believe that rapid exploration and development of the outer continental shelf areas of the United States are essential to reduce the growing dependency upon imported liquid fuels, there is an understandable temptation to look to the North Sea experience for a solution to our own difficulties. Yet one must be rather cautious about accepting the proposition that the North Sea provides some kind of offshore development model for the United States. That is not to say that there may not be lessons to be learned—merely that the North Sea experience, and particularly the efforts of planners to promote development while minimizing community and environmental disruption, must be subjected to careful scrutiny rather than superficial acceptance.

In this context, it is important to recognize that while exploration for and development of offshore oil are indeed a continuous and ongoing process, distinct phases exist—usually referred to as the *exploration*, the *development*, the *production*, and the *depletion* phases—each of which needs to be considered separately in terms of potential impacts and appropriate planning strategies. These phases, of course, apply to individual fields as well as to the entire producing area of the North Sea and hence overlap, with the impacts of exploration, development, and production being experienced conjointly.

Nevertheless, it seems justifiable to claim that at the macro scale, the United Kingdom is entering the production phase. Whether at the macro or micro scale, each of these phases tends to be characterized by particular sets of activities. During the exploratory phase, operations are speculative and the elements of risk and uncertainty influence both government and company policies. The focus of attention (apart from the issue of supply bases) is essentially offshore, and leasing policy is likely to be the major source of controversy.

With the discovery and verification of commercially exploitable deposits, additional support facilities are required to bring the fields into production. These include production platforms, pipelines, storage terminals, and processing facilities. Decisions made during the development phase will be crucial in determining the range and magnitude of onshore impacts. The focus of most planning activity has shifted to the coastal zone, although the safety of installations and the prevention of pollution remain major offshore concerns. As oil and gas begin to flow ashore during the production phase, onshore impacts and priorities are likely to change as planners and communities begin to grapple with the long-term implications of a finite resource. The growing debate over the adequacy of environmental and land use planning procedures as these relate to the development of North Sea oil may well reflect the onset of the production phase and the greater visibility of onshore impacts at a time when the less tangible economic benefits have already been discounted.

What seems to be called for, therefore, in any assessment of North Sea development is a critical review of the planning process with particular attention to the manner in which the various issues that arise during these phases are resolved. These issues may be thought of as critical linkages in the development process, the "successful" handling of which will determine the character and pace of oil development. Critical factors in the way in which these issues are resolved will include (1) the nature and timing of public participation; (2) the adequacy, timing, and quality of information available to both planners and the public; (3) the relative inputs from national, regional, and local planning authorities; (4) the scope, sophistication, and policy effect of impact assessment; and (5) the opportunity for forward or contingency planning intended to reduce locational conflicts and minimize the risk of marine pollution.

Initially however it is necessary to consider U.K. licensing policies which provide the context within which environmental and land use planning decisions have traditionally been made. In particular, exploration and production will be critically affected by the financial terms under which companies must operate, by administrative requirements such as specified work programs, and by the extent and nature of government participation in offshore planning. As Chapman (1976, p. 76) points out, from the perspective of those companies involved in offshore drilling a decision to start exploration in one area rather than in another with similar potential will be based upon a comparative assessment of the general regulatory framework, as this affects the economics of resource development. In general U.K. offshore policy, as reflected in the terms of production licensing, has sought to encourage exploration and development while at the same time ensuring that the nation receives an appropriate share of benefits, either directly in the form of royalties and taxes or indirectly through the involvement of British companies. Yet all decisions affecting offshore activity have profound implications for both the marine environment and the coastal zone. And it is precisely the more visible impacts of offshore oil development that bring home to the general public the tradeoffs that are involved when the national "need" for energy development confronts what are usually referred to as quality-of-life aspirations (which may embrace issues of cultural identity, social welfare, and amenity value). The major emphasis in this book therefore will be on the planning framework as it evolved during resource exploration and development. Planning for the marine environment and for the coastal zone will be treated separately, although it must be stressed that this is a highly arbitrary and artificial distinction. Incidents, such as oil spills, may extend beyond the immediate physical impact on the marine environment to influence local community attitudes towards offshore oil development. Similarly, onshore development—particularly controversial locational decisions for support facilities—will affect the pace of offshore activity.

2. Offshore Concessionary Terms

Introduction

All stages in the exploitation of offshore oil and gas resources are subject to government regulation. The offshore regulatory framework includes the financial terms under which companies may operate, administrative requirements including specific work programs and the nature of government participation, and operating restrictions intended to minimize the risk of marine pollution.

Clearly the regulatory framework is likely to vary over time and from one country to another. Legislation affecting offshore exploration and production is intended to implement specific government policies and objectives. Policies that are appropriate for India, a country heavily dependent upon imported oil and lacking the financial and technical resources to carry out its own offshore exploration program, would not necessarily be in the best interests of Mexico, a country which is already an oil exporter and which possesses both an experienced national oil company and substantial proven onshore reserves. Similarly Norway, with its small domestic requirements and sizable offshore reserves, initially felt able to impose more restrictive terms on North Sea operators than did the United Kingdom.

The very success of exploration activities, however, is likely to result in a reordering of government objectives and priorities and a shift in national policies. Thus as the United Kingdom has begun to anticipate a greater degree of energy self-sufficiency, successive governments have come under pressure to modify the terms and requirements of offshore concessions to ensure that "the best interests of the nation are served." Translated into human terms of self-interest, once oil is discovered, the demand is for a larger piece of the pie.

Irvin L. White et al. go so far as to suggest that "to the extent that North Sea policy is debated [within the United Kingdom], the issue seems to focus on whether the U.K. is getting a proper share of the benefits either directly in fees, royalties, and bonuses or indirectly by the participation of British companies in various development activities" (1973, p. 6). Since White et al. deal essentially with that period immediately following the discovery of exploitable reserves, it may be surmised that in part at least this perception of the debate over North Sea oil policy in the United Kingdom reflects a particular phase in the development process.

As development moves into the production phase—involving substantial flows of oil to the shore for processing, refining, and marketing—the onshore impacts of offshore production will acquire greater visibility, and the focus of the debate may well shift accordingly. In any case, in terms of government policy affecting offshore exploration and production activities, the United Kingdom has gradually moved much closer to the Norwegian position, requiring a greater degree of state participation and imposing higher levels of taxation. Needless to say, the extent and form of government involvement as well as the rate at which offshore reserves should be developed have proved to be highly contentious issues.

The impact of changes in offshore restrictions and regulations and of the uncertainty that may arise as a result of shifts in government policy varies enormously from one company to another. "The organisations involved in the offshore search are far from uniform in corporate circumstances and long-term goals and their reactions to political pressures are correspondingly diverse" (Chapman, 1976, p. 98). Some offshore operators will be prepared to accept higher financial and political risks; for others even the threat of state participation may be sufficient to deter exploration.

In short, a complex interaction exists between offshore activity and the regulatory and financial terms of offshore legislation. Thus the initial licensing terms influence the pattern and rate of exploration, while the results of exploration feed back and affect those terms (Chapman, 1976, p. 81). In the same way policies affecting the location and pace of offshore exploration and development have major implications for land use planning in the coastal zone. In turn, the extent to which onshore planners can locate essential facilities and provide critical services without incurring significant social or environmental costs will feed back and affect the timetable for North Sea development.

Exploration Licenses

The British, Dutch, and Norwegian governments differentiate between exploration and development in terms of licensing arrangements. While the activities permitted under these various licenses vary from one country to another, exploration or prospecting licenses merely allow an operator to search for oil and gas; any further activity requires a production license. The exploration licenses issued by the British and Norwegian governments cover virtually the entire continental shelf area and do not convey exclusive exploration rights. Exploratory drilling is not permitted. Nondrilling exploration methods may include (1) identification of local variations in the earth's magnetic and gravitational fields, (2) seismic surveying, and (3) bottom sampling and coring (White et al., 1973, p. 56). All subsequent activity (including exploratory as well as development drilling) requires a production license.

In contrast to exploration licenses, production licenses are exclusive and apply only to the area (or block) specifically licensed. As production licenses have been issued, the area covered by exploration licensing has gradually been reduced. A slightly different situation exists in the Netherlands, where an exploration license (termed a prospecting license) conveys exclusive exploration rights and allows drilling. Once the government is convinced that a commercial deposit has been discovered, a production license is issued to the exploration licensee.

Exploration licenses may specify various operating procedures intended to ensure safety and prevent pollution. A major emphasis of these regulations is to minimize conflict between exploration activities and other "uses" of the North Sea, particularly fishing (White et al., 1973, p. 56). An additional requirement is that licensees are required to furnish the respective governments with all raw exploratory data and interpretations of those data. As the governments themselves are also actively involved in collecting and evaluating offshore data, their information base upon which to determine subsequent production licensing policy is significantly broader than is the case in the United States. Neither the data nor the assessment may be publicly disclosed, thereby ensuring that each company's commercial advantage is maintained. Yet this restriction in turn means that the advantage of the government's information base and its foreknowledge of where offshore activities are likely to be concentrated is not available to planners at the local level. They continue to complain of an "information gap" and of the difficulty of forecasting likely social impacts in the absence of adequate information about company and government intentions.[1] In referring to criticism in Scotland that the oil companies "withhold information," John Francis and Norman Swan observe:

> Decisions, involving millions of pounds and many people, have to be made with incomplete data about the total North

> Sea potential and its location. Inevitably, there are periods of policy gestation when premature publicity would be of benefit mainly to rivals and speculators. It is unfortunate, but perhaps inevitable, that this periodic reticence is sometimes regarded as prevarication. In order to counteract this impression, it seems important that oil companies and their contractors should take the public and not least the local community, into their confidence at the earliest practicable moment. (1973, p. 59)

Production Licenses

In the most general sense, a government's attitude and priorities with respect to offshore resources is apparent in the method of production licensing that is adopted and in the financial terms that are applied to the recovery of oil and gas. Production licensing is central to any attempt to formulate and implement a coherent offshore oil and gas policy. The licenses contain the terms that define the pace and character of all offshore activity; subsequent impacts—whether offshore or onshore, national or local, financial or social—will be influenced by these terms. D. I. MacKay and G. A. Mackay aptly summarize this situation in the context of British North Sea oil policy: "On this all else depends. If these aspects of policy are at sea, metaphorically as well as literally, then the situation cannot be retrieved by any other combination of policies" (1975, p. 18).

Licensing policy considerations may include a desire to accelerate exploration and development activity, to direct exploration to particular offshore areas, to maximize the economic rent accruing to the nation, to provide for full involvement by national as opposed to foreign-based concerns, to minimize the risks involved in offshore operations, and/or to mitigate the social and environmental impacts of energy development. Priorities with respect to these varying objectives are reflected in the criteria used to select licensees and in the operating and financial terms imposed on the licensee.

Arriving at a "balanced" offshore policy is likely to be more difficult for a government than for a private company, where the objective is comparatively straightforward—a reasonable return on investment. Moreover a government's efforts to formulate and implement an appropriate exploration and development strategy must take into account the uncertainties that inevitably surround drilling in a previously unexplored area. While the assessment of risk can be reduced by preliminary seismic surveys and comparison with known areas of similar geological character, and while a statutory obligation may exist to share such information with the national government involved, initial offshore policy and associated legislation must frequently be formulated on the basis of very limited and incomplete data. A. P. H. Van Meurs (1971, pp. 170–177) notes that in these circumstances initial concession terms and codes are more often than not based upon the assumption of a low probability of success. The primary concern in such circumstances is to stimulate interest in outer continental shelf exploration.

Offshore licensing policy for the U.K. sector of the North Sea certainly conforms to this general pattern. A parliamentary report prepared by the Committee of Public Accounts is particularly revealing with respect to the considerations influencing initial offshore policy and the legislation (Continental Shelf Act of 1964) intended to give effect to that policy:

> Your Committee were told that in determining basic policy the following factors were taken into account:
>
> (i) The North Sea was completely unproven as an area in which petroleum existed. The considerable interest of the companies pointed to the existence of structures that might contain petroleum. Whether or not they did could only be determined by deep drilling, a risky and expensive operation, much more so offshore than on land. Conditions even in the Southern Basin of the North Sea presented new challenges in technology.
>
> (ii) The U.K. could gain substantially from the production of indigenous oil or gas, providing an additional and secure source of primary energy and benefiting our balance of payments. Retained oil imports were then costing the U.K. about £300 million a year in foreign exchange.
>
> (iii) Most of the countries from whom we imported our oil supplies were members of the Organisation of Petroleum Exporting Countries (OPEC) and were pressing for increased revenue from the oil company concessionaries in their countries . . . The Department [of Trade and Industry] thought that if the U.K. were to impose onerous financial terms it might have incited OPEC countries to follow suit, to the detriment of our overseas oil interests and balance of payments.
>
> (iv) The Departments view was that the greater the effort put into the exploration of the U.K. shelf, the greater would be

> the potential benefit to British firms supplying equipment and services to the licensees—whether the results were successful or not. (Committee of Public Accounts, 1973, p. x)

In essence, therefore, the British government's offshore policy was to encourage the most rapid exploration for and development of North Sea hydrocarbons, to secure the most favorable financial terms without discouraging maximum effort by oil companies, to protect British oil industry interests overseas, and to ensure adequate involvement in North Sea activities by British companies. This basic policy, the Committee of Public Accounts noted, was arrived at by the Department of Trade and Industry in consultation with other government departments and "consultations . . . on legal matters, and on technical and operational subjects, with two committees set up by interested oil companies" (ibid., p. ix).

Many consequences stem from these policy decisions and from the assessment of high risk, not least the method of licensing, the financial terms offered, and (other policy objectives notwithstanding) reliance on foreign capital and expertise.

> It was judged in 1964 that, in the unproven North Sea, competitive bidding would be unlikely to lead to full and thorough exploration, and it was thought that bids might well have been small, and confined to strictly limited areas, and that British participation might well have been less than was possible to achieve under a discretionary system. Although it was recognised that American oil companies would inevitably play a major part because of their numerical superiority, technical expertise and ownership of or ready access to the equipment needed for off-shore drilling, it was considered essential that British interests should be well represented. Under a discretionary system the Department felt that they would be able to insist, as a condition of a production license, that the licensee carry out an effective work programme and also to persuade him to buy British goods and services where they were readily available. In the light of these considerations it was decided that the method of allocation should be left to Ministerial discretion. (Committee of Public Accounts, 1973, p. xi)

Production licenses for the U.K. sector of the North Sea have so far been offered on seven different occasions since 1964 (Table 13). A striking, and somewhat surprising, feature of these licensing rounds is that the basic considerations influencing British offshore policy, and hence the method and terms of licensing, remained essentially

Table 13. Continental Shelf: Licensing Rounds, 1964–1981

	Blocks Offered	
	Areas Covered	*Number of Blocks*
First round (1964)	North Sea	960
Second round (1965)	North Sea	1,102
	Irish Sea	
	English Channel	
Third round (1969–1970)	North Sea	157
	Irish Sea	
Fourth round (1971–1972)		
Discretionary	North Sea	421 (North Sea, Irish Sea, Celtic Sea)
	Irish Sea	
	Celtic Sea	
Competitive bidding	North Sea	15
Fifth round (1976–1977)	North Sea	71
	Irish Sea	
	Arctic Sea	
	Orkney/Shetland Basin	
	English Channel	
	West of Scotland	
Sixth round (1978–1979)	North Sea	46
	West of Scotland	
	Cardigan Bay	
	Bristol Channel	
	Southwest Approaches	
Seventh round (1980–1981)		
Discretionary	North Sea	80
	North & West of Shetland	
	West of Scotland	
	Southwest Approaches	
	English Channel	
Nominated[b]	North Sea	—

[a]Preliminary awards only.

[b]Blocks nominated by companies; a premium of £5 million was to be paid by the licensee on award of a production license.

Source: Department of Energy, 1980.

Applications		*Licenses Granted*	
Number Received	*Number of Blocks Applied for*	*Number of Licenses Granted*	*Number of Blocks Awarded*
31	394	53	348
21	127	37	127
34	117	37	106
92 }	271 }	}	267
		118 }	
31	15 }		15
53	51	28	44
55	46	26	42
125	71	37[a]	—
—	54	42	—

unchanged throughout the first four rounds of licensing despite the fact that this period saw the discovery of oil and gas in sufficient quantities to establish the North Sea as a major hydrocarbons province. The fifth and sixth rounds of production licensing (held during 1976–1977 and 1978–1979) did see a significant change in licensing strategy, with the government attempting to achieve a more stable level of offshore activity through the licensing of smaller amounts of offshore acreage at more frequent intervals. Whether or not this approach had succeeded in sustaining an adequate rate of exploration drilling was to become a major political issue as the decade drew to a close.[2]

FIRST LICENSING ROUND (1964)

With the exception of Denmark, all governments involved in North Sea development have opted for a system of block licensing. This approach, it is argued, has the particular advantage of allowing the state to control the location and rate of exploration and to involve itself in the organization of offshore exploration and production. Thus a government may choose to offer for license large numbers of blocks at frequent intervals if it wishes to stimulate activity. Initially concerned with promoting rapid exploration, most nations opted for this approach.[3] The U.K. government, for example, placed virtually the entire North Sea area on offer during the first round of production licensing in 1964. Applications were invited for 960 blocks, although only 394 blocks (concentrated in the southern sector of the North Sea) actually attracted any interest.

The criteria for allocating production licenses were neither specified in the Continental Shelf Act of 1964 nor promulgated in any formal set of regulations. Instead, criteria were announced by the Minister of Power on the floor of the House of Commons:

> First, the need to encourage the most rapid and thorough exploration and economical exploitation of petroleum resources on the continental shelf. Second, the requirement that the applicant for a license shall be incorporated in the U.K. and the profits of the operation shall be taxable here. Thirdly, in cases where the applicant is a foreign-owned concern, how far British oil companies receive equitable treatment in that country. Fourthly, we shall look at the programme of work of the applicant and also at the ability and resources to implement it. Fifthly, we shall look at the contribution the applicant has already made or is making towards the development of resources of our continental shelf and the development of our fuel economy generally. (Committee of Public Accounts, 1973, p. xii)

As noted by Kenneth W. Dam, "these criteria were anything but precise . . . [They] were merely statements of the preferences and predilections that were to guide the ministry in making its awards" (1976, p. 25).

Implicit in the minister's statement, however, was the rejection of any form of competitive bidding (in the sense of an auction system where the applicant willing to pay the largest cash bonus obtains the license) comparable to that used in the United States. In this respect, the terms of the production license were fixed: royalty rate of 12.5 percent, an initial license fee of £6,250 per block, and the surrender of one-half of each block after six years. Unsurrendered portions of each block could be retained for forty years at an annual rental of £10,000, rising in annual increments of £6,250 to a maximum of £72,500 per year. In retrospect, the annual rent for each block seems ludicrously low; moreover, despite subsequent discoveries first of oil and then of gas, this rate was only marginally raised in the third and fourth licensing rounds. At the time, however, it was argued that these financial terms would ensure the most efficient investment of resources, that instead of advancing (essentially nonproductive) cash payments to the government, companies would be able to utilize their working capital in financing rapid exploration. A secondary consideration, somewhat ironic in view of the scale of North Sea discoveries, was the desire to protect British oil interests in the Middle East. Thus the Committee of Public Accounts report records the ministry view that at the time the financial package was "as favourable to the Government as any in completely unproved areas throughout the world, while not providing ammunition to OPEC in their demands for higher revenues from British oil companies operating there" (Committee of Public Accounts, 1973, p. xi).

In the absence of an auction system, the government was confronted by the problem of selecting licensees for those blocks (such as those lying closest to the Dutch coastline) that had attracted several applications. In these circumstances, the government was able to promote its primary objective of "the most rapid and thorough exploration" of the North Sea by initiating what Dam described as "a form of competitive bidding in work programs" (1976, p. 26).

> It came to be known that the ministry expected much more active drilling programs in areas that were widely sought after than in less coveted areas. Indeed, by a process that is none too clear to the outsider looking in after the fact, a "going price" came to be known for each area. This going price was denominated in such things as holes drilled and exploration work undertaken . . . Moreover, where an applicant's work program for

> a particular block seemed insufficient to the ministry, he was informed that unless he increased the extent of exploration and drilling activity he could not expect to receive a final allocation of that area. By means of this kind of direct negotiation, the ministry was able to introduce an element of competition into the work programs. (Dam, 1976, p. 28)

The overriding emphasis on rapid development is further evident in the relinquishment provisions of the license. Each licensee was required to surrender one-half of the licensed area after six years, thereby providing an obvious incentive for quickly determining which areas offered the best prospects. This emphasis in licensing policy was constrained only to the extent that political considerations also demanded active involvement in North Sea oil activities by British companies. "The two principal domestic oil interests, British Petroleum and Shell, probably could not alone exploit this vast area with the speed and efficiency desired even if foreign policy considerations were to permit the exclusion of American and other foreign-owned oil interests. On the other hand, it was thought desirable that these two companies should be somewhat favored in the distribution of blocks" (Dam, 1976, p. 25).

As a result about 30 percent of those blocks allocated in the first licensing round were awarded to British companies. Moreover, it would appear that British companies were well represented in the most promising areas. "In what geologists have always considered the most 'prospective' area of the North Sea-bed . . . British Petroleum has licenses for perhaps the largest number of blocks . . . of any single company" (*Economist*, September 25, 1965, pp. 1237–1238).

SECOND LICENSING ROUND (1965)

Applications for production licenses were again invited in 1965. In all some 1,102 blocks were placed on offer, including a number of North Sea blocks that either had not attracted interest or for which applications had been rejected in the first round, as well as blocks in the Irish Sea and the English Channel (Table 13). The response was disappointing; only 21 applications were received for 127 blocks. Moreover, little interest was shown in any area other than the southern North Sea, which by the end of the round was almost fully allocated.

Despite a change of government following the Labour Party's victory in the October 1964 general election, the criteria and procedures utilized in the first round of licensing remained essentially unchanged in the second. The Committee of Public Accounts report notes that after a careful review of offshore policy the new administration concluded that "there was no good case to change either the method of allocation or the financial terms and conditions" (Committee of Public Accounts, 1973, p. xii). In allocating licenses, it was announced that the minister would take into account three considerations additional to those used as a basis for allocating licenses in the first round.

> These were:
>
> (i) any exploration work already done by or on behalf of the applicant which was relevant to the areas applied for, and his facilities for disposing, in the U.K., of any oil or gas won;
>
> (ii) the contribution the applicant had made or was planning to make to our economic prosperity, including the strengthening of the U.K. balance of payments and the growth of industry and employment in the U.K., with particular reference to regional considerations; and
>
> (iii) any proposals which may be made for facilitating participation by public enterprise in the development and exploitation of the resources of the continental shelf. (Committee of Public Accounts, 1973, p. xii)

In practice, only the last criterion significantly affected licensing policy. As a result of increased participation by the Gas Council (a 50 percent interest in a group that included Amoco, Amerada, and Texas Eastern) and of the involvement of the National Coal Board (in a group that included Gulf and Allied Chemical), total British participation rose from 22.7 percent (9.2 percent public sector interest) in the first round to 33.6 percent (15.5 percent public sector interest) in the second round (Committee of Public Accounts, 1973, p. xiii).

THIRD LICENSING ROUND (1969–1970)

The overall situation in the North Sea had changed quite dramatically by the time of the third licensing round in 1969. The discovery of the West Sole, Leman Bank, Indefatigable, Hewett and Viking gas fields between 1965 and 1968 had demonstrated the significance of the southern North Sea as a major gas-producing province. In these changed circumstances, government officials (assisted by industry professionals) undertook a major review of licensing policy. Crucial to this review appear to have been the assessments (1) that no more large gas fields were in prospect in the Southern Basin, and (2) that no assessment of the potential of the Northern Basin could be made in view of the lack of exploratory data (Committee of Public Accounts, 1973, pp. xiv–xv). In these circumstances it was concluded "that the prospect of further big finds could not be relied on to attract licenses, and that it was necessary to retain their interest in developing smaller, higher cost, gas fields. Since the basic policy was still to encourage the rapid and thorough exploration and development of the petroleum resources of the United Kingdom shelf, Ministers decided to continue with the discretionary system of allocating licenses in the third round of licensing" (ibid.).

As a result, despite substantial finds of gas, similar criteria and guidelines for selecting licensees were employed during the third licensing round as during the preceding two rounds. "Plus ça change, plus c'est la même chose." Even the financial terms remained intact with only minor adjustments in annual rental payments "to reflect the changing value of money" (Committee of Public Accounts, 1973, p. xv). A greater emphasis on participation by nationalized industries is discernible, particularly in the Irish Sea, where it was decided that licenses would be granted only to applicants who provided for participation by the Gas Council or the National Coal Board. Dam interprets this as a conciliatory gesture to the left wing of the Labour Party and to the growing sentiment on the Labour Party National Executive Committee for some form of nationalized hydrocarbons authority. "Viewed against such a background, the new criteria, and particularly the decision to allocate a license to a newly created Hydrocarbons subsidiary of the Gas Council, appeared to be a partial concession to views held by important segments of the Labour Party" (Dam, 1976, p. 31).

One influential factor in maintaining that status quo may well have been the government's expectation that offshore discoveries would continue to be in the form of natural gas (Committee of Public Accounts, 1973, p. xxvi). In this context the position of the Gas Council as a monopoly buyer was crucial to government offshore leasing policy, since it could be argued that any possible loss in economic rent arising from unduly favorable concessionary terms was more than offset by the advantage of abundant, low-priced supplies of natural gas. Indeed, gas prices and what constituted a "reasonable price" for offshore gas had become a controversial negotiating point between the government and the oil companies. Thus, the government argued that oil companies were entitled to development costs plus a reasonable return on capital invested in the field in question. "The heart of the case for a cost-based price was that the oil companies would make unconscionably large profits if they were permitted to sell the gas at any higher price" (Dam, 1976, p. 75). Oil companies, understandably, demanded market prices, arguing that anything less would reduce the incentive for further exploration.

While these various considerations all influenced the government's decision to continue with existing licensing arrangements and concessionary terms, it is perhaps worth commenting on the government's extremely cautious assessment of the potential of both the Northern and Southern Basins of the North Sea. With the advantage of hindsight it is easy to be critical, yet it must be noted that the first significant strike in the Northern Basin, the Cod gas condensate field, had been made in May 1968, nearly eighteen months before the announcement of the third licensing round. Ekofisk, the major discovery in the Norwegian Sector, was announced in November 1969. It was clear that the U.K. sector held equal promise from the Montrose discovery, made in one of the most northerly blocks allocated in the first round and announced in September 1969. It must be acknowledged that announcement of the Montrose and Ekofisk strikes followed both the selection of blocks and the announcement of criteria for the third licensing round, thereby somewhat constraining the government's options.[4] Nevertheless, there clearly was considerable exploratory activity and growing interest in the Northern Basin prior to the third licensing round; the significance of these factors appears to have been underestimated in the government's review of leasing policy.

FOURTH LICENSING ROUND (1971–1972)

The U.K. government has indeed been unfortunate in the timing of its licensing rounds. In the same way that the West Sole gas strike in October 1965 came shortly after the announcement (August 1965) of the terms for the second licensing round, so the announcement of the third round was immediately followed by the Montrose and Ekofisk strikes—indeed, both these discoveries were made before the actual granting of third round licenses in June 1970. In November 1970, the Forties discovery confirmed the potential of the U.K. sector. The discovery of oil in commercial quantities again significantly altered the circumstances of North Sea development, particularly in the fact that the onshore flow of oil would be handled by private companies rather than by a nationalized industry.

Despite these changed circumstances, the Department of Trade and Industry (successor to the Ministry of Power) decided to press ahead with a further round of production licensing in 1971–1972. In subsequent testimony to the Committee of Public Accounts, the department argued that a further round was necessary in view of (1) a marked decrease in exploratory drilling activity in the U.K.

sector and (2) an increase in the price of OPEC oil following the success of the Libyan government in September 1970 in raising its take from oil company concessionnaires (Committee of Public Accounts, 1973, p. xv). Against this background there was a continued need "to press ahead quickly with a further major licensing round, to open up further territory for exploration and to attract to the U.K. shelf more of the technical and financial resources—which might otherwise go elsewhere—of existing operators as well as those of companies not so far represented" (ibid., p. xvi). To maintain interest, the department placed on offer a total of 436 blocks. This represented a significant increase over the area offered in the third round, which the department argued had proved insufficient to satisfy the demands of applicants, a disturbing and seemingly illogical argument in view of the simultaneous claim that exploratory activity was declining and that there was a need to stimulate interest.

Equally surprising perhaps was the department's decision to continue to give priority to the discretionary method of allocating production licenses, arguing that such an approach (1) had served well in opening up the U.K. continental shelf, (2) ensured the involvement of a large number of companies, (3) enabled British-owned companies to be awarded a substantial interest, and (4) allowed flexibility in selecting the most appropriate work-program. Regardless of the merits of these arguments, one might reasonably have anticipated a review of licensing policy following the discovery of oil in commercial quantities in the U.K. sector. The Committee of Public Accounts was clearly taken aback by the admission that, in marked contrast to the first three rounds of licensing, no such interdepartmental review was undertaken prior to the crucial fourth round (Committee of Public Accounts, 1973, p. xxvi).

Despite its continued preference for discretionary licensing, the Department of Trade and Industry did agree to experiment with an auction of fifteen selected blocks. Applicants were invited to submit a straightforward cash bid for these blocks, the only requirement being that applicants should possess the necessary technical as well as financial resources to undertake development work. The sums bid for these blocks far exceeded the department's expectations and raised a storm of controversy. Tenders totaling £135 million were submitted for the fifteen blocks from thirty applicants, the high bids for the blocks totaling £37 million (Committee of Public Accounts, 1973, p. xviii). Although one block received a high bid of nearly £21 million, the average amount bid on each of the remaining blocks still exceeded £1 million each.

In order to place these figures in perspective, the cash bonuses for the auctioned blocks must be compared with initial rental payments of only £2.69 million on the 267 blocks allocated by the discretionary method (Table 13). In defense of the Department of Trade and Industry, MacKay and Mackay (1975, pp. 29–30) note that these higher bids could not possibly have been sustained over the entire area offered for licensing and that the sums involved were of minor significance compared with the revenues that will ultimately be derived by the government through taxation, royalty, and participation agreements. Nevertheless, it seems reasonable to assume that the areas allocated by discretionary licensing in the fourth round included at least some quality blocks comparable to those auctioned. Moreover, the government proceeded with its discretionary allocation despite its knowledge of the very high bids that were being submitted in the auction experiment.

These considerations prompted the following interesting exchange between members of the Committee of Public Accounts and Sir Robert Marshall, Industrial Secretary at the Department of Trade and Industry (DTI):

> MR. HAROLD LEVER (Chairman): So you allocated most of these 267 blocks after you knew that 15 blocks had fetched £37 million?
>
> SIR ROBERT MARSHALL: They were offered at the same time.
>
> LEVER: You tell me you received the tenders by August 1971?
>
> MARSHALL: Yes.
>
> LEVER: So by that time you were aware that you might get £21 million for one block and £6 million for another block and perhaps a few thousand for another?
>
> MARSHALL: Yes.
>
> LEVER: That was then known to the Department?
>
> MARSHALL: Yes.
>
> LEVER: After you had received that information you still went on with the allocation of 267 blocks to 210 companies for a total of £3 million?
>
> MARSHALL: Yes.
>
> LEVER: How do you explain that?

MARSHALL: Because they had been offered at the same time as the auction in the summer of 1971. The applications had been invited.

MR. PAGET (Committee Member): But they were not allocated until the winter. Why did you not say to them that you were not taking any of these tenders because of what happened at the auction, and they would have to think again if they wanted to get them?

LEVER: May I put another question to you: are you telling us that in allocating one of these 267 blocks, the Department allocated for £2,000, knowing that that very block might be one for which someone would be prepared to pay £20 million or £6 million, or whatever a block would fetch?

MARSHALL: Yes. We allowed the applications to go ahead.

LEVER: Notwithstanding the fact that you were under no kind of contractual obligation to do so?

MARSHALL: I think that is correct. (Committee of Public Accounts, 1973, p. 15)

As the committee concluded, "We are surprised that, when the results of the tender competition . . . were known on 20th August 1971, the questions of reconsidering or withdrawing the invitation for the discretionary allocation were not discussed interdepartmentally or put to Ministers for a policy decision" (Committee of Public Accounts, 1973, p. xxxiii).

As is apparent in the preceding discussion, both the discretionary method of allocating licenses and the financial terms of those licenses came under fire as the pace of discovery quickened. The very success of the auction experiment served to underscore the fact that the bulk of the production licenses (covering by far the most promising areas of the North Sea) had been allocated on terms that had to all intents and purposes been fixed in 1964, long before any discoveries had been made and when the potential of the offshore area was unknown. Whether the nation was receiving a fair and adequate return from its offshore resources became a matter of public debate. Moreover, the Committee of Public Accounts, appointed in large measure in response to public expressions of concern, revealed to the public for perhaps the first time the profitability of North Sea operations.[5] In these circumstances, significant changes in licensing policy and concessionary terms were almost inevitable.

Discretionary Licensing Reexamined

A major argument advanced in support of discretionary licensing is that it promotes rapid exploration by allowing the relevant authority to favor those companies willing to pursue an intensive search program. Such an argument would understandably have appealed to the U.K. government. Equally compelling, however, was the fear that competitive licensing might handicap the interests of British companies, a fear that MacKay and Mackay (1975, p. 29) suggest seems to have been borne out by the relatively low share of auctioned blocks obtained by British interests in the fourth licensing round. Thus from the U.K. government perspective, discretionary licensing offered a satisfactory way of balancing two major political considerations—the need for rapid exploitation and the involvement of British companies.

It is important to note that further arguments against an auction system, more usually advanced by those companies actually involved in exploration, are that such an approach results in a highly inefficient investment of capital and that the payment of large cash bonuses places an intolerable strain on the working capital of companies, thereby curtailing drilling activity and the pace of exploitation. In this respect one may discern

> a certain congruity of interest between the government and the oil companies in rejecting an auction system. To the oil companies an auction system threatened to increase costs and thereby to reduce prospective returns. The government sought low-cost, secure, and foreign-exchange-saving energy supplies as soon as possible and in effect paid a subsidy for more rapid development of those resources. This subsidy was not only undisclosed and hence shielded from criticism but also had the politically attractive quality of not coming out of tax revenues but rather of being merely a reduction of monies that would otherwise have flowed from the oil companies to the Treasury. (Dam, 1976, p. 36)

Criticism of U.K. offshore policy between 1964 and 1972 has focused not only on the method but also on the scale and terms of production licensing. Thus, the large number of blocks placed on offer in the first and second rounds has been condemned in that "it made it impossible for the government to modify the terms in this area to take account of changed circumstances arising from the discovery of several major gas fields without resorting to the morally dubious practice of retrospective legislation" (MacKay and Mackay, 1976, p. 88).

Such criticism has the advantage of hindsight. Given the unknown potential of the North Sea in the early 1960s, it must be recognized that the blanket allocation of production licenses did achieve the primary objective of stimulating interest and activity. Whether an auction system of allocating blocks would have produced a similar result is, of course, very much a matter for debate, although the high level of interest shown even in the first licensing round in the areas closest to the Dutch coastline is significant. The offer of large numbers of blocks in the fourth round of licensing (which included blocks in what was to prove to be the prolific East Shetland Basin) is more difficult to understand. The reasons behind this offer have already been noted, particularly the fear that unless additional areas were made available, drilling rigs would be moved out of the North Sea. In retrospect, this fear was undoubtedly exaggerated; Sir Robert Marshall, in testimony to the Committee of Public Accounts, acknowledged that fewer blocks could have been offered for licensing without any risk of slowing the momentum of exploration (Committee of Public Accounts, 1973, p. 153).

Others have criticized discretionary licensing on grounds of the inefficiencies that arise from "concealed discrimination." MacKay and Mackay suggest that "there are strong a priori grounds for preferring a system of allocation which places more weight on the market solution and less on administrative discretion" (1975, p. 29). They also challenge the validity of the premise underlying discretionary licensing, arguing that

> there is no reason to believe, as the oil companies claim, that a system of auctioning licenses would prohibit rapid exploration because it would significantly reduce their working capital. After all, if the oil companies were concerned about this issue it would surely be reflected in the lower prices they would be prepared to bid . . . Indeed, auctioning might lead to faster exploitation. After all, having invested capital in buying a license the rational licensee will wish to obtain a return on that capital as quickly as possible. There is no reason to suppose that this elementary point has escaped the oil companies. (Ibid., p. 30)

Dam's analysis of the results of the fourth licensing round (1976, pp. 38–41) represents a convincing rebuttal of the argument that discretionary licensing (in conjunction with a work program requirement) induces a higher level of exploratory activity. Indeed, as

Table 14. Details of Licenses Granted, 1964–1972

	British Participation (as % of licensed territory) as of 1/1/73			*Minimum Work Programs*	
	Total British Interests	*Public Sector Interests*	*Total Initial Payment (£ million)*	*No. of Wells*	*Approx. Value (£ million)*[a]
First round	22.7	9.2	2.02	98	80
Second round	33.6	15.5	0.66	44	30
Third round	36.5	20.0	0.63	48	34
Fourth round					
Discretionary	34.7	9.6	2.69	225	200+
Competitive bidding	20.0	10.0	37.37[b]	—	—

[a]Includes the value of seismic and other geophysical work.
[b]Includes £37.21 million in premium payments.
Source: House of Commons, 1973, pp. 45, 89.

Tables 13 and 14 demonstrate, discretionary licensing in the United Kingdom has in effect provided a subsidy for exploration:

> The cost of the subsidy is represented by the payments foregone—here £37 million for fifteen blocks. How much exploration activity could have been purchased for £37 million? The answer is presumably the amount that would have been expended on those fifteen blocks under a discretionary system. Using averages from the previous three rounds, the amount of exploration would have been less than £12 million. Even taking the fourth round discretionary licenses issued the following year, the amount would have been less than £15 million. In other words, it costs at least twice as much as the companies will actually spend to induce exploratory expenditure.
>
> In this light, the discretionary system turns out to be a most expensive subsidy. Of course, it is a subsidy that . . . also subsidized participation in the North Sea by British petroleum interests. And the ministry may have had a point in arguing that one cannot extrapolate from what was bid on fifteen blocks to what would be bid on hundreds of blocks. But one must still recognize that the auction experiment raises a serious question

> about the efficacy of the discretionary method in accomplishing its avowed purposes. (Dam, 1976, p. 39)

Despite such criticism, it seemed clear from testimony before the Committee of Public Accounts that there was little official enthusiasm for any repetition of the auction experiment. In the final analysis, the decisive factors appeared to be the opportunity offered by discretionary licensing for (1) protecting British interests and (2) bringing informal pressure to bear on oil companies (for example, in terms of concessions in the siting of onshore plants or drilling in unexplored areas) through the threat of future sanctions (Dam, 1976, p. 43).

Growing reservations outside official circles regarding licensing policy were reinforced by the Committee of Public Accounts revelations about the profitability of North Sea oil operations. Thus a large oil field such as Forties was likely to show a rate of return of well over 30 percent on invested capital even after payment of rent, royalties, interest charges, and taxes. Moreover, under the existing tax regime, oil companies were able to shelter their North Sea profits from U.K. corporation taxes by offsetting them against losses from overseas operations: "Because of the operation of double taxation relief, liability to U.K. Corporation Tax was extinguished by credits for tax paid elsewhere. Moreover, the size of this relief has led to the accumulation of huge tax losses of the order of £1,500 million for nine major companies which could be carried forward indefinitely to be set against future taxable profits. And tax losses were continuing to accumulate" (Committee of Public Accounts, 1973, p. xxi). Testimony to the committee from the deputy chairman of the Board of Inland Revenue indicated that U.K. tax receipts from oil companies amounted "in recent years [to] about £½ million" and that these were "really quite special" (Committee of Public Accounts, 1973, p. 102).

Leaving aside those licensing rounds which preceded the discovery of oil and gas, any critique of British offshore policy during the third and fourth licensing rounds cannot avoid reexamining the basic considerations underlying that policy. Many of the deficiencies of British licensing methods and terms arise from (or perhaps were required by) the perceived need for rapid exploration. Andrew Graham poses the crucial questions:

> Why exactly? What are the benefits? Cheap oil? Well, no because as we have said, it will be sold at the world market price . . .
>
> What of the prospects for UK suppliers of equipment—will

> rapid exploration help them? In fact, in the short run, as a direct result of the policy of rapid development, UK firms have been unable to meet specifications and have received under 30 per cent of all the orders placed. But then, of course, there are the exciting long-term prospects (says the DTI). But how on earth are British firms to get in on this long-term market and learn about the technology if they don't get the orders in the first place? (1973, pp. 724–725)

Significant foreign-exchange savings could not occur in a situation where (1) North Sea oil was being developed by foreign companies, (2) development costs (services, supplies) were being incurred in foreign currency, and (3) all profits were being repatriated. "Yet this nightmare is exactly what the policy of rapid development has made likely" (Graham, 1973, p. 724). Unfortunately, the force of Graham's argument was to be diminished by the quadrupling of oil prices and the subsequent embargo, events that Graham could not have foreseen but which he dismissed as "hard to imagine." It is equally evident, however, that any chance of a slower rate of exploration and development in the U.K. sector of the North Sea was simply out of the question in the post-1973 oil environment and the renewed emphasis on "energy-independence" and balance-of-payments savings.

Finally, the continued emphasis on the need to stimulate interest in North Sea exploration and development (up to and including the fourth licensing round) cannot be divorced from the government's very conservative appraisal of the potential of the North Sea. Throughout this period, Lord Balogh and others repeatedly drew attention to the government's low estimate of reserves and future production levels (at least when compared with industry and independent estimates) and its refusal to amend estimates "in spite of the fact that they were successively contradicted by events" (Balogh, 1972, p. 3).[6] In criticizing the government's offshore policy, Balogh even went as far as suggesting that "the attitude which explains this approach cannot be understood unless we assume that the Government was misled into thinking that the quantity of oil and gas likely to be found in the North Sea was of minor importance. Only in that case would it have been reasonable to be so cautious—small gains in the North Sea might have been more than offset by the adverse treatment of British companies elsewhere by way of government's discriminating against them in granting licenses or by extracting a greater share of profits" (ibid., p. 2).

How were such assumptions, considerations, and decisions determined? Balogh refers to "governmental incompetence and company excellence" (1972, p. 10). Graham raises similar issues. "Why did the civil servants treat the oil companies so lightly? Why was policy not reviewed interdepartmentally before the fourth round of licenses in 1971? Was information from companies about costs really not thought necessary, or was it thought to be counterproductive to obtain it? And why wasn't the tax situation examined more closely?" (1973, p. 725).

Graham's conclusions are particularly interesting in view of his participatory role as economic advisor to the Cabinet Office during the Harold Wilson administration:

> We probably need to look for answers somewhere in the twilight zone of the relationships between industry and department, and one department and another. The DTI not only regulates, but is the sponsoring department for the oil industry—consultations take place almost daily, and the department is heavily dependent on the industry for information. But how close should such a relationship be, and how far can it be combined with impartial advice to ministers? In addition, what steps are taken to obtain independent outside advice—particularly when negotiating with the industry? The DTI witnesses [before the Committee of Public Accounts] stressed that neither they nor the companies were intentionally underestimating the magnitude of the finds in the North Sea, but did they examine successive forecasts from other oil- and gas-producing areas which show that early estimates are typically revised upwards by as much as four and five times? Their actions hardly suggest so. (Graham, 1973, p. 725)

The British National Oil Corporation

The Committee of Public Accounts report indicated broad, bipartisan support for changes in U.K. offshore policy. In view of the committee's revelations it was perhaps not surprising that the primary concern was to ensure that the nation received its share of the profits of North Sea oil and gas development. More specifically "attention turned from methods of allocating licenses to methods for recouping the economic rent that had already passed to the companies under existing licenses" (Dam, 1976, p. 43). The nature of such methods was to be strongly influenced by the return of a Labour government in the February 1974 general election.

In July 1974 the Labour government issued a White Paper entitled "U.K. Offshore Oil and Gas Policy" that called for government "participation" in North Sea development through the establishment of a British National Oil Corporation (BNOC). In discussing the form of participation, the White Paper made at least a formal distinction between past and future licensees. In the case of the former, the White Paper proposed "negotiating" participatory arrangements through BNOC. Future licensees posed less of a problem. The White Paper stated that a condition of all future licenses would be that "the licensees shall, if the Government so require, grant majority participation to the state in all fields discovered under those licenses." Such a distinction was perhaps politic, yet as Dam points out, "since there was no intention to offer new licenses before 1976 at the earliest and since in fact all of the measures specifically proposed would be applicable to existing as well as future licenses, the White Paper must be viewed as an exercise in changing the terms surrounding existing licenses. The only instrument that would be different for future licenses than for existing licenses would be government participation, and there the difference lay only in how it would be imposed: participation would be negotiated with existing licenses but imposed as a condition on future licenses" (1976, p. 105).

The other major proposals in the White Paper focused on reform of the existing tax regime and were intended to redress the balance between oil company profits and government revenues (Dam, 1976, pp. 108–111). Particularly, the White Paper sought to eliminate what the government regarded as tax loopholes, notably the sheltering of North Sea profits against losses and/or capital investment elsewhere. The change was to be achieved (1) by separating the North Sea operations of a company from its other activities; (2) by refusing to allow accumulated losses incurred elsewhere to be carried forward against future North Sea profits. The latter change was justified on the grounds that such losses did not represent any real commercial loss (on North Sea development) and that there was therefore "no equitable reason why they should be used to eliminate future tax liabilities" (House of Commons, 1974a, p. 14). Despite the high costs of exploration and development it was already clear that the first generation of North Sea fields would be extremely profitable. In addition therefore to eliminating various tax loopholes, the White Paper proposed a new barrelage tax on oil production as a way of ensuring that the nation as a whole would receive the "lion's share" of the profits of North Sea oil development.

The White Paper's various recommendations formed the basis for two major pieces of legislation—the Oil Taxation Act and the Petroleum and Submarine Pipe-lines Act—passed by Parliament in 1975. The Oil Taxation Act sought to reform the tax regime (1) by disallowing credit for losses and capital allowances from non–North Sea operations against North Sea profits, and (2) by introducing a new petroleum revenue tax (PRT).[7] This was to be a flat-rate tax assessed on the revenue from individual fields in order to prevent averaging. As enacted in 1975, PRT was payable at a 45 percent rate, liability being assessed after the deduction of operating costs, royalty payments, and a "free allowance" of ten million tons, from gross revenue. Moreover offshore producers did not become liable for PRT until they had recouped 175 percent of their capital expenditures on a field (Dam, 1976, pp. 124–130). As calculated by the Treasury, the revised tax structure was intended to ensure that around 70 percent of North Sea oil profits accrued directly to the nation (or, perhaps more accurately, flowed into the government's coffers).

The White Paper's recommendations with respect to the creation of a British National Oil Corporation were carried into effect in the Petroleum and Submarine Pipe-lines Act. As defined in this act and as elaborated in its subsequent actions, BNOC's role has embraced (1) majority participation in all production licenses; (2) direct involvement in exploration and development through its equity interest in offshore fields and as a licensed operator; and (3) serving as advisor to the Department of Energy on questions of oil policy. In view of the findings of the Committee of Public Accounts, the changes in the tax system appeared reasonable, even generous, and most controversy surrounded the ambiguous role of BNOC as an operator (and therefore a direct competitor to the private companies) and an adviser to the government: "Not unnaturally, offshore operators did not welcome what they saw as the BNOC cuckoo in their North Sea nest. The specter loomed large of the young state company using its political muscle to oust companies that had taken exploration risks and pioneered development of the new breed of platforms needed to contend with North Sea conditions" (Vielvoye, 1979, p. 48).

BNOC AND PARTICIPATION

The Petroleum and Submarine Pipe-lines Act provided the government with the necessary authority to regulate North Sea production rates. Under the act all producers had to submit development programs to the Secretary of State for Energy, who could reject or modify any program judged not in accordance with "good oilfield practise" or contrary to "the national interest." The scope of these powers, and particularly what was perceived as a lack of protection against arbitrary ministerial decisions, created considerable anxiety among offshore companies (Robinson and Morgan, 1978, pp. 29–34). Of more immediate concern, however, were the terms on which the government sought to "participate" with companies holding pre–fifth round production licenses in the development and operation of proven or potentially commercial fields. And in the course of negotiating agreements with these companies, the form and substance of state participation (at least as had been envisioned in the 1974 White Paper) were significantly altered.

At first the government was able to "coerce" a few licensees who were in financial difficulties or seeking work capital by underwriting development costs in return for majority shareholding for the state. In such circumstances "participation began to look more like a technique for the government to provide capital and thereby assure the commercial success of a venture than to use participation as a device for recapturing economic rent" (Dam, 1976, p. 117). This impression was reinforced by subsequent negotiations between the government and the major oil companies. Clearly "opportunity tactics" of the type described above were unlikely to succeed with the majors—indeed, as Dam observed, "with the escalating British inflation and the weakness of the pound during 1974–75 at least one of them—Esso—could borrow in world capital markets on more attractive terms than could the British government" (1976, p. 118). A breakthrough in the negotiations appears to have come early in 1975 when the head of the government's negotiating team pledged that companies would not be either better off or worse off financially as a result of participation. Ambiguous as this pledge appeared to be, it is difficult to reconcile with the participatory strategy outlined in the 1974 White Paper—or indeed with the same White Paper's angry condemnation of "enormous and uncovenanted profits" to be reaped by oil companies from their North Sea activities.

Indeed, as the negotiations proceeded, the government appears to have backed away from any real form of participation in favor of an oil purchase agreement. As described by the Department of Energy, participation agreements "give BNOC the right to take at market price up to 51 percent of each company's share of petroleum from the producing fields, together with full membership, with an independent vote, of the operating committees which manage fields, pipelines and terminals" (1979, p. 22).

Dam, writing before negotiations were completed, suggested that the government's role would likely be confined to that of a "sleeping partner" whereby "the government would essentially be acting as a banker, lending development capital and receiving a return based on risk—a return that might in high-risk cases include an equity 'kicker'" (1976, p. 121). In actuality the government's retreat on participation extended even to its proposed role in financing field development. The arrangements for financing development of the Ninian field, for example, illustrated the extent of the government's "retreat" (*Economist*, February 7, 1976, p. 75). The initial financial prospectus (issued by the joint consortia in November 1975) indicated that the government would provide 51 percent of the development costs in line with the 1974 White Paper. In February 1976, however, it was announced that the consortia developing the field would raise *all* of the required development capital. As the *Economist* commented with some acerbity: ". . . so Britain will follow neither the sheikhs (once cited favourably by ministers as a precedent) who swiped companies' assets at a knock-down price, and so got an equity interest in the oil, nor Norway (held up by White Papers as a model) which will take a specified proportion of oil for a matching state contribution to costs . . ." (ibid.). In effect, majority participation as negotiated meant that the government acquired a supply of oil, not in return for a contribution to development costs as proposed in the White Paper, but through an option to purchase up to 51 percent of the oil produced at market prices. And even here certain of the agreements provide for a buy-back by the companies of up to 100 percent of the BNOC allotment. Thus the participation agreement negotiated by both Shell and Esso provided for the repurchase of up to 100 percent of BNOC's oil entitlement if output failed to meet the companies' refining requirements.

> It would perhaps not be too cynical to say that, by an implicit promise of favorable future treatment, the British government obtained an oil purchase agreement that they chose to call a participation agreement. In this way, the White Paper promise of participation was carried out in form rather than substance, and the task of attempting to recapture the rent transferred to the companies under the discretionary allocation system was left to the tax system. (Dam, 1976, p. 123)

As noted in the *Economist* (February 7, 1976), "among the tatters of participation, two strands remain, which Whitehall now likes to pretend were really the object of the exercise." One was that "BNOC staff, through membership of the operating committees that manage fields, will gain experience of the industry and later influence decisions in the North Sea"; the second that BNOC "will have direct access to oil." In part, the growing emphasis on "direct access to oil" (rather than recapturing economic rent) as the major objective of the participation exercise reflected a new priority, namely, the government's desire to ensure that the bulk of North Sea oil production would be directed toward U.K. rather than European refineries. From this perspective it was argued that BNOC's option to buy 51 percent of the oil produced in the U.K. sector would (1) avoid the need, in the event of a disruption in world oil supplies, for emergency legislation that might prove to be offensive to the European community, and (2) enable the government to pressure oil companies into refining two-thirds of their North Sea crude in the United Kingdom. In practice the government's rationale appeared less than persuasive in view of surplus refining capacity and the depressed state of product prices in Europe.

> The two-thirds goal is less a policy than the guideline to what would be a sensible policy if the market were the way the government would like it to be . . . The government is not really expecting two thirds of North Sea crude to be refined in Britain in the early 1980s any more than it expects those who are granted long-delayed planning permission for new refineries necessarily to put them up. Of course, as with the petroleum revenue tax or participation, the government is not advertising its retreat. But retreating it is. (*Economist*, April 3, 1976, p. 98)

Although the negotiation of participation agreements with those companies holding pre–fifth round production licenses took nearly four years, opposition to the government's policy gradually became less strident.[8] Undoubtedly offshore operators recognized that in the absence of a participation agreement they would be at a serious disadvantage in any future licensing round. Indeed, one of the criteria for the fifth round of licensing announced in 1976 was "the record of the companies in the voluntary compliance negotiations" (*Offshore*, vol 36, no. 10, 1976, p. 126). Any thought that the government was less than serious in its intentions was dispelled by the exclusion of Amoco, one of the few larger companies that had failed to reach a participation agreement, from all fifth round awards. Although the government was inevitably accused of attempting to "blackmail" companies into accepting BNOC as a partner, it must also have become apparent to the companies that the diluted terms of participation meant little change in the status quo. The implications of the buy-back provision negotiated by Shell-Esso were not lost on the oil industry. An article in *Offshore* observed that the tough negotiating battle fought by Shell and Esso had resulted in a "hollow" success for the government.[9] The agreement whereby Shell and Esso could repurchase up to 100 percent of BNOC's oil entitlement provided "a sort of 'double take' buy that, excluding all the official jargon, means that the status quo is maintained. Nothing has been altered, except on paper, and Shell/Esso still have control over their North Sea oil" (B. Smith, 1977, p. 61).

This view of course was not accepted by the government, which argued that its objectives of acquiring first-hand experience of North Sea operations and securing access to an assured supply of oil had been achieved. As a result of the participation policy, Dr. J. Dickson Mabon, Undersecretary for Energy, stated, "BNOC is now firmly established where it should be at the heart of Britain's offshore industry. That could not have been the case had we not given the corporation a direct involvement in commercial fields which have been and will be developed under pre–Fifth Round licenses . . . [The agreements] give BNOC invaluable experience in oil trading and will give the country a direct influence over the disposal of North Sea crude, a crucially important consideration if we are to make the most of our oil wealth" (*Offshore*, vol. 38, no. 11, 1978, p. 182). And while this may not have been the original objective of participation as envisioned in the 1974 White Paper, there can be no disagreement that the option to purchase 51 percent of the oil produced from fields in the U.K. sector of the North Sea will allow BNOC to play an increasingly significant role in the world market for light, low-sulfur crude. According to one estimate, when existing fields reach peak production sometime in the early 1980s, BNOC will be able to supply in excess of one million barrels per day, the bulk of which will be crude sold to BNOC under the participation agreements.[10]

BNOC'S OPERATING ROLE

Even as offshore operators came to terms with participation, new arguments arose over BNOC's role as a North Sea operator. The direct involvement of the national corporation in offshore exploration and development is clearly anticipated in the Petroleum and Submarine Pipe-lines Act. Such involvement was regarded as essential if the government was to acquire the necessary financial and technical expertise with which to protect the national interest. Yet it may well be that in the evolution of the government's offshore strategy, direct involvement (through BNOC) took on added importance and significance in light of the diluted version of participation as finally negotiated by the companies.

As a result of its acquisition of the offshore oil and gas interests of the National Coal Board (NCB), BNOC became an immediate participant in North Sea operations.[11] Through its NCB interests the corporation obtained a direct stake in the Viking gas field and in four oil fields then in production or under development—Thistle, Dunlin, and the U.K. portions of Statfjord and Murchison. Subsequently the corporation purchased most of the North Sea interests of the financially troubled Burmah Oil Company, including its holding in the Ninian field and the bulk of its stake in the Thistle field, and also took over responsibility from Burmah for developing and operating the Thistle field. Significantly, the state oil company recruited Burmah's engineering and production groups, thereby acquiring the necessary expertise to sustain an expanded exploration and development role in the North Sea. Early in 1978 the government announced that licensees seeking to dispose of all or part of their holdings must give BNOC first right of refusal, thereby providing the opportunity for BNOC to further add to its equity portfolio. In this way BNOC acquired a portion of Ashland's share in the Thistle field (raising its equity holding in that field to 18.93 percent) and later purchased a 10 percent stake in the Mesa group's small, nearshore Beatrice field. By mid-1979, BNOC had an equity interest in one producing gas field, in three producing oil fields, and in three oil fields under development. The corporation's share in the reserves of these six fields is estimated to be 662 million barrels; potentially the fields at peak production could yield BNOC around 220,000 barrels per day of crude oil (Vielvoye, 1979, p. 47).

Quite apart from these interests are the awards made to BNOC in the fifth and sixth rounds of production licensing held during 1976–1977 and 1978–1979. In these rounds BNOC was the designated operator on twelve of the eighty-six blocks for which licenses were awarded. Moreover, in April 1978, outside normal licensing rounds, the government awarded BNOC exclusive production licenses for a further eleven blocks.

It was on one of the blocks awarded during the fifth round, 30/17b located approximately two hundred miles east of Dundee, that BNOC made its first major strike as an operator. With recoverable reserves of between 17 and 21 million tonnes (120–150 million barrels) this discovery, subsequently named the Clyde field, was viewed by BNOC as a particularly important step in its evolution as a major oil company.[12] However, the circumstances of its discovery gave rise to considerable controversy and to suggestions that the corporation was abusing its position as advisor to the Department of Energy. More specifically, it was alleged that BNOC had acquired operating responsibility on the basis of confidential information supplied to the Department of Energy by the Shell-Esso group. It was claimed that exploration drilling by this group had identified the Fulmar field on an adjacent block *prior* to the call for fifth round applications (although the discovery had not been publicly announced) and that the group's regional surveys indicated similar geological conditions extending into block 30/17b. In an irate letter to the *Times*, David Singleton claimed that BNOC had had access to the exploration data submitted by Shell/Esso in support of its application for the production license. "Any junior technical person could have seen the potential after a review of the Shell/Esso information. All that was left was for BNOC to contract a rig and drill at a proper location, probably already determined by the applicants" (Letter to the *Times*, July 5, 1978, p. 20). The two companies had been informed by the Department of Energy that "they would be awarded the relinquished part of the block but only if the BNOC were operator for the group" (ibid.).[13]

By way of rebuttal BNOC claimed that its own geologists were quite capable of evaluating the block's potential without any help from Shell-Esso and noted that several other applicants in the fifth round had identified this as prime acreage on the basis of their own surveys and interpretation. This explanation did little to reassure those in the offshore industry who chose to see in BNOC's emergence as a North Sea operator a direct threat to their own established position. The ill-concealed fear that BNOC would receive preferential treatment appeared to be confirmed by (1) the government's award of exclusive production licenses outside normal rounds, and (2) the proposed conditions for the sixth round of licensing as announced in August 1978. The latter indicated that BNOC would be the designated operator on six of the forty blocks to be offered, thereby resurrecting the specter of BNOC using the confidential information submitted to the Department of Energy to select the six best prospects for itself. In order to ensure that BNOC's integrity as an advisor was not further compromised, the Department of Energy subsequently agreed to indicate in the formal invitation for applications those blocks on which BNOC would be the licensed operator. This change in announced licensing and policy procedure came in response to representations by the United Kingdom Offshore Operators Association (UKOOA) and amid press speculation that the major companies were planning to boycott the sixth round.

From the government's perspective, the preference shown to BNOC—and indeed the whole strategy of direct involvement in North Sea exploration and development—was intended not to displace the private sector from offshore development but to strengthen the government's own body of offshore expertise. First-hand experience of the real problems, costs, and revenues of oil operations, it was argued, would strengthen the government's hand in setting licensing terms, in "auditing" oil companies, in evaluating the effectiveness and equity of the tax regime, and in formulating a depletion policy. This rationale was clearly expressed by the Secretary of State for Energy, Tony Benn, in an invited article published in *Offshore*, where he noted that the award of exclusive licenses was primarily intended

> to give BNOC the opportunity to increase as rapidly as possible the number of blocks for which it acts as operator and so the speed with which it can gain experience and expertise.
>
> This is vital for the U.K., since the whole purpose of creating BNOC was to create a technically expert and commercially-oriented body, both to hold the public stake in the oil fields and to advise the government on questions of policy. Without such expertise our interests will be at a fatal disadvantage. It is also in the interests of our private-sector licensees that the corporation with which they deal should gain expertise and should be able to make a real contribution—as I know BNOC will—to the development of the fields. (Benn, 1978, p. 81)

One suspects that the inadequacies and deficiencies of the government's "information base," as exposed by the Committee of Public Accounts, had been very much taken to heart. To strengthen government expertise, Benn acknowledged that there would be a "tilt in the balance" and that BNOC's share of the licensed interest would rise.

It might be argued that the government's objectives could equally well have been achieved through its participation agreements which gave BNOC full membership of all operating committees. In this context a major government concern (though one that was rarely openly acknowledged) appears to have been the fear that once it had worked out a *modus vivendi* with the offshore operators on development of the more profitable fields, companies might be less interested in exploring for and developing the smaller, geologically more complex, and financially less attractive marginal fields. In the event of any future withdrawal by major operators, BNOC would have to assume responsibility for such development, a task that would require the sort of engineering, exploration, and development expertise that could be acquired only through direct field (as opposed to boardroom) experience. Thus the importance attached by the government to BNOC's acquiring experience as an operator was related as much to future production concerns as it was to monitoring offshore activities. Whether, in its effort to protect the national interest and safeguard against the consequences of any future decline in interest in the North Sea by private operators, the government was creating the very conditions that would precipitate such a withdrawal, is open to debate.

PRODUCTION AND DEPLETION POLICIES

In the early stages of North Sea exploration and development very little thought appears to have been given to "the very vexed question of whether there would in certain circumstances be advantage in delaying the exploitation of our own resources of oil rather than expediting their development" (Committee of Public Accounts, 1973, p. 304). As already noted, the policy of successive governments was to stimulate exploration and to extract both oil and gas reserves as quickly as possible.[14] Even prior to the dramatic price increases imposed by OPEC producers during 1973–1974, the realization that North Sea reserves might well be sufficient to eliminate net imports of crude oil (and perhaps even provide a surplus for export) had already led to a questioning of government policies and priorities. The actions of OPEC served to reinforce the view that there might be significant national advantages to be gained from deferring production over a longer period of time. "Just as the previous popular view had been that oil prices would decline into the indefinite future, the idea now took hold that oil prices would rise for ever—or rather, until the exhaustion of resources in the not far distant future. Whether or not such a turnaround in expectations was justified, in these conditions there must have appeared to be some case for making one's resource savings in the more distant (higher-price) future" (Robinson and Morgan, 1978, pp. 24–25).

Under the 1964 Continental Shelf Act the government already enjoyed substantial discretion with respect to controlling North Sea oil depletion rates through its ability to determine the extent and pace of production licensing. However the physical uncertainties that surround exploration and the length of time needed to bring discoveries on stream render the periodic process of issuing licenses a rather imprecise tool with which to implement a depletion policy. The Petroleum and Submarine Pipe-lines Act of 1975 therefore significantly broadened the government's powers by providing for controls over field development and limitations on production. Since any depletion policy designed to stretch out indigenous supplies by deferring production directly impacts the profitability of a field, and thereby the basis for all cash flow and investment calculations, exactly how the government proposed to exercise its powers was a matter of concern to all companies involved in the North Sea.

An indication that the government intended to make use of

these new powers came early in 1978 when the Department of Energy announced that companies seeking approval for field development programs would henceforth receive only "phased permission" rather than the lifetime approval granted to the first generation of fields.[15] In the case of Fulmar, the first major development affected by the government's new policy, approval of the development program extends only to the end of 1985, at which time production plans will be reviewed in the light of operating experience and national needs. As noted by Tony Benn, "[A] major preoccupation has been developing procedures for our exercise of the new powers which we took in 1975 to control the physical development of oil and gas fields. The U.K. government cannot and will not abdicate its responsibility for seeing that development is carried out in a way which will secure the maximum recovery of an irreplaceable resource" (1978, p. 81).

Quite apart from "developing procedures," as the United Kingdom has moved closer to self-sufficiency, the government has begun to use its depletion powers to intervene more directly in production decisions. The terms of production licensing, for example, require the Secretary of State's approval before any associated gas can be flared or reinjected. The aim is to minimize wastage and encourage operators to arrange for transporting the gas ashore or install reinjection equipment. Until recently the government has more or less routinely consented to some gas flaring in order to allow fields to be brought quickly into production.[16] In 1977, however, the government halted production at Brent "in the interests of maximizing total hydrocarbon recovery," noting that the field lacked gas reinjection facilities and that the pipeline to the St. Fergus terminal was not scheduled for completion until 1979 (Department of Energy, 1978, p. 6). As noted in the *Economist*, "This is an important change in British policy towards the North Sea . . . Until now, with balance of payments worries uppermost, the government has been concerned to get as much oil ashore as possible. So Mr. Tony Benn, the Energy Secretary, agreed to the flaring of gas as a necessary evil. But now, with seven fields on stream and North Sea oil production equivalent to half Britain's oil needs, Mr. Benn feels he can re-emphasise his policy of conserving as much gas as possible for future use . . . For the government, the decision marks a new self-confidence about its ability to endure minor interruptions to oil production . . ." (July 2, 1977, p. 85).

Broadly speaking a depletion policy is concerned with the rate at which oil reserves should be exploited. In the debate over North Sea depletion policies the central issue has been whether or not the government should intervene once national self-sufficiency has been achieved in an effort to "flatten" the production peak and thereby extend the life span of indigenous oil supplies. "What happens when Britain hits its own self-sufficiency mark, which it is now doing occasionally. Or more importantly, what happens when the North Sea touches peak output, believed now to be 1982–1986. Will all the fields be forced to reduce their flow while Britain conserves its supply, or can the companies expect to continue producing its oil unimpeded?" (*Offshore*, vol. 39, no. 12, 1979, p. 41).

Robinson and Morgan (1978, pp. 18–49) argue against any government intervention to regulate production rates. In their judgment it is unrealistic to expect public employees or politicians to be any more benevolent, any more farsighted, or any less self-interested than any other group. There is therefore "no more reason to believe that the results of their actions will necessarily benefit society than there is for maintaining that the market will always achieve the social optimum" (Robinson and Morgan, 1978, p. 43). They conclude that the exercise of government regulatory powers is unlikely to improve on the market outcome. Undoubtedly there are formidable difficulties confronting any government that attempts to formulate and implement a depletion policy for the North Sea over the next fifteen to twenty years. "There is uncertainty about the actual extent of the oil reserves, about the course of oil prices, about the macro-economic framework within which the depletion policy will need to evolve, and about how energy demand will develop" (Department of Energy, 1978, p. 55). Thus while it is comparatively easy to identify the factors which taken together should be considered in determining future production rates, evaluating and weighing such factors as the likely increase in the real price of oil, the impact on employment in the U.K. offshore supply and service industry, and the comparative economic benefits of rapid depletion (eliminating the current account deficit or even creating a surplus in the early 1980s when the burden of repaying the United Kingdom's overseas debt will be greatest) as opposed to deferred production (prolonging the benefit to the balance of payments towards the end of the century when oil supplies will be scarcer and more expensive) is much more problematic.

In such a situation, whose judgment should the nation rely upon? Perhaps all that can be demanded of a depletion policy is that it be flexible and capable of subsequent adjustment, and that it be arrived at as openly as possible with adequate safeguards against arbitrary ministerial decisions. But there is always the possibility that the interests of the major companies and the nation will diverge over the long term. As the government's Green Paper on Energy Policy noted: "An operator is likely to prefer deferment only if the expected rate of increase in the real price [of oil] is likely to be higher than the discounting rate he employs. Commercial organizations tend to seek higher rates of return and more rapid pay-off periods than governments, are often subject to pressures on capital and management resources, set a higher value than government on the risks attached to delay and have to budget for reinvestment elsewhere" (Department of Energy, 1978, p. 54). By way of contrast, a depletion policy pursued in the national interest must take into account not only the macro-economic consequences of alternative policies but also the strategic advantages of conserving indigenous supplies and of providing more time in which to adjust to alternative energy supplies. In these circumstances it seems highly desirable that the government should have the power to control depletion. Not, of course, that government policy will necessarily be guided by the long-term national interest: as noted in Chapter 1, short-term political expediency may well encourage a continuing production build-up beyond immediate self-sufficiency levels. Since its election in 1979, for example, the Conservative government has indicated its desire "on strategic and security of supply grounds . . . to prolong high levels of United Kingdom Continental Shelf production to the end of the century" (*Economist*, July 26, 1980). Yet the measures to be employed in implementing such a policy (at least as announced in mid-1980) amounted to little more than a willingness to delay development of one or two small fields. "If the government takes seriously its rhetoric [to prolong the period of self-sufficiency for as long as possible] . . . it will have to consider cutting production directly by (1) using its statutory power to control output . . . or (2) by postponing its entitlement to 12½% of output as a royalty. Either will cost the government money in revenue foregone, which helps explain why the government has placed the cheap policy (delays in development) at the top of the list of options" (*Economist*, July 26, 1980, p. 60).

While companies have continued to express apprehension over future U.K. depletion policies that may interrupt the "natural" growth of production, the extent to which any government can actually restrict output in the immediate future appears to be severely limited by the assurances given to the oil companies in 1974 by the then Energy Secretary, Eric Varley. These assurances were intended to dispel uncertainty about depletion policy and to encourage a continued, rapid build-up toward self-sufficiency. The Varley "guarantees" include: (1) for discoveries made before the end of 1975, there would be no delays on field development and no cutback in production until 1982 or for four years from the start of production, whichever came later; (2) for finds made after 1975, there would be no cutback in production until 150 percent of investment had been recovered; (3) exercise of the powers would be made with full regard for the technical and commercial aspect of the field in question and would not generally involve a cut in production of more than 20 percent; and (4) the needs of the offshore supply industry for a stable and continuing market would be taken into account (Department of Energy, 1978, p. 53).

According to Dam, these reassurances did not entirely dispel the uncertainty created by the government's powers over depletion rates. Moreover, "the reference to self-sufficiency levels implied that the government would probably not encourage production beyond those levels and thus this power was expected by the industry to discourage future exploration" (1976, p. 115). However, as the Department of Energy has noted, "the effect of these assurances is that no delay can be imposed on the development of fields accounting for between a half and two-thirds of our estimated total reserves, and that no cut-backs can be made from such discoveries before 1982 at the earliest—in some cases it may be considerably later, depending on when production starts" (Department of Energy, 1978, p. 53). Thus, to a very considerable degree, the government's ability to limit production rates in the early 1980s has already been significantly constrained.[17]

The Producers' Rebellion

By finally accepting majority public sector participation in offshore licenses, private operators may have been under the impression that they had worked out a satisfactory *modus vivendi* with the Department of Energy and the state-owned oil corporation. However it soon became apparent that the participation agreements per se represented only one element in the government's evolving strategy for the North Sea. As negotiated, participation agreements provided access to oil and to information about the nature and operation of the world oil market. From the government's perspective such agreements could not, by themselves, guarantee that the national interest would be fully taken into account in determining the manner and timing of oil recovery. Hence the use and refinement of other means to implement government policies—acquisition of field operating experience through BNOC; adjustment of the tax regime to ensure that the nation received at least 70 percent of the profits of North Sea oil; and the exercise of licensing and depletion powers to match the rate of exploitation to the needs of national policy.

Increasingly, therefore, private operators saw the major threat to their position in the North Sea as deriving from the direct involvement and advisory powers of BNOC and from the manner in which the government chose to exercise its power to regulate output. Despite occasional public outbursts, such as the vice-chairman of Standard Oil of California's description of BNOC as "a growing albatross that you have chosen to hang about our necks" (from a British T.V. interview with David Frost, April 4, 1978, quoted by Mutch, 1978b, p. 98), criticism by the offshore companies initially tended to be somewhat muted. "Behind the public silence was the belief that if they were seen to openly criticize BNOC and official policy, they might be penalized in the next round of licensing at least. To remind them of the possibility, they had the example of Amoco" (Mutch, 1978b, p. 98).

However the tenor of the editorials in the *Oil and Gas Journal* and *Offshore* reflects the companies' dislike of the changing political climate and their desire to return to the "good old days" of the late 1960s:

> It may be hard to believe, but less than a decade ago the U.K. sector of the North Sea was considered a bargain basement for oil producers. With a teetering economy, the mere prospect of more energy for less money led Britain to offer initial license fees at the modest rate of $480 for up to 10 blocks offshore; a paltry $12 secured each additional block and a square kilometer could be rented for just $108 for the first six years of licensing.
>
> After the first wave of successful discoveries, however, there developed a political climate as perilous as the waters in which the drillers operated. Over the ensuing years, oil companies have been inundated with price wrangles, competitive bidding, currency problems and finally a whole new set of rules as producers are nudged into signing participation agreements with the newly formed British National Oil Corporation. (*Offshore*, vol. 37, no. 12, 1977, p. 55)

As 1978 progressed, however, criticism of the government's North Sea policies became increasingly vocal. This change of mood, and more particularly the hostile reaction to new licensing and taxation proposals, must be set in the broader context of continued uncertainty over the political future of the minority Labour government. Collectively the protests were described by *Offshore* (vol. 38, no. 8, 1978, p. 41) as marking "the opening salvoes of what may come to be called the Producers' Rebellion . . . After years of being forced to take whatever contract terms the host government decided it would offer, the offshore producers now want to be heard."

ADJUSTING THE TAX REGIME

The "producers' rebellion" gathered momentum as it became known that the Treasury was planning to revise the basic rate and allowances of the petroleum revenue tax (PRT). As announced in the House of Commons in August 1978, the changes included (1) an increase in the basic rate at which PRT was payable (from 45 percent to 60 percent on post-royalty payments after operating costs and allowances); (2) a reduction in the percentage of investment costs that could be recovered before incurring PRT (from 175 percent to 135 percent); and (3) a decrease in the amount of oil exempted from PRT payments (from 1 million tons a year to 500,000 tons a year).[18] In introducing the changes, Joel Barnett, the Chief Secretary to the Treasury, noted that "when we fixed the rates and allowances for PRT at the beginning of 1975 we deliberately adopted a cautious approach. We set the rate of PRT no higher than 45% and we gave generous allowances and reliefs because of the great uncertainties at the time. Now . . . we are in a position to take stock and it is apparent that companies are obtaining very large profits from the natural resources of the nation. We believe that the public share of these profits can and should be increased without endangering the exploitation of the less well-placed fields (Parliamentary Report, *Times*, August 3, 1978, p. 7). In making the original PRT calculations, the Treasury had estimated that total government revenues from North Sea oil between 1976 and 1980 would be around £5 billion (at 1976 prices). In practice revenues had fallen far short of expectations (Table 15). In part this shortfall reflected delays in bringing the fields into production, but a more significant factor was the structure of the tax itself. This allowed for recovery of 175 percent of capital costs plus other oil allowances before payment of PRT. With escalating development costs, tax revenues were taking far longer to materialize than originally anticipated. In 1978–1979, for example, the total contribution to government revenues from PRT was only £183 million. In this situation the changes in PRT were intended to provide an additional £150 million in the 1979–1980 financial year, while in the long term (i.e., up to 1985) total government revenues from the North Sea would be increased by £2 billion to approximately £25 billion.

Table 15. U.K. Government Revenues from North Sea Oil (£ million)

Financial Year	*Royalties*	*Petroleum Revenue Tax*	*Corporation Tax*[a]	*Total*[b]
1976–1977	71	—	10	81
1977–1978	228	—	10	238
1978–1979	288	183	50	521
1979–1980	628	1,435[c]	166	2,230

[a]Estimated proportion of corporation tax receipts attributable to North Sea oil and gas.

[b]For comparison and to place North Sea tax revenues in perspective, the 1979–1980 estimated value added tax (VAT) yield was £8.2 billion and the income tax yield £20.6 billion.

[c]The change in collection procedures for PRT meant that three payments fell due in 1979–1980 rather than the normal two.

Source: Department of Energy, 1980, p. 18.

These revisions were far less drastic than had been anticipated in some quarters. The financial correspondent of the *Times* described the terms as "undemanding," noting that "under the new rates and slightly changed allowances, the total take of North Sea company profits is estimated ultimately to increase from 70 to 75 percent; a reduction of a sixth in after-tax profits which will hardly be welcomed, but with fields like Forties proving more profitable than expected, this is not excessive" (*Times*, August 8, 1978, p. 19). Indeed considering (1) that the government was willing to reduce royalties where necessary to encourage the development of marginal fields; (2) that the 175 percent allowance on capital expenditures remained unaltered for fields with approved development plans; and (3) that the increased rate of PRT was allowable against corporation tax, the tax regime (even revised) hardly seemed onerous.

However reaction from the multinational companies was predictably harsh. An editorial in the *Oil and Gas Journal* (vol. 76, no. 35, 1978, p. 29) suggested that the United Kingdom was in the process of "committing energy suicide." Pointing to the many benefits to the British economy arising from the increasing flow of North Sea oil (balance of payments, employment in supply and servicing industries) the editorial concluded that "the government's short sighted impatience on taxes and fixation on increasing them is incredible when weighed against what the surging flow of oil and gas from present leases is doing for the British economy." Many North Sea operators indicated that the changes in the tax regime (combined with the proposed terms for the sixth round of offshore licensing announced earlier in 1978) would force them to review their future plans for the North Sea:

> Companies have threatened to move out of the North Sea before, but an accommodation was always reached, mainly because commercial fields had been found and in many cases were under development.
>
> The government feels these are idle threats and all part of the negotiating procedure.
>
> This time they could find that the companies mean what they say. No one is going to leave a commercial proposition, but there could certainly be less interest in new exploration. (Quoted by Vielvoye, 1978, p. 112)

Under U.K. parliamentary procedures no change in the tax structure could be enacted until the following session, a "crumb of comfort for the oil companies" as the *Economist* (July 29, 1978, p. 125) noted, since this would be preceded by a general election. However any hope the oil companies may have had that a Conservative administration would be more sympathetic to their position proved unfounded. Indeed, even in opposition, the Conservative Party had given the proposals a qualified welcome, stating that there was "clear scope for adjustments in the tax regime" (Statement by Tom King, Opposition Spokesman on Energy, Parliamentary Report, *Times*, August 3, 1978, p. 7). And in mid-1979 the newly-elected Conservative government announced its intention to implement the tax changes proposed by the previous administration virtually unchanged, a decision that "surprised and disappointed the oil industry" (*Offshore*, vol. 39, no. 8, 1979, p. 25).[19]

LICENSING STRATEGY

The fifth and sixth rounds of production licensing (held during 1976–1977 and 1978–1979) involved a significant change in the government's offshore strategy. In contrast to earlier rounds, when the objective had been to encourage rapid exploration, the intention now was to achieve a stable level of offshore activity by licensing small amounts of OCS acreage at more frequent intervals. More specifically it was hoped to provide for a rate of exploration drilling that would maintain national self-sufficiency yet avoid the extreme fluctuations in demand for support facilities and structures (drilling rigs, supply vessels, production platforms, fabrication yards) such as had occurred as a result of the very large number of licenses issued in the third and fourth rounds. There were, for example, only forty-six blocks placed on offer in the sixth round, far fewer than in any previous licensing round. Moreover, most of the prospective licenses comprised acreage in the Atlantic approaches to the English Channel and the Shetland Islands. Only fifteen of the blocks for which applications were invited consisted of prime acreage in the North Sea contiguous to areas already licensed for exploration.

Reaction to the government's new licensing strategy, and particularly to the consultative document (published by the Department of Energy in May 1978) outlining the terms that would govern the sixth round, was extremely hostile. The Department of Energy's proposal that the state-owned oil corporation should have *at least* a majority interest in all licenses, interpreted to mean that operators offering only the minimum 51 percent would find themselves at a disadvantage, was regarded as a particularly undesirable auction element at a time when offshore companies were confronted by escalating development costs. So was the suggestion that private sector partners might consider carrying all or part of BNOC's exploration and appraisal costs:

> If the government insists on taking part in oil production operations, then it must pay its way. As a matter of fact, it should also pay its proportionate share of risk costs. Why shouldn't government, if it is a partner in the producing operation, pay its share of exploration money? And take its losses when the hole is dry? (*Offshore*, vol. 38, no. 8, 1978, p. 41)

Finally, BNOC's role as advisor to the Department of Energy, and possible use of confidential information to select the most promising blocks for its own operatorship, remained a matter of concern to many operators.[20]

Timothy O'Riordan (1976, p. 237) argues that compromise through informal consultation between regulators and regulated is a central element in British policy-making. Yet the distinction between consultation and bargaining is often a fine one. And the initial sixth round terms outlined by the Department of Energy in its consultative document, as well as the critical response of offshore operators, undoubtedly represented initial bargaining positions. The Department of Energy is charged with protecting the national interest; the offshore companies are responsible to their corporate shareholders; both have an ultimate interest in accommodation (through compromise or concession trading) and resolution of any controversy. Thus the Department of Energy could agree to identify those blocks on which BNOC would be the designated operator before issuing the formal invitation for applications, yet remain firm in its intention that operators should consider carrying BNOC's exploration costs. The latter had been anticipated in the 1974 White Paper and hence it had been the fifth round that had been out of line with government policy rather than the sixth round. As private companies, the offshore operators could not be too strident in their criticism of the "competitive elements." As noted by the energy correspondent of the *Times*, "as with the suggestion that applicants might wish to offer BNOC more than 51 percent participation in the blocks they receive, the 'carried interest' element amounts to a tender style of application. If the potential is high, the oil companies will be prepared to offer more to the Government, if it is low they are surely capable of making a suitably revised application . . . all the companies are really being asked to do is to make a commercial judgement" (Hirst, 1978a, p. 23). Moreover, it did not go unnoticed that carrying the whole of a national company's exploration and operating costs had been a fairly standard practice in the Middle East since the mid-1960s.[21]

It is not possible of course to separate criticism of the specific terms for the sixth round from the general hostility of offshore operators toward the whole philosophy of government involvement in the development of North Sea oil. From the offshore operator's viewpoint the United Kingdom had "a North Sea success story because the private sector—not the government—took nearly all the risk in the first place and still continues to do so" (statement by C. C. Pocock, chairman, Shell Transport and Trading Company, quoted in *Offshore*, vol. 38, no. 11, 1978, p. 72). The multinational oil companies in particular resented the preferential treatment shown to BNOC and questioned the corporation's judgment, integrity, and operating ability. In the words of one oil executive, BNOC "makes no contribution . . . to your economy or your control of the oil situation and it does slow down and complicate our attempts to run a very professional business. I can think of nothing which has been accomplished so far in the North Sea which couldn't have been accomplished faster at lower cost if there were no BNOC" (George Keller, vice-chairman of Standard Oil of California, quoted in *Offshore*, vol. 38, no. 7, 1978, p. 98). However, BNOC was by no means the only source of friction and other forms of indirect government pressure—for example to award development contracts to United Kingdom–based industries—were equally resented.[22]

In this situation, offshore operators were often hard-pressed to conceal their desire for a change of government. After it had become apparent that there would be no autumn general election, the *Econ-*

omist observed that "the oil companies, unhappy at the way Mr. Benn was making them 'bid' for the blocks they want by offering bigger stakes to the British National Oil Corporation, were hoping they would have a Conservative minister to deal with—one who would take a very different view of BNOC's empire-building. Now the companies have to think again" (September 11, 1978, p. 112). However as the deadline for sixth round applications approached, it became apparent that Esso and Shell, two of the companies most heavily involved in the North Sea, had found neither the geological prospects nor the financial terms to be sufficiently attractive. In these circumstances Esso decided against participating in the sixth round, although Shell announced that it was applying for a small number of blocks.

The decision by Esso in particular raised serious doubts about the government's overall licensing strategy and timetable for North Sea oil development once self-sufficiency had been achieved. Not surprisingly, despite the discouraging attitude of Esso and Shell, the government claimed that the response to the sixth round more than vindicated its licensing policies. According to the Department of Energy a total of one hundred companies were involved in bidding for all of the forty-six blocks on offer, suggesting a far higher level of interest than in the fifth round, when several blocks attracted no bid.[23] Nevertheless the government was undoubtedly concerned at the decline in exploration activity during 1978. Department of Energy forecasts had indicated a rate of exploration drilling similar to that achieved in 1977, when sixty-seven wildcats were drilled. Yet in the first six months of 1978 only nineteen exploration wells had been drilled. Oil companies inevitably saw in this sharp decline, as in Esso's decision not to participate in the sixth round, proof that government licensing and taxation policies were inhibiting exploration. In discussions with the Department of Energy during 1978, the United Kingdom Offshore Operators Association (UKOOA) argued that only a "mammoth effort" by international oil companies, involving between sixty and ninety-five exploration wells each year for a decade, could ensure continued self-sufficiency in the second half of the 1980s as production from proven fields began to decline (*Offshore*, vol. 39, no. 1, 1979, p. 41).[24] To accomplish this the government would need to provide financial incentives to encourage development of marginal fields; additional prime acreage for exploration (including tracts in deeper offshore waters); assurances that production would not be controlled in the interests of depletion; and, finally, an element of continuity and predictability with respect to future fiscal and development policies.

Such arguments appeared to be somewhat self-serving, and many observers questioned whether such a high rate of exploration activity was really necessary.

> The oil companies believe that . . . unless the pace is kept up, interest will fade, the infrastructure will decline and the nation ultimately will receive less oil than it otherwise would.
> This argument suits the oil companies' case. They wish to avoid depletion controls to maximise cash flows, but the faster oil is . . . used, the more needs to be found. If that oil needs to be found in more difficult fields and at greater depths, the more incentives the oil companies will need to search for it.
> Theoretically, there is no reason why a low, but steady, rate of drilling should not find as much oil as rapid exploration. (Hirst, 1978b, p. 27)

Moreover it seemed unlikely that government policies could be held entirely accountable for the slowdown in exploration drilling. Oil company critics noted that the proposed changes in the tax regime could hardly have affected the drilling rate in the first half of the year and that many companies, already heavily involved in developing earlier finds, lacked the capital and crews necessary to mount a major new exploration program. Certainly there appeared to be little interest in moving quickly to explore the acreage that had been made available in the fifth round. As the chairman of BNOC, Lord Keaton, observed, the demand by oil companies for additional acreage was not always matched by a willingness to fulfill existing drilling obligations (*Offshore*, vol. 39, no. 1, 1979, p. 81). From this perspective it seemed possible that the slowdown in exploration activity in 1978 represented a short-term phenomenon rather than a permanent withdrawal.

Thus the North Sea decade drew to a close much as it had begun, on a note of uncertainty and controversy. Lord Balogh writes of "a sort of cyclic rhythm" to oil company protests whenever new measures are thought to be impending. "They have wailed about the hardships already inflicted on these valiant heroes, fighting to conquer horrible storms in the icy seas; these hardships would in the end force them to give up and sail away. It was never explained where they could find better conditions" (Balogh, 1978, p. 8).

Clearly private operators have been the prime movers in North Sea exploration and development. It is not unreasonable that these companies should seek assurances as to their future role in the North Sea. And their plea for a satisfactory return on investment and a stable basis on which to make development decisions is both understandable and legitimate. At the same time it is important to recall the findings of the Committee of Public Accounts and the reasons why the government became involved in the first place. As with the oil companies, so there are legitimate national interests to be considered and protected. It was the belief that the two would not necessarily coincide that led to enactment of the Petroleum and Submarine Pipe-lines Act in 1975. And, provided private companies are assured of a reasonable return, it is surely appropriate that the government, rather than the multinational oil companies, should determine the rate at which U.K. petroleum reserves should be exploited. But to effectively discharge its responsibilities, a government needs access to advice, information, and expertise that is independent of that available from the oil companies.

> The crude oil market is *sui generis*. Hardly any genuine arms-length sales take place. Swaps and special relations to suppliers enable the giants quite legally to avoid taxation by shifting profits. Much the most important function of a national oil corporation is the first-hand knowledge of real costs and real prices it acquires through its membership of operating committees. It is naive in the extreme to think that in the absence of that knowledge adequate taxation can be exacted. (Balogh, 1978, p. 8)

Similarly, a national oil company without field operating experience lacks credibility. In this respect a major objective of Labour policy in the second half of the 1970s was to ensure that BNOC had the necessary ability not only to explore for oil (any discoveries could always be banked in the interests of conservation for future recovery) but to develop fields that might be important in a national context but of little interest to oil companies involved in supplying the world market.[25]

Given the changing circumstances of North Sea oil development, it should not be surprising that government offshore policies have required frequent revision and refinement. The prospect of self-sufficiency raises rather different (and far more complex) sets of problems for both government and industry than does the task of stimulating exploration in an unproven area. Moreover, despite all the pleas by private operators for stable fiscal policies, North Sea development cannot be isolated from global events over which the U.K. government has little direct control. In retrospect, for example, many of the arguments advanced by UKOOA during 1978 were undermined by the significant increase in oil prices that followed the revolution in Iran. As outlined in *Offshore* in 1980, there appeared to be every likelihood of a "boom" in North Sea exploration. "Oil prices have surged upwards, and major North Sea producers such as Shell and BP have been able to raise rates to the $30-a-barrel mark. Field economics have improved and, quite simply, future profits appear to merit the investment involved in another round of exploration. In the light of economic and political developments, what were previously considered marginal oil bearing structures are also now being reassessed for possible future development" (*Offshore*, vol. 40, no. 2, 1980, p. 41-35).

The election of a Conservative government in 1979 clearly presaged major changes in offshore policy. While the full extent of these changes still remains unclear, the new administration is committed to encouraging North Sea exploration and development through greater incentives to the private sector. The terms of the seventh round of production licensing, for example, permitted companies to apply for any unlicensed block or blocks within the well-explored parts of the northern North Sea, these blocks being subject to a premium payment of £5 million instead of the normal £6,250 annual licensing fee. At the same time it was announced that BNOC would no longer receive a mandatory majority equity interest in offshore licenses but would have to apply for production licenses on the same terms as private sector companies.[26] However BNOC will retain its option to purchase at market prices up to 51 percent of the oil and gas produced from existing and future licenses (Department of Energy, 1980, p. 6).

At least initially therefore it seemed likely that the state-owned corporation would continue its oil-trading activities, but subject to the same taxes and restraints as any other commercial operator. An early decision, for example, was to remove BNOC's exemption from paying PRT, while in another policy reversal licensees seeking to dispose of all or part of their offshore interests are no longer required to offer BNOC or BGC the first option. Although early plans to sell off some of the corporation's assets to private operators aroused such opposition that they were quickly shelved, there seems little likelihood that BNOC will be permitted to extend its North Sea activities, for example, into downstream processing using associated gas as petrochemical feedstock. However the recent decisions (announced by the Secretary of State for Energy in December 1980) (1) that BNOC would exercise its rights to all associated gas available under existing participation agreements, and (2) that petrochemical companies seeking to acquire feedstock through the proposed gas-gathering network must negotiate supply contracts with BNOC, suggests that even under a Conservative administration the corporation will continue to function not only as a commercial concern but as an instrument of state policy. In effect the decision will allow BNOC to play a key role in the allocation of natural gas liquids and appears to have been forced on a reluctant government by the difficulty of coordinating supply contracts with some forty potential sellers of gas and by the increasingly acrimonious debate between petrochemical companies over the merits of rival proposals for new or expanded complexes at Nigg Bay, Peterhead, Mossmorran, Grangemouth, and Teesside. At the same time the government has indicated on various occasions that it will intervene to hold down production in the interests of prolonging self-sufficiency and seek an increase in the British stake in the North Sea, though this is to be accomplished through "encouraging" smaller British operators rather than through awards to BNOC. In all probability therefore the changes in policy will not go as far as many offshore operators had anticipated in reducing direct government involvement in North Sea oil development.

This chapter has outlined some of the difficulties involved in reconciling public and private interests in the North Sea and in arriving at a balanced and equitable set of offshore policies. The next two chapters will examine the implications of these policies for the marine environment and for coastal zone management.

3. Offshore Oil: The Risk of Marine Pollution

Introduction

Few subjects carry heavier emotional overtones than marine oil pollution. A recent report on the potential threat to the British coastline, for example, observes that "not infrequently, the only visual evidence that an oil slick has passed by offshore is the appearance of oiled birds on our beaches—sometimes dead and so heavily encased in oil as to be scarcely recognisable for what they are, sometimes still alive, struggling ashore unable to fly or dive but making valiant, futile attempts to preen their contaminated plumage" (Royal Society for the Protection of Birds, 1979, p. v).

Concern over the potential impact of oil pollution on valuable biotic and recreational resources is by no means new. Indeed the first major spill in British coastal waters occurred as early as 1907, when the *Thomas W. Lawson* broke up and sank off the Scilly Isles. During the past two decades, however, events such as the blow-out in the Santa Barbara Channel and the grounding of the *Amoco Cadiz* have provided a forceful reminder of the risks associated with the offshore production and transportation of oil (Table 16). Such incidents, frequently described in the media as "environmental catastrophes," have led to a questioning of current priorities with respect to our use of the marine environment. Does the present rate of input of petroleum-derived hydrocarbons into the oceans pose a significant threat to human health and welfare? Is offshore drilling and production a major contributor to marine oil pollution? Are the risks acceptable given the benefits to be derived from exploiting offshore oil and gas reserves? Should the search for oil be restricted in order to protect other environmental resources and attributes? Have those involved in producing and shipping oil adopted all available pollution control measures? Given the range and complexity of the issues that are involved, such questions are not easily answered. Yet they reflect legitimate public concerns and raise major policy questions with respect to the most appropriate balance between alternative uses of the marine environment.

With the exception of the Ekofisk *Bravo* blow-out on the Norwegian continental shelf, no serious spills have occurred during drilling and production operations in the North Sea. However, past experience is not always an accurate or helpful guide to the future. The possibility of equipment failure and of human error, whether due to negligence or misjudgment, means that the risk of a major spill can never be entirely eliminated. Moreover, the unexpected is always possible. Offshore operators, for example, have long argued that the drilling record does not support the contention that exploration per se involves exceptional risks.[1] Yet the *Ixtoc I* blow-out in the Bay of Campeche during 1979–1980, involving the release at one time of around thirty thousand barrels of oil a day into the Gulf of Mexico, demonstrated that mishaps can occur even during exploration drilling.

Understandably it is these spectacular incidents that attract public attention, yet a greater danger in terms of marine oil pollution may result from the increase in tanker traffic that is required to ship crude oil to national and international markets. From a strictly economic perspective the use of pipelines may not be justified in the case of smaller fields such as Auk and Argyll. At present around 20 percent of the oil produced from the U.K. sector is offloaded onto tankers, a practice that greatly increases the chances of chronic discharges during routine handling and shipping operations as well as of a major spill in the event of a tanker accident. In addition, the location of the Sullom Voe terminal, which upon completion will handle storage and transshipment of around 50 percent of U.K. offshore production, will contribute to a significant increase in tanker traffic in waters to the north of Scotland which have hitherto been free of the risk of oil pollution.

Exploitation of the oil and gas resources of the North Sea inevitably involves some degree of risk to the marine environment. A basic question is whether these risks have been reduced to an acceptable level through careful contingency planning and the adoption of available pollution control measures. As Gerald Moore has pointed out, what is essentially involved is achieving a balanced use of the environment, a balance "weighted according to the values set upon each use or activity by society, and taking into account the ecological dangers of overstepping certain critical levels. In practical

Table 16. Selected Oil Pollution Incidents

Date	*Vessel or Other Source (Location)*	*Estimated Spillage (barrels)*	*Impact*
3/47	*Tampico Maru* (Baja California, Mexico)	60,000 Diesel fuel	Devastation of marine flora and fauna; 90% biota restored after 3 to 4 years but certain species still showing population changes 12 years after spill.
7/62	*Argea Prima* (Guayanilla, Puerto Rico)	70,000 Crude	High mortalities among shallow water and shore-dwelling organisms including vertebrates.
3/67	*Torrey Canyon* (Southwest England)	860,000 Crude	Very high mortalities among seabirds and intertidal shore life primarily caused by misuse of highly toxic chemical detergents and emulsifiers.
10/67	Pipeline (Louisiana, Gulf of Mexico)	160,000 Crude	Ruptured pipeline due to anchor dragging.
3/68	*Ocean Eagle* (Puerto Rico)	83,000 Crude	Oil and chemical emulsifiers resulted in significant losses among subtidal and intertidal organisms.
1/69	Platform A21 (Santa Barbara Channel, California)	33,000–78,000 Crude	Significant contamination of coastline. Claims for damages totaled $1.8 billion; direct clean-up costs ca. $10 million. Recovery well advanced after one year.
9/69	*Florida* (Buzzard's Bay, Massachusetts)	4,500 No. 2 fuel oil	Oil incorporated into coastal marsh sediments; restrictions on shellfish harvesting still in effect in 1977.
2/70	*Arrow* (Chedabucto Bay, Canada)	108,000 Bunker C	Local, short-term effects.
3/70	Platform C (Main Pass Field, Gulf of Mexico)	30,000–65,000 Crude	Much of the spilled oil burned off; remaining oil contained by mechanical equipment.
12/70	Platform B (Marchand Field, Gulf of Mexico)	52,000–95,000 Crude	Blow-out and fire. Spill allowed to burn to minimize oil pollution.

2/71	*Wafra* (Cape Agulhas, South Africa)	445,000 Crude	Little damage to intertidal life.
3/73	*Zoe Colocotroni* (Bahia Sucia, Southwest Puerto Rico)	48,000 Crude	Deliberate discharge in an effort to refloat the vessel. Significant amounts of spilled oil penetrated coastal mangrove swamps.
8/74	*Metula* (Strait of Magellan, Chile)	381,000 Crude	Oil contamination along 75 miles of coastline; high winds and intense cold impeded efforts to offload oil. No clean-up effort. Spilled oil still present in large amounts along 250 km of coastline 12 months later.
10/74	*Universe Leader* (Bantry Bay, Ireland)	ca. 17,000 Crude	Spill occurred when a valve failed to close during off-loading. Clean-up operations hampered by lack of trained manpower and suitable equipment.
6/76	*Urquiola* (La Coruña, Spain)	770,000 Crude	Severe impact on local fishing and shellfish industry due to tainting. Dispersants failed to prevent oil reaching coastline.
12/76	*Argo Merchant* (Nantucket, Massachusetts)	175,000–200,000 Fuel oil	Spilled oil congealed into large lenses that drifted in the Gulf Stream for several months; very little oil incorporated into coastal sediments.
4–5/77	Platform *Bravo* (Ekofisk field, North Sea)	140,000 Crude	Blow-out; no oil reached the coastline.
9/77	*Adrian Maersk* (Hong Kong)	7,000 Bunker C	Fish farming losses estimated to exceed $2 million.
3/78	*Amoco Cadiz* (Brittany)	1,561,000 Crude	Clean-up costs estimated to be ca. $50 million.
5/78	*Eleni V* (East Anglia)	~38,500 Heavy fuel oil	Mechanical containment and chemical agents proved ineffective in controlling and recovering spilled oil.

Table 16. Continued

Date	Vessel or Other Source (Location)	Estimated Spillage (barrels)	Impact
10/78	*Christos Bitas* (Dyfed, South Wales)	17,500 Crude	Tanker grounding off Welsh coast. Despite intensive spraying, oil eventually reached the North Devon coast and the Gower peninsula of South Wales. Inquiry revealed numerous defects in navigational equipment, compounded by negligence on the part of the ship's master in failing to reduce speed when visibility restricted by fog.
12/78–1/79	*Esso Bernica* (Sullom Voe, Shetlands)	~8,500 Bunker C	Equipment failed to contain oil; local mortalities among seabirds and otters.
1/79	*Betelguese* (Bantry Bay, Ireland)	Unknown Crude	Tanker exploded following offloading of 120,000 tons of crude oil. Inquiry indicted operators of terminal for failing to observe safety standards and procedures and shipowners for consciously ignoring the corroded state of the vessel and failing to install an inert gas system to reduce the risk of explosion.
2/79	*Antonio Gramsci* (Ventspils, U.S.S.R., Baltic)	35,000 Crude	Tanker grounding. Released oil, mixed with slushy pack ice, formed a slick that drifted among islands in the Stockholm archipelago causing extensive seabird mortalities. Total clean-up costs estimated at $12 million.
3/79	*Kurdistan* (Nova Scotia)	56,000 Bunker C	Vessel broke in two and sank near Cape Breton Island.
6/79–3/80	*Ixtoc I* (Bay of Campeche, Gulf of Mexico)	~3,000,000 Crude	Largest spill yet recorded. Claims filed in U.S. courts total $365 million. Petroleos Mexicanos (PEMEX) reportedly spent $133 million in capping the well and in con-

			taining the environmental damage while losing $87 million in oil revenues.
7/79	*Atlantic Empress/Aegean Captain* (Trinidad-Tobago)	Unknown	Collision of supertankers; early estimate of 1.6 million bbls spilled appears exaggerated. (Inquiry found that the officer on watch on the *Aegean Captain* had no proper radar training and failed to take usual seamanlike precautions during a squall which reduced visibility. Watch officer on *Atlantic Empress* was a radio officer without qualifications as a deck officer.)
12/79	*Andros Patria* (Cape Finisterre)	Unknown Crude	Explosion following development of a crack in tanker hull. Slick 20 km long treated with dispersants. Some oil came ashore near La Coruña, site of rich oyster and shellfish grounds affected by *Urquiola* spill in 1976.
1/80	*Funiwa 5* (Nigeria)	200,000 Crude	Blow-out of production well. Approximately 50% of spillage reached shore. Local fisheries reported to be seriously affected in the Niger delta and drinking water supplies contaminated by oil.
2/80	*Irenes Serenade* (Navarino Bay, Greece)	Unknown Crude	Tanker exploded while at anchor. Burning oil ignited vegetation on Sfaktiria Island, reported slick 32 km long treated with skimmers and dispersants.
3/80	*Tanio* (Brittany)	Unknown Fuel oil	Vessel broke in two about 40 km off the Brittany coast. Heavy fuel oil from ruptured tanks formed a slick over 100 km long. Dispersants not used in view of ineffectiveness against heavy oil. Limited success in collecting oil using nets and recovery devices. Much of the spilled oil polluted the same stretch of the Brittany coastline as had been affected by the *Amoco Cadiz*.

Sources: National Academy of Sciences, 1975, pp. 74–75; U.S. Environmental Protection Agency, 1976; *Marine Pollution Bulletin*; *Annual Reports* of Council on Environmental Quality; *Annual Reports* of Advisory Committee on Oil Pollution of the Sea; *Oil Spill Intelligence Reports*.

terms, this means setting a limit to the amount of contamination of marine waters that is acceptable and ensuring that the limit is not overstepped" (1976, p. 590).

Unfortunately, establishing such thresholds, let alone obtaining international agreement to respect and enforce standards, is problematic. In part the sheer scope and pervasiveness of the problem complicates the search for an acceptable solution. Any effort to monitor and control marine oil pollution must deal with spillages of both crude oil and refined products (exhibiting widely different chemical properties and toxicities) from a multitude of sources (including indirect discharges from diffuse sources such as the atmosphere and urban runoff as well as direct discharges of an unpredictable nature from point sources such as production platforms and tankers) under extremely variable environmental conditions. Moreover, "marine ecology cannot as yet claim to be a science beyond a stone's throw from the fringes of the sea" (Johnston, 1976a, p. viii), and there remains considerable scientific uncertainty over the fate and effect of petroleum hydrocarbons in the marine environment. The recent National Academy of Sciences (NAS) study *Petroleum in the Marine Environment* emphasizes in its conclusion that

> A basic question that remains unanswered is, "At what level of petroleum hydrocarbon input to the ocean might we find irreversible damage occurring?" The sea is an enormously complex system about which our knowledge is very imperfect. The ocean may be able to accommodate petroleum hydrocarbon inputs far above those occurring today. On the other hand, the damage level may be within an order of magnitude of present inputs to the sea. Until we can come closer to answering this basic question, it seems wisest to continue our efforts in the international control of inputs and to push forward research to reduce our current level of uncertainty. (National Academy of Sciences, 1975, p. 107)

Given these limitations, it is perhaps not surprising that there exist widely divergent assessments with respect to the nature, impact, and severity of oil pollution. Controversy surrounds virtually every stage in the sequence of events whereby oil enters and is removed from the ocean. Uncertainty and disagreement over the relative contribution from different sources, the need for and effectiveness of control measures, the adequacy of contingency plans, and the safety of clean-up techniques greatly complicate the task of evaluating the costs and benefits of alternative courses of action. In these circumstances there is an unfortunate tendency to draw only those inferences from the data that appear to support the views of the protagonist. As Robert J. Stewart observes with respect to oil spills from tankers, "because the problem is as obscure as it is large, it serves as a ready vehicle for special interest groups. The technique is to state the problem so that the obvious solution incorporates one or more of the program elements desired by the definer of the problem" (1977, p. 74).

In dealing at some length in this chapter with efforts to control oil pollution on the U.K. continental shelf, my intention is to provide a basis for more informed decision-making. In this way not only the public but those involved in formulating and implementing policy will be better equipped to evaluate the nature and scale of action required to deal with the problem. If it is indeed the case that public attitudes toward environmental hazards are strongly influenced by the way in which relevant information is presented, it is disturbing to read reports that "insufficient official information on oil pollution is made available to the public" (Royal Society for the Protection of Birds, 1979, p. 11). Without public access to such information, neither affected coastal communities nor government agencies can deal with the oil and shipping industries from a position of strength.

The Sources of Marine Oil Pollution

In those coastal communities most exposed to offshore exploration and development, an understandable concern is the extent to which such activities will have a deleterious effect on other resources that contribute to the local economy. Concern over the incidence and potential impact of oil spills is likely to be expressed most forcefully where the leased area includes valuable fishing grounds or borders a coastline of outstanding recreational and amenity value.

Such fears, perhaps reinforced by major oil spills that may not be in any way related to offshore drilling, are usually depicted as exaggerated by offshore operators. Referring to the offshore industry's actual drilling record, an article in the magazine *Exxon* argues that "oil spills from platforms are rare. During the 20-year history of modern offshore petroleum operations, including the drilling of over 21,000 wells and the production of over 7.5 billion barrels of oil and over 41 trillion cubic feet of natural gas, the industry has experienced no more than a half-dozen major spills" (Dedera, 1977, p. 13). William B. Travers and Percy R. Luney suggest that the public's perception of the problem has been distorted by "the occasional, but spectacular, oil spills that have occurred during drilling and production" (1976, p. 791). They note that although the dramatic and highly publicized blow-out on Union Oil's Platform A in the Santa Barbara Channel in January 1969 may have released up to 78,000 barrels of crude oil into the channel, this represented less than 15 percent of the estimated amount of oil spilled that year in the course of tanker operations in U.S. waters. Nevertheless the fears persist and may, as in the extended and bitter legal campaign over the sale of oil and gas leases on the Georges Bank off the Massachusetts coast, significantly affect the rate and location of offshore drilling.[2]

Although offshore operators might prefer to see the issue of oil pollution from offshore facilities treated as an independent variable, others see marine oil pollution as a single, interrelated problem, a view encouraged by the difficulty that exists in identifying the source of many spills and by the indirect way in which offshore production may contribute to the overall problem. In this respect, while the most spectacular influx may be in the form of oil spills, petroleum-derived hydrocarbons may be introduced into the environment at any stage in the production-transportation-refining-combustion sequence. Robert W. Holcomb (1969, p. 204) somewhat wistfully observes that in the best of all possible worlds oil produced from offshore environments or transported across the ocean surface would be confined within the "industrial system"—wells, pipelines, tankers, refineries, and ultimately boilers, furnaces, and internal combustion engines. However, for all our ingenuity in "managing" the environment, we are perhaps finally recognizing that we cannot entirely control events to suit our purposes. Once crude oil is pumped out of the ground for human use, it is inevitable that petroleum-derived hydrocarbons will eventually enter the marine environment, where they become subject to normal ecological processes and cycles. Thus the environment must assimilate a wide range of petroleum waste products, including lubricating oils utilized in industrial operations and hydrocarbon emissions from the internal combustion engine, as well as the crude oil spilled in the course of offshore production and transportation.

Most discussions of marine pollution emphasize the range of pathways whereby hydrocarbons can reach the sea, stressing particularly the role of rivers in transporting urban runoff into estuarine environments and indirect injection via the atmosphere. The National Academy of Sciences (NAS) study *Petroleum in the Marine Environment*, for example, estimated that there is a total world-wide influx of petroleum-derived hydrocarbons into the atmospheric environment amounting to around 68 million tonnes a year (National Academy of Sciences, 1975, p. 10). Approximately two-thirds of this

Table 17. Estimates of Petroleum Hydrocarbons Entering the Ocean

Source	MIT/SCEP (1970)	USCG (1973)	Kash (1973)	NAS (1975)
	(millions of tons/year)			
Marine transportation (including tanker discharge, accidents, terminal operations, bunkering, etc.)	1.13	1.72	2.30	2.133
Offshore oil production	0.20	0.12	0.20	0.08
Coastal oil refineries	0.30	—	0.30	0.20
Industrial waste	—	1.98	—	0.30
Municipal waste	0.45	—	2.19	0.30
Urban runoff	—	—	—	0.30
River runoff	—	—	—	1.60[a]
Subtotal	2.08	3.82	4.99	4.913
Natural seeps	?	?	0.10	0.6
Atmospheric rainout	9.0[b]	?	9.0[b]	0.6
Total	11.08	—	14.09	6.113

[a]Input of petroleum hydrocarbons from recreational boating incorporated in river runoff value.

[b]Atmospheric fallout estimated at between 9 million and 90 million tons, latter representing *total* estimated emission of petroleum hydrocarbons into atmosphere (see MIT/SCEP study). Kash study assumed atmospheric input would double between 1970 and 1980. These figures have been criticized in some quarters as grossly inflated (see, for example, Organization for Economic Co-operation and Development, 1977a, p. 13).

Sources: Massachusetts Institute of Technology, 1970 (MIT/SCEP); U.S. Coast Guard, 1973 (USCG); Kash et al., 1973 (Kash); National Academy of Sciences, 1975 (NAS).

input comes from automobiles and aircraft, with fuel combustion in stationary sources, industrial processes, and solvent and gasoline evaporation accounting for the remainder. According to the NAS study nearly 10 percent of this atmospheric flux (0.6 million tonnes per year) reaches the ocean through direct fall-out, rainfall, and other atmospheric-ocean exchanges (Table 17). Other studies attach even greater significance to the atmosphere as a source of petroleum input into the ocean; a study by the Massachusetts Institute of Technology, for example, puts the influx of petroleum hydrocarbons into the marine environment at around nine million tons a year (Massachusetts Institute of Technology, 1970, pp. 140–141). Thus in the unlikely event that it proved possible to control the more visible and direct sources of marine pollution, oil would continue to enter the ocean ecosystem from more diffuse sources for as long as industrialized nations rely upon oil and natural gas. In this respect the fate of petroleum hydrocarbons in the environment illustrates Barry Commoner's ecological dicta that there is "no such thing as a free lunch" and that "everything must go somewhere" (Commoner, 1971).

Estimates of the total influx of petroleum hydrocarbons into the oceans as a whole range from six to fourteen million tons a year (Table 17). The major direct contributors may be identified as:

1. Maritime traffic accidents, particularly those involving large tankers.

2. Offshore exploration and production (including accidental discharges during tanker loading operations at offshore platforms or mooring buoys, effluent disposal from routine production and processing operations, as well as losses due to equipment failure, blowout, or pipeline rupture).

3. Routine shipping operations (including the deliberate or accidental discharge of ballast water, tank washing, and bilge wastes, as well as the spillage of oil during bunkering).

4. Land-based sources (including point sources such as coastal refineries and non-point sources such as urban and highway runoff).

According to C. C. Bates and E. Pearson (1975) the study prepared for the National Academy of Sciences provides the most reliable data base (Table 18). The NAS budget is noteworthy in two particular respects. First, 1.2 million tonnes (or nearly 20 percent) of the petroleum hydrocarbons introduced into the oceans is attributed to natural seeps and atmospheric rainout. An additional 2.7 million tonnes enters the ocean either in the form of municipal and industrial waste or through urban and river runoff. Offshore oil production and marine transportation thus account for only 36 percent of the total volume of oil entering the marine environment. Second, the contribution from offshore drilling and production is minimal when compared to the introduction of oil and petroleum products during shipping operations. In assessing the reliability of the data, the panel expressed greatest confidence in their estimate for marine transportation, concluding that "this input represents the major source of visible accumulations of petroleum hydrocarbons both in the open oceans and along coastlines" (National Academy of Sciences, 1975, p. 104).

Such data must be interpreted with great care. The most immediate problems arise as a result of (1) weaknesses in the background data base, and (2) the extreme level of aggregation involved in the global budget approach. U.S. background data provided the basis for

Table 18. National Academy of Sciences Budget for Petroleum Hydrocarbons Entering the Ocean

	Best Estimate	*Probable Range*
	(millions of tonnes/annum)	
Artificial		
Marine transportation		
LOT tankers[a]	0.31	0.15–0.4
Non-LOT tankers	0.77	0.65–1.0
Dry docking	0.25	0.2–0.3
Terminal operations	0.003	0.0015–0.0005
Bilges/bunkering	0.5	0.4–0.7
Tanker accidents	0.2	0.12–0.25
Non-tanker accidents	0.1	0.0002–0.15
Subtotal	2.133	
River runoff	1.6	—
Urban runoff	0.3	0.1–0.5
Coastal municipal wastes	0.3	—
Coastal (non-refining) industrial wastes	0.3	—
Coastal refineries	0.2	0.2–0.3
Offshore production	0.08	0.08–0.15
Atmospheric rainout	0.6	0.4–0.8
Natural		
Offshore seeps	0.6	0.2–1.0
Total	6.113	

[a]LOT: Load-on-top tankers equipped with slop tanks. This procedure requires that crude oil carriers transfer water used to clean tanks to a settling tank where the oil and water can separate out over time. The "clean" water is discharged and the crude oil residue retained, the next cargo being "loaded on top."

Source: National Academy of Sciences, 1975, Table 1.5.

much of the NAS assessment and there must remain uncertainties about the validity of extrapolating from this data base to a global level, even when allowance is made for differences in consumption patterns, operating procedures, and environmental regulations in countries other than the United States. Second, the lack of geographical specificity is likely to conceal the severity of the problem in those offshore areas where oil production is presently concentrated or along coastlines bordering the major shipping lanes. Quite apart from the limitations of the estimates per se, the global budget approach is of little help in defining or clarifying the problem of marine oil pollution. Is it the rare, massive spill of crude oil that produces the most adverse biological impacts, or smaller but repeated spills of refined products? Under what circumstances will spilled oil be dispersed or concentrated? Who pays for the costs and damages incurred in the course of an oil spill if the source cannot be identified? It is in these areas that difficulties in interpretation and differences of opinion are likely to be most pronounced. The gross global budget approach cannot distinguish between the volume, characteristics, circumstances, or frequency of particular spills, yet such distinctions are critical in terms of the long-term biological impact and the ability of marine ecosystems to recover from oil pollution. Nor does this approach shed much light on the possible causes of pollution incidents—human error, equipment failure, deliberate discharge, or natural catastrophe. As a recent U.K. study notes, "these exercises [in determining global inputs] appear to have restricted value, since calculations of quantity alone have little significance in assessing priorities for remedial action. In seeking to prevent environmental damage, other important factors must be taken into account including location, season, frequency of occurrence and types of oil spilled" (Royal Society for the Protection of Birds, 1979, p. 61).

Given the emotional nature of the issue, it is perhaps not surprising that the precise impact of the major sources and their relative contribution to oiled beaches and seabird mortalities remains highly controversial. Nevertheless, from the oil industry's perspective, the National Academy of Science study is confirmation (1) that sources other than offshore production and shipping are the major contributors to marine oil pollution, and (2) that offshore drilling does not constitute a major pollution threat to the marine environment. An article in *Exxon* applauds the NAS assessment as a landmark study that "refutes popularly held beliefs regarding sources of oil in the oceans" (Dedera, 1977, p. 13).

Other authors have arrived at essentially the same conclusion, although it should be emphasized that efforts to identify those particular activities and practices most likely to result in an oil spill rely heavily on U.S. data.[3] In their 1976 analysis of the risk of oil spills on the Atlantic outer continental shelf, for example, Travers and Luney conclude that blow-outs involving major spills have been comparatively rare on the U.S. continental shelf. "Since 1953, more than 18,000 wells have been drilled offshore in U.S. waters with only 11 major oil spills [defined as over 5,000 barrels spilled]. Since 1972 no major oil spills have occurred. In the past 5 years, 102 wells have been drilled on the Canadian Atlantic outer continental shelf without an incident." In their judgment, "offshore production and pipelines invariably introduce less crude and petroleum products into the environment than do tankers and sources of automotive waste oil" (Travers and Luney, 1976, p. 791). On the basis of the less hazardous geological conditions off the Atlantic seaboard, the relative infrequency of blow-outs leading to major spills, and the dangers inherent in surface shipments of imported oil, Travers and Luney concluded that the exploration and development of petroleum resources on the Atlantic OCS is environmentally preferable to increasing oil imports. "Without petroleum production from the Atlantic outer continental shelf, imports of crude oil and petroleum products will increase and the concomitant use of more tankers will increase the number of collisions and accidental and deliberate spills occurring in Atlantic coastal waters" (ibid., p. 794).

Records for the U.S. outer continental shelf suggest that irrespective of the actual amount of oil spilled as a result of offshore exploration and production, few major spills actually occur in the course of drilling.

> Particularly noteworthy is that no spills of more than 50 barrels resulted from drilling operations [in the Gulf of Mexico] during 1971–75, even though 4,105 new wells were started. No such spill has occurred since July 19, 1965, when 1,688 barrels of crude oil was discharged into the Gulf during a blowout in the Ship Shoal area off the coast of Louisiana. (Danenberger, 1976, p. 9)

Inputs from OCS activity are primarily transportation-related (including pipeline damage), are secondly caused by failure of production platform equipment, and only thirdly related to actual drilling operations. Thus fear of major spills from blow-outs during drilling is judged to be exaggerated.

> During 1971–1975, 22 blow-outs occurred during drilling, workover or production. Nine of the 22 caused significant pollution or property damage. Of the nine, five were associated with drilling operations. However the 4 non-drilling incidents were responsible for all 725 barrels of "Blow-out spillage." According to OCS blow-out records, which go back to 1956, only two drilling-related blow-outs caused significant oil spills—the Santa Barbara blow-out . . . and the previously mentioned Gulf of Mexico blow-out . . . (ibid., p. 10)

The U.S. experience suggests that whatever the precise contribution from offshore drilling and production, it represents a small proportion of the total amount of oil entering the marine environment. As an OECD report observes, "transport of oil and nonmarine operations are by far the major polluters, and an improvement there would be far more important in reducing oil pollution of the sea" (Organization for Economic Co-operation and Development, 1977a, p. 12). Such a conclusion does not, of course, mean that efforts to reduce oil pollution from offshore activities should be relaxed. Nothing in the U.S. experience diminishes the need for government-industry cooperation in ensuring that good working practices are rigorously observed, for thorough and frequent inspections of offshore facilities, or for carefully rehearsed contingency plans to make certain that essential equipment and information are readily available in the event of a spill. Nor, Travers and Luney's assertion to the contrary, does the U.S. offshore record *necessarily* support a decision to proceed with exploration, though it is clear that any delay in leasing cannot be justified *solely* on the basis of the risks associated with drilling. If anything, their analysis demonstrates the need to consider a much broader array of socioeconomic, political, and environmental factors in any decision affecting offshore leasing. As Richard E. Chaisson, Lester B. Smith, Jr., and Jamie M. Fay suggest, not only does the Atlantic outer continental shelf contain valuable fisheries resources (not considered in Travers and Luney's analysis) but there is also at least the possibility that tanker traffic in coastal waters will increase rather than decrease as a result of any oil and gas discoveries. "The estimate of oil and gas resources recoverable from that area is too low to economically justify the use of pipelines for transport of oil ashore, making small tankers (20,000 to 30,000 dead weight tons) the most probable mode of transport. Use of these smaller and generally older tankers could result in chronic discharge of oil and more accidents" (Chaisson, Smith and Fay, 1978, pp. 128–129).

The previous discussion is intended to give some background to the issue of marine oil pollution in the North Sea. There is, of course, always a danger in extrapolating from such a spatially and temporally restricted data set to the deeper offshore waters of the North Sea, where operational conditions, structural problems, and environmental circumstances will be very different from those of the U.S. outer continental shelf. However, information on marine oil pollution in the North Sea is extremely restricted in terms of scope and completeness. A recent report by the Royal Society for the Protection of Birds (RSPB) claims that "no detailed systematic investigation of the causes and sources of oil pollution has been undertaken by any government department" (1979, p. 61). The RSPB's own survey of oil pollution, based on a systematic national inventory of "beached birds" begun in 1971, provides one long-term source of information on the incidence and impact of oil spills along the British coastline. Additional data are compiled by the Advisory Committee on Oil Pollution of the Sea (ACOPS), which distributes an annual questionnaire to local authorities requesting information on oil pollution incidents (whether or not birds are involved) occurring within their jurisdiction. For all their limitations, both surveys suggest that routine shipping operations pose a far greater threat to U.K. coastal waters than does offshore drilling. According to the RSPB investigation, out of a total of forty-six major oil pollution incidents (defined as those involving fifty or more oiled birds found within a limited stretch of the coast) occurring between January 1975 and June 1979, twenty could be attributed to deliberate or accidental spillage of oil from routine shipping operations or to land-based sources. Five of these incidents involved offshore collisions or groundings. The remaining incidents could not be attributed to any source but were assumed to be the result of unreported accidental or deliberate discharges: ". . . most spillage which occurs in ports or at supervised anchorages can be correctly assigned to source. Similarly, spills from marine traffic accidents, offshore exploration/pollution or land-based developments can normally be attributed. However for the majority of spills outside ports sources are not or cannot be specifically identified, and it is concluded that these result from unreported accidental or deliberate discharges" (Royal Society for the Protection of Birds, 1979, p. 65).

Not all incidents, of course, were of equal severity. Particularly serious oiling incidents (in terms of bird casualties) included the spill of fuel oil from the *Esso Bernica* during docking at the Sullom Voe terminal on December 30, 1978, loss of crude oil as a result of the grounding of the *Christos Bitas* off South Wales in October 1978, and an unattributed spill of fuel off the Northumberland coastline in January 1978. Clearly the potential for large-scale pollution is much greater from collisions or strandings involving loaded oil tankers. Yet in the RSPB's judgment such incidents have been responsible for less damage than routine shipping operations. "Although major ship disasters have the potential to kill enormous numbers of birds at a time, the unreported and often illegal dumping of oil by ships has been the most important cause of bird deaths by oiling in recent years" (Royal Society for the Protection of Birds, 1979, p. 6). Moreover, investigation of "the sources of oil pollution, the types of oil spilled, and of individual incidents involving bird mortality shows that there are two main sources of such spillage. One is the discharge of crude oil from tankers. The other is the discharge of black fuel oils from ships of all types; hitherto the serious harm caused by this practice has not been fully recognized" (ibid., p. 5).

The RSPB assessment must be interpreted with some caution. Both the definition of major oiling incidents and the actual data base are problematic. Yet the society contends that its investigation may well underestimate the full dimensions of the problem. And there is some evidence to suggest that true seabird mortality rates as a result of oil spills may be as much as three times that recorded in the beached bird survey. Moreover the responses of local authorities to inquiries by ACOPS tend to confirm (1) that such official statistics as do exist seriously underestimate the number of pollution incidents in U.K. waters; (2) that fuel oil spills are now occurring with greater frequency; and (3) that most incidents involve unreported discharges, either accidental or deliberate, during routine shipping operations.

In view of the number of oil spill incidents occurring each year in U.K. waters ACOPS contends that "the complacency which is still common in certain Government departments is unjustified" (Advisory Committee on Oil Pollution of the Sea, 1978, p. 4). In 1977 the committee's chairman, Baroness White, noted in the introduction to the *Annual Report* that "our survey of oil pollution incidents around the coasts of the United Kingdom has revealed some alarming facts. The number of spills in 1976 increased by almost 20% by comparison with 1975. . . . It is not surprising, but it gives cause for concern that offshore related spills are becoming almost commonplace in Eastern Scotland. Suggestions sometimes heard in authoritative circles that pollution is becoming controllable are therefore unsubstantiated" (Advisory Committee on Oil Pollution of the Sea, 1977, p. 4). This disturbing trend continued during 1977, when the ACOPS survey revealed a total of 642 incidents in U.K. waters, compared with 595 the previous year and 500 in 1975. Moreover, as the committee noted, "If one bears in mind that there were no major tanker accidents and no blow-outs or pipe-line fractures resulting in massive pollution, the increase in incidents is particularly worrying. Chronic pollution is more harmful to the marine environment than less frequent spills, and we were thus alarmed to receive a fair number of questionnaires reporting pollution over large sectors of the coast lasting for several months or even the whole year. Eastern and Western Scotland were particularly affected" (Advisory Committee on Oil Pollution of the Sea, 1978, p. 13). Since 1977 there does not appear to have been any further deterioration in the overall picture. Indeed the ACOPS survey recorded only 507 oil pollution incidents in

U.K. waters during 1978 and 558 incidents during 1979 (Advisory Committee on Oil Pollution of the Sea, 1979, p. 46; 1980, p. 57). Nevertheless over 500 incidents a year must still be regarded as representing a serious situation. And while the number of incidents may have declined, 1978 saw accidents involving the *Eleni V*, the *Christos Bitas*, the *Litiopa*, the *Esso Bernica*, and the grounding of the *Amoco Cadiz* off the Brittany coastline. As a result the total quantity of oil discharged into U.K. waters was greater than in any previous year (Advisory Committee on Oil Pollution of the Sea, 1979, p. 38). Quite apart from their ecological impact, these incidents demonstrated how misplaced had been Baroness White's hope, expressed in the 1976 report of the Advisory Committee, that Britain would not have to experience a major pollution incident "before we realise that another *Torrey Canyon*, or worse still, chronic pollution in various parts of our coasts, is a real possibility" (Advisory Committee on Oil Pollution of the Sea, 1977, pp. 4–5).

Thus, although there is some general consensus that offshore exploration and production do not represent the primary source of oil pollution in the North Sea, there remains concern about the overall dimensions of the problem and particularly the indirect role of offshore activities in contributing to more frequent tanker movements through previously unpolluted waters. The RSPB report, for example, notes that the chronic pollution apparent in the vicinity of major ports and busy coastal shipping lanes is beginning to appear off the north coast of Scotland, an area not notably affected in the past (Royal Society for the Protection of Birds, 1979, p. 67). The spill of nearly 1,200 tons of Bunker C fuel oil into the waters of Sullom Voe and Yell Sound from the tanker *Esso Bernica*, less than two months after the North Sea terminal became operational, illustrates the indirect threat from offshore activities:

> At first the oil was contained by Vikoma booms (one flown up especially from Orkney), but there were no pumps available capable of pumping it ashore when it became viscous at low temperatures during a cold spell. After 4 days, mechanical failures of different types caused first one and then the other boom to fail and to deflate, leaving the oil free to spread over the whole of Sullom Voe and the adjacent coasts of Yell Sound for up to 16 km around . . . [This incident] appears to have made a fairly clean sweep of seabirds in Yell Sound and Sullom Voe, but these are not their most important habitats in Shetland. Fortunately it failed to reach the main seabird concentra-

> tions a few miles further away on the far side of the island of Yell, where it might have created a world record with the first well-documented six-figure birdkill. (Bourne, 1979, pp. 93–94)

Yet even prior to the incident the Shetland Islands Council oil pollution officer had warned that "from an ecological point of view Sullom Voe is a disaster before it starts," predicting that with 700–800 tankers of up to 300,000 tons using the terminal each year there would be a minimum of at least one minor spill a week, one over 100 tons annually, and one over 2,000 tons once in a decade (quoted in Bourne, 1979, p. 93). Data compiled by ACOPS confirm the very real threat posed by the transshipment of North Sea oil at onshore terminals. There were thirty-five oil pollution incidents recorded during 1979 in the waters surrounding the Orkney and Shetland Islands, compared with only fifteen during 1978. In addition to oil spilled from the *Esso Bernica*, Orkney beaches were reported to have "suffered chronic pollution from February through to April, and this [is] believed to be due to a number of vessels washing their tanks, dumping waste oil fuel, or discharging dirty ballast while en route to Sullom Voe" (Advisory Committee on Oil Pollution of the Sea, 1980, p. 62).

Quite apart from the indirect contribution of offshore activities, it is inevitable that oil will be spilled either during tanker loading operations offshore, through pipeline failure, or as a result of a blow-out or other equipment malfunction. There is always the potential for significant environmental damage should a large spill occur under conditions that inhibit physical dispersal or biological degradation of the oil before the slick is washed ashore. R. Johnston concludes that with current offshore production, storage, and transportation techniques, "one might expect a 100,000 ton spill somewhere in the North Sea once in 25 years and one of 400,000 tons over a 50-year period. No refinement of statistics can tell whether well-head, platform, shipping or terminal would be involved or where the accident might happen" (1976b, p. 119). Hitherto, with the exception of the Ekofisk blow-out in 1977, there have been few problems associated with drilling in the North Sea. And even the Ekofisk incident, occurring in the open waters of the North Sea, posed far less of a threat to the environment than did the much smaller spill of fuel oil from the *Esso Bernica* in the confined waters of Sullom Voe.

Yet, despite the excellent drilling record to date, it is possible that the geographical location of offshore activity has encouraged complacency, at least among British authorities. It is frequently noted that any oil spilled during North Sea production will tend to move away from the British coastline toward Norway and to be dispersed by natural processes long before it reaches coastal waters, as indeed happened during the Ekofisk blow-out. As the RSPB study observes, this may not always be the case. If the focus of oil exploration continues to shift to the west of the Shetlands and to the approaches to the English Channel, the tendency would be for oil to move toward the British coastline. Moreover, for all the preoccupation with blowouts, there will be routine operational discharges at offshore facilities as well as other types of accidents involving oil spills.[4] The latter have occurred in the U.K. sector of the North Sea more frequently than is usually appreciated, although on most occasions the spillage has been limited. In 1976, for example, ACOPS (1977, pp. 16–17) reported seven spills from offshore facilities, including two from the Beatrice field, where the operator was threatened with revocation of the production license. The development of North Sea fields has inevitably been accompanied by an increase in the number of spills from the offshore installations. In 1979, for example, ACOPS recorded a total of thirty such incidents and expressed concern "that the number of spills originating on offshore installations appears to have doubled since 1978 (when 15 incidents were recorded), but [ACOPS] accepts that the number of offshore fields now in production, and better reporting procedures, may have contributed to the higher figure" (Advisory Committee on Oil Pollution of the Sea, 1980, p. 25). Most incidents appear to have occurred during the loading of tankers at single buoy moorings (SBM). Simple mechanical or procedural failures resulting in a break in the connection between the tanker and the buoy have been the primary cause. According to one study, tanker offloading operations were responsible for at least two spills a month in the first quarter of 1975 (Organization for Economic Co-operation and Development, 1977a, p. 31). One of the most serious incidents of this nature occurred at the Montrose field in 1976 as a result of a mechanical failure of the SBM system used to load tankers. Between three and four thousand barrels of oil were spilled, creating a six-mile slick, before the platform crew was alerted and loading operations halted (Advisory Committee on Oil Pollution of the Sea, 1978, p. 19). However pipelines do not completely eliminate the possibility of a spill, as shown by the loss of around thirteen thousand barrels of crude oil from the pipeline connecting the Thistle field with the Dunlin field in April 1980 as a result of damage caused by a ship's anchor (Advisory Committee on Oil Pollution of the Sea, 1980, p. 15).

The likelihood of serious environmental damage due to accidental spills and/or chronic discharges will be greatly increased when oil is discovered in commercial quantities close inshore. At the present time the Beatrice field has the distinction of being the closest to the U.K. coastline—fourteen miles off Caithness.

> It is within 15 miles of Britain's largest guillemot colony (Inver Hill), within 35 miles of five other major auk colonies and Britain's top scoter and long-tailed duck wintering area, and is located in the middle of a rich offshore seabird feeding ground. Furthermore, its associated onshore terminal shares a common boundary with a National Nature Reserve of international importance for estuarine birds [Nigg Bay on Cromarty Firth]. Tanker routes into the terminal pass very close to two other major auk colonies and the above mentioned seaduck site. The oil pollution risk attached to the commercial development of this oilfield is a matter of grave concern. (Royal Society for the Protection of Birds, 1979, p. 74)

In other areas oil pollution risks may need to be weighed against not only the "costs" to wildlife but other competing uses of the coastline as well. Thus one notes, for example, the recent onshore discoveries at Wytch Farm in Dorset and the interest in the possible offshore extension of this field into the English Channel, an area that is immediately adjacent to major resort communities.

The Fate and Effect of Oil in the Ocean

Evaluating the potential impact of energy development on the marine environment involves two basic questions—first, how frequently oil will be spilled and in what amounts; second, whether the introduction of oil (directly or indirectly) will have an adverse impact on human health and/or welfare.[5] Both questions, however, defy a very precise answer—there are simply too many unknowns. In a recent review of the effects of petroleum on marine organisms, A. D. Michael comments that "there is perhaps a tendency to assume that we know a good deal about oil spills now since there is a considerable volume of literature. This attitude emanates from the supposition that if one cannot see effects, they do not exist. I hold the view that we still have much to learn about the consequences of spills" (1977, p. 132). The chronic discharge of oil into estuaries and waters in particular has received little attention even within the scientific community (Kerr, 1977). At the present time it seems safe to say only that there is some general understanding both of the movement of oil through the marine ecosystem and of the processes that contribute to the dispersal and degradation of oil. However, oil spills tend to be highly individualistic phenomena. Crude oil exhibits marked differences in chemical composition (even on occasion from one well to another within the same field), while the external conditions—location, time of year, wind and water conditions—will vary enormously from spill to spill. In this situation no two spills are likely to behave in exactly the same way. In effect "circumstances" determine the nature of the impact (Kerr, 1977). Despite the difficulty of generalizing about the fate and effect of petroleum in the marine environment, an understanding of the variables identified in Figure 6 is critical in order to assess the pollution risk and to ensure that contingency planning is based on sound scientific and ecologic principles.

CHARACTERISTICS OF SPILLED OIL

Crude oil is not a single chemical but a complex mixture of hydrocarbon compounds with widely differing properties and toxicities. Although most of these compounds contain only hydrogen and carbon (arranged in varying combinations), a limited number may contain oxygen and nitrogen; sulfur is also likely to be present as a separate element, as hydrogen sulfide, or in various organic sulfur compounds, from trace amounts to as much as 5 percent by weight (National Academy of Sciences, 1975, pp. 42–43; Ryan, 1977, p. 4). Obviously the particular combination of hydrocarbon compounds present will determine the physical as well as the chemical properties of the crude oil (Table 19). The National Academy of Sciences report, for example, describes an "average" crude from Venezuela as containing 10 percent gasoline (hydrocarbon compounds in the C_5–C_{10} molecular weight range), 5 percent kerosene (hydrocarbon compounds in the C_{10}–C_{12} molecular weight range), 20 percent light distillate oils (hydrocarbon compounds in the C_{12}–C_{20} molecular weight range), 30 percent heavy distillate oils (hydrocarbon compounds in the C_{20}–C_{40} molecular weight range), and 35 percent residuum oil (hydrocarbon compounds of $> C_{40}$ molecular weight).[6] Yet this is a considerably "heavier" crude than the world-wide average, which might contain up to 40 percent gasoline and kerosene (National Academy of Sciences, 1975, p. 43). By way of contrast, North Sea crude is both "lighter" than the global average (i.e., it contains a high proportion of the smaller molecular size compounds) and "sweeter" (i.e., it contains only very small amounts of sulfur).

Of particular importance is the proportion of paraffinic to aro-

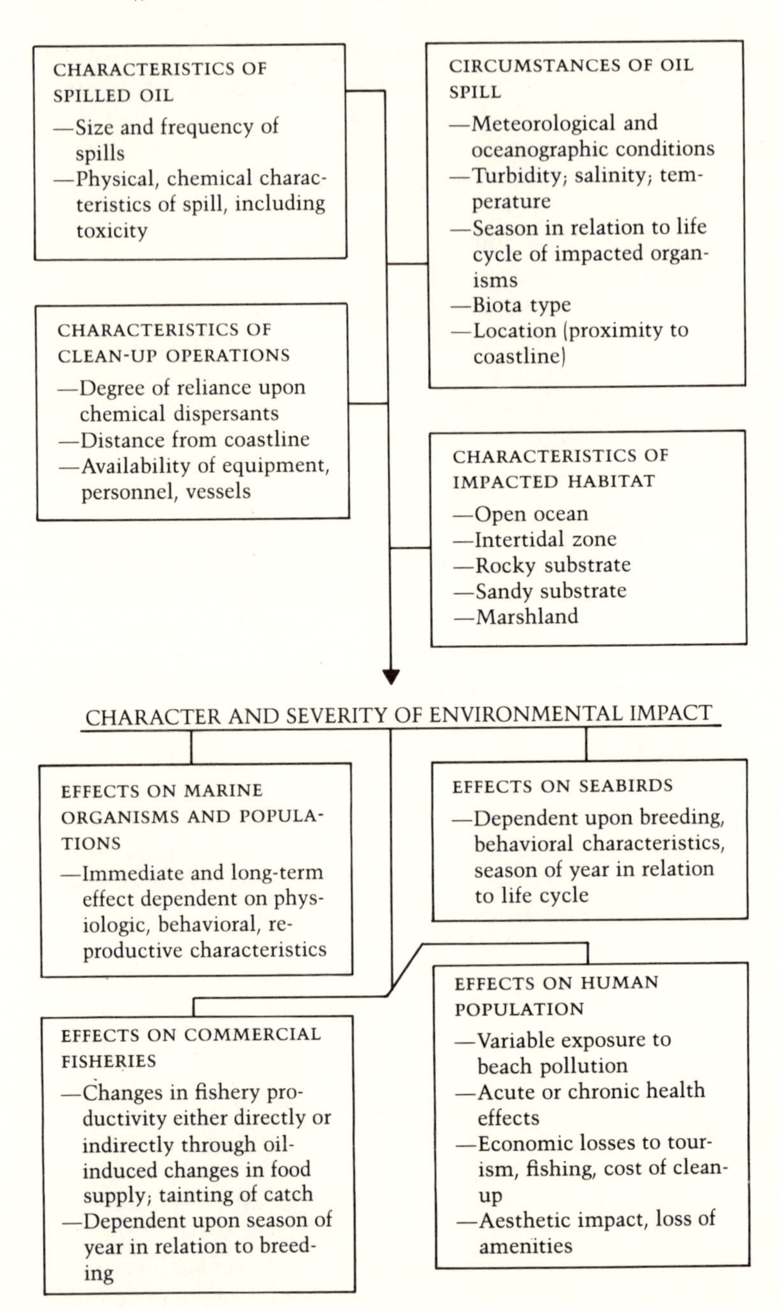

Figure 6. The Circumstances and Impact of an Oil Spill

matic hydrocarbons present in crude oil, since this will affect both the rate of dispersal (spreading, evaporation, solution) and the biological impact of a spill.[7] The lower paraffins (such as methane and ethane), for example, are quickly lost through evaporation, while the aromatic hydrocarbons include the most soluble and toxic components present in crude oil. In this respect there is general agreement that the toxicity of crude oil increases along the series paraffins, naphthenes, olefins, to aromatics, while within each series of hydrocarbons the smaller molecules are more toxic than the larger (Cowell, 1976, p. 369). Since the smaller molecules in each fraction also tend to be the most volatile, a weathered slick will be considerably less toxic than freshly spilled crude oil.[8] Subsequent investigation of the *Torrey Canyon* incident, for example, indicated that the oil stranded on the Cornish beaches was almost biologically inert; the extensive damage inflicted on marine flora and fauna resulted from the use of inappropriate clean-up methods by teams of inexperienced workers anxious to placate public feeling (Cowell, 1976, pp. 370–371). In general therefore the lighter crudes that contain a high proportion of aromatic hydrocarbons of low molecular weight are more toxic but will evaporate more rapidly. According to most estimates, North Sea crudes may lose up to 40 percent of their mass through evaporation within a few hours. However the residue of heavier hydrocarbon compounds may persist for some considerable period of time, posing a serious threat (through physical coating rather than direct toxicity) to marine flora and fauna. Depending upon the location of the spill, this weathered oil will ultimately enter coastal waters and accumulate on exposed shores or gradually sink to the ocean floor through sedimentation.

As already discussed, there is particular concern over the increasing number of incidents around the U.K. coastline that involve spills of fuel oil. While not directly related to offshore activity, the variety of refined products (whose chemical composition will reflect the character of the crude oil from which they were distilled as well as the nature of the distillation process) further complicates oil spill contingency planning. In general the heavy "black" fuel oils (such as Bunker C or No. 6 fuel oil) that are used for industrial heating, power generation, and marine fuels are among the most persistent of all petroleum products. Chemically the Bunker C or No. 6 fuel oils are composed almost entirely of hydrocarbons in the C_{30^+} range. In the event of a spill, very little oil will be lost through evaporation. Slicks of heavy fuel oil are likely to persist on the ocean surface and may well travel considerable distances. As the *Eleni V* incident demonstrated, existing clean-up techniques (whether involving chemical

Table 19. Crude Oil: Chemical Characteristics

Field	*Gravity*[a] *(°API)*	*Sulfur (% weight)*	*Hydrocarbon C_4 (% volume)*
Brent	38.16	0.26	3.2
Ekofisk	36.3	0.21	1.0
Magnus	39.3	0.28	3.15
Murchison	38.8	0.28	3.61
Ninian	35.1	0.41	2.07
Abu Al Khoosh (Abu Dhabi)	31.6	2.0	—
Kuwait	31.2	2.5	2.46
Lagomedio (Venezuela)	32.6	1.23	—
North Slope (Alaska)	26.8	1.04	—
Pennington (Nigeria)	37.7	0.076	1.1

[a]A formula developed by the American Petroleum Institute to determine the weight of crude oil or other petroleum products. The higher the °API Gravity, the lighter the oil. North Sea and other crudes in the mid- to high 30's are considered very light, while North Slope oil with an °API Gravity of 26.8 is heavier but still well within the range of desirable crude oil weight.

Sources: Oil and Gas Journal, vol. 77, no. 48 (1979), pp. 47–53; *International Petroleum Encyclopedia*, 1979, pp. 296–316.

dispersants or mechanical systems) are not very effective in coping with spills of heavy fuel oils.

Light distillates include kerosene, gasoline, chemical feedstocks, and light solvents. In contrast to the heavier distillates, these products contain hydrocarbons in the C_4–C_{12} molecular weight range. Although highly toxic, they present a less serious hazard than crude or fuel oils in open waters. Studies undertaken both in the open sea and under laboratory conditions indicate that virtually all hydrocarbons below C_{12} are lost from oil slicks within twenty-four hours (McAuliffe, 1977a, p. 23). Again, however, the importance of carefully discriminating between the properties of different crude or refined products is critical. Although the lighter hydrocarbon compounds such as gasoline are regarded as having the highest acute toxicity, there are exceptions. Some American No. 2 fuel oils, for example, contain over 45 percent low aromatics (Nelson-Smith, 1977, p. 57). As might be expected, spills of refined products of this type in confined environments are likely to have a devastating impact. The wreck of the barge *Florida* near West Falmouth on the Massachusetts coast in 1969, for example, resulted in a spill of No. 2 fuel which was driven by onshore winds onto the beaches and into the estuaries and marshes of Buzzards Bay. The local fishing industry was immediately affected; restrictions on clam harvesting remained in effect eight years after the spill, at which time the No. 2 fuel oil could still be found in marsh sediments (Sanders, 1977). John W. Farrington points out the obvious lesson: ". . . if you do not see an oil slick on the water, it does not mean that the oil has disappeared from the sediments under the water. Since the sediments and associated bottom dwelling organisms are an integral part of the coastal marine ecosystem, there is continued chronic exposure of these ecosystems to components of fuel oil" (1977, pp. 7–8).[9]

In summary, once petroleum is introduced into the marine environment, weathering processes immediately begin to transform the spilled material as it spreads out across the ocean surface. This transformation involves a wide range of physical, chemical, and biological processes—evaporation, emulsification, oxidation, degradation by microorganisms. The form and rate of weathering will be at least in part determined by the chemical composition of the petroleum fraction. "Since the physical and chemical processes that such petroleum products undergo in the marine environment must depend on this composition, a knowledge of at least the major component types and their properties is an essential prerequisite for predicting the fate of petroleum in a more general sense" (National Academy of Sciences, 1975, p. 42). The physical and chemical changes determine the subsequent behavior of the spill, its toxicity, and the most appropriate clean-up treatment. In effect petroleum-derived hydrocarbon compounds in widely varying combinations are being introduced into a dynamic ecological system and will actively interact with the other elements and organisms within the system.

THE DISPERSAL AND DEGRADATION OF SPILLED OIL

Dispersal of spilled oil by physical processes (spreading, evaporation, emulsification, solution, sedimentation, and direct sea-air exchange through wave action and bubble breaking) is accompanied by alterations in the character of the slick as a result of chemical oxidation and biological degradation. Biological processes (including uptake by larger organisms as well as degradation by microorganisms) are most effective after the physical processes have been initiated (McAuliffe, 1977a). As a result the relative importance of the different weathering agents varies through time, the more rapid physical processes that facilitate dispersal giving way to slower chemical and biological modification as the spill ages. Ultimately most of the oil spilled on the ocean surface will be transferred to the atmosphere, dispersed in the water column (in both particulate and dissolved form), incorporated into the bottom sediments (through adsorption or absorption onto particulate matter), or oxidized by chemical or biological means to carbon dioxide.

The fate of spilled oil is summarized in schematic form in Figure 7, although it should be emphasized that the dearth of quantitative information makes it difficult to assess the relative importance of the pathways whereby oil is removed from the marine environment except in the most general of terms. Many intermediary products will be formed in the process of weathering. The lumps and balls of tarry oil that are to be found floating on the ocean surface, for example, are formed from the heavier fractions of spilled oil.[10] According to the NAS report (1975, p. 105) the presence of paraffinic wax in many of these tar balls similar to that formed on the compartment walls of oil tankers, together with their high iron concentration, is evidence that most of these materials originate from tanker washing and bilge discharges. These tarry oil residues are a major focus of public concern, particularly when washed ashore on exposed coastlines, but they do not represent the ultimate fate of the spilled oil. For all that, the increasing amount of tar found both in the open oceans and along formerly unpolluted coastlines does raise serious questions about the *rate* of input of petroleum products in relation to the capacity of natural dispersal and degradative processes.[11] As the NAS study notes, "When oil becomes incorporated in coastal sands protected from the weathering effects of sun and oxygen, its residence time may be measured in years or decades. Unless steps are taken to reduce the input to a level that can be assimilated through natural degradation processes, we will all have to reconcile ourselves to oil contaminated beaches," (National Academy of Sciences, 1975, p. 105).

The weathering of oil is an even more complex and variable phenomenon than the above description suggests. As already noted, the physical and chemical properties of the oil as well as the external circumstances of the spill (i.e., wind and sea conditions, water temperature, salinity, the presence of organisms, and the availability of nutrients) greatly affect the rate and nature of the changes that will occur within the oil slick. Of particular concern, given the expansion of offshore exploration and development into Arctic and sub-Arctic regions, is the fate of any oil spilled in colder or even ice-covered waters.[12] Dispersive and degradative processes are likely to operate much more slowly in such environments; the viscosity of oil is increased in cold water, evaporation and solution rates are reduced, and photochemical oxidation and microbial degradation will

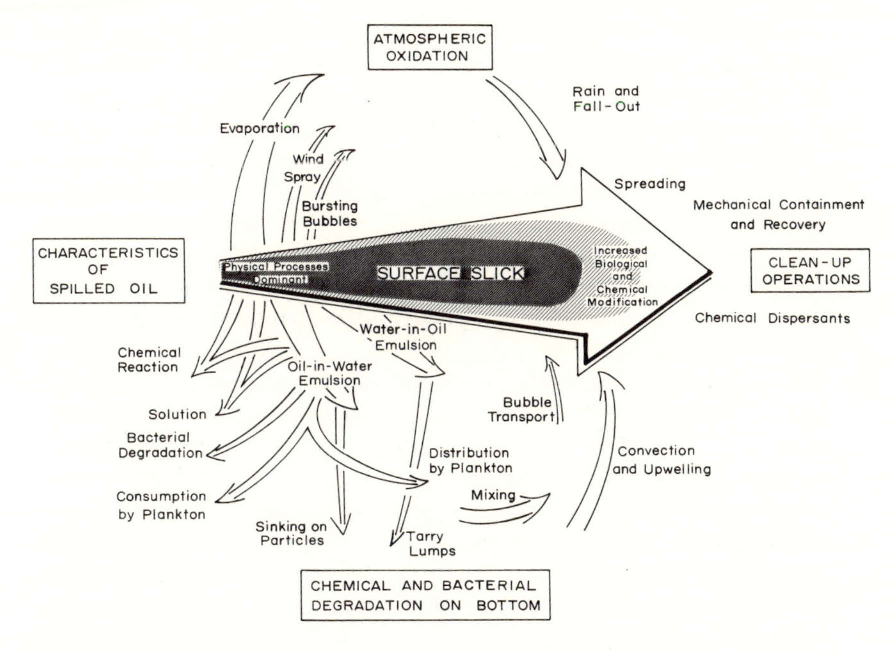

Figure 7. The Fate of Spilled Oil (after Nelson-Smith, 1977, p. 51)

exhibit marked seasonal variations (McAuliffe, 1977a, pp. 28–29). Additional complications in determining the fate of spilled oil arise from the fact that weathering processes do not operate independently of each other but interact in ways that are both complementary and competitive. Less viscous oils will spread more rapidly; this in turn will facilitate evaporation of the volatile components. Evaporation and solution will be among the first compositional changes to occur, yet these are competitive processes, with actual rates dependent upon the vapor pressure and solubility of each volatile hydrocarbon (McAuliffe, 1977a, p. 20). Rough seas tend to increase the sea-air transfer of petroleum, since sea spray and bursting bubbles will eject both volatile and nonvolatile components into the atmosphere. Yet, closer inshore, rough seas may contribute significantly to the process of sedimentation as sand, silt, clay, and shell fragments stirred up in the shallow water provide particles for adsorption of dissolved hydrocarbons. Finally the application of chemical dispersants can significantly alter the behavior of spilled oil. Such dispersants are intended to accelerate the dispersal and degradation of a slick by exposing a larger surface area to evaporation, photo-oxidation and biodegradation; however there remains disagreement over the risk of synergistic toxic effects resulting from the addition of the chemical dispersant (McAuliffe, 1977a, pp. 27–28).

As noted in the NAS report, despite the apparent complexity of the processes affecting an oil spill, "it is possible to identify stages in the life time of a spill and to assign priorities to the processes acting to modify it" (National Academy of Sciences, 1975, p. 51). Initially the spilled petroleum spreads outward under the influence of gravitational and surface chemical effects, although the density and viscosity of the spilled material as well as wind, wave, and current action will all influence the extent, shape, and rate of slick growth. Some of the more viscous heavy fuel oils, such as Bunker C, may be subject to very little spreading. More characteristically, an expanding slick will consist of a distinct central core composed of the heavier fractions that spread relatively slowly, while the outer fringes take the form of a rapidly expanding film of lighter distillates. This provides the characteristic rainbow-colored sheen that approaches monomolecular dimensions (a faint silvery sheen) at the extreme margins of the slick.

The spreading of the oil slick facilitates other dispersive processes, since a much larger surface area is exposed to air, sea, and sunlight. Evaporation of the more volatile components is the predominant process, and its rate increases with wind speed and the further spreading of the spill.[13] However, all crude oils will leave a residue of heavier distillates; indeed, as evaporation proceeds the spilled crude will more closely resemble Bunker C fuel oil in carbon number distribution, having lost those hydrocarbons containing less than C_{12} through evaporation.

Since most of the residual components of the spill are relatively insoluble in water, an important mode of dispersion for those fractions not lost through evaporation is emulsification. Water-in-oil emulsions, frequently described as "chocolate mousse," contain as much as 80 percent water along with the more coherent, semi-solid lumps of oil. Their formation appears to be related to the presence of nitrogen-, sulfur-, and oxygen-containing compounds in the oil, and hence the rate of formation is likely to vary dramatically with the type of material spilled. Most mousses appear to be highly resistant even to chemical dispersants and may persist as floating masses for a considerable period of time. Water-in-oil emulsions have frequently been identified as the source of pelagic tar balls. However, McAuliffe (1977a, p. 20) suggests that mousses derived from crude oil spills contribute only a small proportion of the oil appearing on beaches as tar balls and that a much more significant source of such tar is bilge washing and Bunker C discharges.

Other processes, including photochemical oxidation, direct transfer to the atmosphere, and sedimentation through adsorption of hydrocarbons onto nonbuoyant particulate matter, all act to disperse the spilled oil. Some of this oil may find its way back into the marine environment. The oil-coated droplets ejected into the atmosphere from sea spray and bursting bubbles, for example, may well fall out downwind of the spill. Such fall out of oil-coated droplets may be significant where chronic pollution exists or if a major spill occurs in coastal waters.

It should be evident that given the range and complexity of the variables involved, any precise estimate of the rate of removal and dispersal by these various processes (even for a hypothetical spill of an "average" crude oil) is virtually impossible. Under particularly favorable circumstances dispersal may be very rapid. In the case of the Ekofisk blow-out, for example, about half of the crude oil is believed to have evaporated before it hit the water. Two months after the blow-out researchers from the Torrey research station in Aberdeen found few indications of any oil in the vicinity of the platform, while none of the spilled material ever reached the shore (Kerr, 1977, p. 1135). Similarly favorable circumstances appear to have existed in the case of the *Argo Merchant* spill involving the release of 175,000–200,000 barrels of heavy fuel oil in shallow waters off Nantucket. Prevailing winds moved the spilled oil, which quickly congealed into large, pancake-like lenses, into the open waters of the Gulf Stream (Kerr, 1977, p. 1135). In the months following the spill very little oil could be detected in sediment samples, suggesting that the bulk of the spilled oil had either been transported to another location or become too dispersed to be traced (Kerr, 1977; Milliman, 1977). Rapid dispersal of course does not mean that all the spilled oil has been removed from the marine environment; rather it contributes to the background level of petroleum products distributed through the water column, in bottom sediments, and incorporated into organic matter.

Under less favorable circumstances dispersal will be neither as rapid nor as complete. However, as a spill ages the more persistent tarry residues are increasingly subjected to chemical and biological processes of weathering. Photochemical oxidation, degradation by adapted microorganisms to carbon dioxide or organic matter, uptake by larger organisms and subsequent metabolism, storage, or discharge all play a role in the ultimate disposal and degradation of the oil. Such processes operate only very slowly in this final stage: "Further degradation, weathering, and interaction with the environment is extremely slow since the surface-to-volume ratio of unspread tar is small, and most dispersive reactions occur at interfaces. More importantly, since the petroleum residue is nonfluid, additional spreading ceases and the internal contents become encapsulated and isolated from effective interaction with dispersive and degradative processes. Microbial degradation becomes important as populations of hydrocarbon-adapted bacteria develop" (National Academy of Sciences, 1975, p. 52).

As the NAS report makes clear, these processes (and particularly microbial degradation) have not been well studied in the field. "On the basis of available information, the most that can be stated is that some microorganisms capable of oxidizing chemicals present in petroleum (under the right conditions) have been found in virtually all parts of the marine environment," that the rate of microbial degradation varies enormously, and that "the time required for substantial decomposition of the most resistant components of petroleum in the marine environment is probably measured in years to decades" (National Academy of Sciences, 1975, p. 66). Yet an understanding of these processes is particularly important for any assessment of longer-term impacts on biological communities. The rate of degradation of the more toxic fractions will be particularly crucial. According to the NAS study, "both laboratory experiments and some field observations have shown that microorganisms consume the least toxic fraction of petroleum (normal alkanes) in a few days or months, depending on temperature and nutrient supply" (1975, p. 106). David T. Gibson points out, however, that comparatively little is known about the microbial degradation of the more toxic aromatic hydrocarbons present in crude oil. "It is possible that many of the polcyclic aromatic hydrocarbons [a group that includes known carcinogens] . . . that are present in crude oil are poor candidates for biodegradation" (Gibson, 1977, p. 43).

Understanding the fate of oil entering the marine environment clearly presents a research challenge of enormous magnitude. Given the range of variables involved, perhaps all that can be safely stated by way of summary is that with luck and under favorable circumstances a major spill of crude oil in open ocean waters may be rapidly dispersed; other spills, however, may remain toxic or persist for much longer periods of time, posing a major threat to biological communities should they occur close inshore or become confined in protected inlets and estuaries.

THE ENVIRONMENTAL IMPACT OF MARINE OIL POLLUTION

Although much of the oil spilled on the ocean surface is fairly rapidly dispersed and degraded, this does not occur without some effect on living systems. Direct toxicity together with the physical coating of flora and fauna are the most obvious ways in which spilled oil inflicts damage on the marine environment.[14] As Eric B. Cowell, Geraldine V. Cox, and George M. Dunnet (1979, p. 5) point out, however, loss of habitat and changes in the available food supply represent a more subtle form of environmental damage—one that may cause as much damage to ecosystem structure and productivity as direct toxicity, if not more. They cite, by way of example, the loss of sediment stability following the destruction of spartina grass in a coastal marsh or estuary. Crab and mussel communities will be affected by the loss of habitat, with implications for other dependent organisms in the food web, but these effects will only become apparent long after the initial spill.

The ecological consequences of a spill are likely to vary enormously. A single factor such as wind direction may be crucial in determining whether the spilled oil is driven onshore with severe effects on biota or dispersed at sea with little observable impact (Michael, 1977, p. 132). Toxic effects may well be intensified on rocky coastlines where trapped oil saturates the small volume of water remaining in tidal pools. Damage may be far more serious at certain times of the year than at others, with populations that congregate in particular locations during feeding, migration, or breeding proving especially vulnerable. The RSPB study suggests that the United Kingdom has been singularly fortunate in the location and timing of oil spills: ". . . few birds were affected by *Eleni V* because few seabirds of vulnerable species are present off East Anglia in May. Similarly, the total affected by *Christos Bitas* off the coast of SW Wales in October 1978 could have been far higher if the incident had occurred in July, when auks and many more gannets would have been in attendance at important colonies nearby" (Royal Society for the Protection of Birds, 1979, p. 69).

Quite apart from the external variables, there will be significant inter- and intra-species differences in response. A single coating of crude oil will cause mortality in certain species, whereas others may survive several coatings. Intertidal organisms, already adapted to alternating periods of immersion and exposure, to sudden infusions of fresh water and heightened salinities due to evaporation, tend to be more resilient to oil-induced stress than subtidal species. Where organisms are already close to their limits of tolerance to temperature or salinity, oil pollution will be more serious than where environmental conditions are optimal. Immobile benthic organisms (such as barnacles) will be more drastically affected than fish and more mobile crustaceans. Suffice it to say that the interactions between spilled oil and the organic environment are extremely complex.[15]

Despite wide variations in the ecological consequences of a spill, the NAS report concluded that a number of generalizations could be made: "In general, where damage was severe, the oil spill was massive relative to the size of the affected area, and the spill was confined naturally or artificially to a limited area of relatively shallow water for a period of several days . . . For a given quantity of oil, the more localized the distribution of the spill, the greater is the mortality" (National Academy of Sciences, 1975, p. 98).

The potential effects of a major spill may include:

1. Permanent or temporary modification of ecosystem structure and productivity as a result of changes in species numbers and composition.

2. Damage to marine fishery resources.

3. Loss of valuable wildlife resources including seabirds and marine mammals.

4. Threat to human health through uptake of petroleum hydrocarbons by marine organisms.

5. Decline in the aesthetic and recreational value of coastal resources due to unsightly slicks and tarred beaches.

These effects are clearly very different in character, ranging from the biological to the aesthetic.[16] Some are of immediate practical significance to society; others initially appear to be unimportant but may prove to be cumulative and potentially significant in the long-term (Michael, 1977, p. 129). In many instances (as in the "loss" of less tangible amenity and aesthetic attributes of the environment) the cost simply cannot be measured in conventional economic terms. Inevitably there will be widely divergent views as to how these various effects should be evaluated and weighed against the benefits of energy development. As the National Academy of Sciences report observed, "at various places in the marine environment and at various times these [effects] will be accorded different priorities in the evaluation of the impact" (1975, p. 73).

From the public's perspective, the most serious consequences of a spill are likely to be those that have a high visibility—the tarring of beaches, the tainting of seafood (particularly shellfish), and the oiling of seabirds. These will undoubtedly be most pronounced in the aftermath of a major spill in coastal waters. Quite apart from the direct cost of cleaning up the shoreline, economic losses will be particularly severe in those communities where tourism and fishing are important activities.[17] These losses may even extend to communities at some distance from the site of the spill, as in the reluctance of the public to visit neighboring beaches or to buy seafood for fear it may prove to be contaminated. Anthony Nelson-Smith (1977, p. 60) describes the drastic impact of the *Torrey Canyon* spill on fish sales in Paris markets even though the supply came from areas unaffected by the spill. Clearly the direct costs to society of a major spill, together with the massive loss of wildlife, more than justify efforts to prevent such spills from occurring. Yet to describe such incidents as "ecological catastrophes" tends to obscure the fact that even where spilled oil is trapped within a coastal marsh or estuary, persisting for considerable periods of time in the sediments, a diverse biological community is eventually re-established. Even in the *Torrey Canyon* incident, which perhaps more than any other spill was responsible for arousing and shaping public concern in the United Kingdom, despite very high mortalities in intertidal shore life and decimation of local seabird populations, recovery of the impacted area was virtually complete within ten years. The combined effect of the spilled crude oil and the toxic chemical dispersant used in clean-up operations was removal of virtually all plants and animals along the affected coastline. Yet the ecosystem recovered through the natural processes of community colonization and succession.[18] Initially an unstable pioneer community became established, dominated by a single species of alga. Gradually the more opportunistic species that characterized the initial stages of recovery gave way to a more diversified autotrophic community, grazing animals returned, and with time a more complex and stable ecological system was re-established.

Obviously recovery time will vary greatly. Length and degree of exposure, the properties of the spilled oil, the ecology of the impacted area all influence the process of recovery. Cowell, Cox, and Dunnet emphasize that resilience (defined as the ability of an ecosystem to regain stability after a disturbance) varies markedly between habitats:

> . . . a rocky shore has a low diversity in types of organisms—few can withstand: periods of salt water and air; temperature ranges from −10°c to 40°c; dessication; and the heavy pounding surf. They are characterized by short-lived opportunistic species with highly mobile planktonic dispersion phases.

> Therefore, such communities are of low persistence but highly resilient. They recolonise and recover rapidly from major disasters.
>
> By contrast, sheltered littoral communities, the shrub stages of tidal marshes, mangrove swamps, and the inner lagoons of coral atolls are characterized by low-diversity, narrow-niche specialists. These communities have relatively high biomass and ecosystem persistence, but they have low resilience. Recovery from an ecological disaster is slow. (Cowell, Cox, and Dunnet, 1979, p. 5)

Moreover *full* recovery (defined as the complete elimination of petroleum hydrocarbons and the re-establishment of all faunal and floral constituents present before the spill with their full complement of age classes) may require decades (Vandermuelen et al., 1978, p. 7).[19]

Despite the preoccupation with major spills, it may well be that chronic pollution, involving repeated or continuous discharges of crude oil, refined distillates, or oily wastes, poses a far more serious long-term ecological problem. In such circumstances biota have insufficient time to recover between inputs resulting in permanent and even irreversible changes in community structure and productivity. Michael (1977, p. 132) estimates, for example, that around 300,000 tonnes of oil and grease enter New York Bight each year, largely through wastewater and runoff. Much of this input must be recycled "or we would have an oilfield in the bight by now," but oil is undoubtedly accumulating in the sediments. "We should be learning the dynamics of the cycling of petroleum hydrocarbons, the rate at which they are accumulating and their biological effects in order to make reasonable predictions about the consequences of increased dosages in such systems" (Michael, 1977, pp. 132–133). Unfortunately a major problem in any such investigation is the difficulty of distinguishing between the effects of the various pollutants usually found in chronically contaminated areas. Moreover, as the National Academy of Sciences report observes, in areas most affected by chronic oil pollution such as the Gulf Coast of Louisiana and Texas it is virtually impossible to separate the effects of marine oil pollution from other activities occurring in the coastal zone. Thus dredging of coastal waterways, spoil disposal, reclamation of wetlands, alterations in tidal flow patterns, and salt water intrusion, often themselves a direct or indirect result of oil-related activity, all result in profound changes in the coastal environment, modifying habitats, disrupting food supplies with lethal or sublethal effects on individual organisms. "The period that the Louisiana fisheries can withstand these alterations to the coastal environment remains unknown. That the fisheries have not seriously declined reflects the great reproductive potential and resiliency of these oyster and shrimp species" (National Academy of Sciences, 1975, p. 89). Nevertheless the report cautions against underestimating the potential effect of chronic oil pollution on resources that are of commercial importance or of extrapolating from Louisiana to other areas.

From the public perspective seabird mortality continues to be one of the more important criteria used in assessing the ecological impact of a spill. "The common tendency is to look at a spill at its darkest hour . . . with oil dripping from a dead duck" (Cowell, Cox, and Dunnet, 1979, p. 5). In view of the emotional response to seabird mortality, there is need for particular care in interpreting the available data and in recognizing the variable degree of risk to particular populations.

On occasions oil spills may result in quite massive destruction of bird life. According to James Eric Smith (1968, p. 14) thousands of seabirds were killed following the *Torrey Canyon* spill. Moreover the count included a significant proportion of the total local population of certain species, notably auks (guillemots, razorbills, and puffins). Among British seabirds auks, along with seaducks (scoter, scaup, eider, long-tailed), divers, and grebes are particularly vulnerable to oil pollution. "These birds spend most of their lives on the surface of the sea, dive to collect their food, and are weak fliers or flightless. They dive rather than fly up in response to disturbances; if they dive on encountering floating oil or if they surface, they become completely coated with oil. Since these birds are also highly gregarious, it is possible for a small oil slick to cause very large casualties" (National Academy of Sciences, 1975, p. 92). All three species of auks found in significant numbers in U.K. coastal waters congregate on the surface in dense flocks close to colonies before and during the breeding season. Susceptibility to oiling is significantly increased during periods of flightlessness. "The young of guillemots and razorbills abandon a colony by leaping from the cliff and plummeting to the sea below, 18–25 days after hatching. Accompanied by one parent, the flightless chicks swim quickly out to sea. In July and August, those reared in Shetland and Orkney and on northern coasts of mainland Britain are believed to swim across the North Sea towards the Norwegian coast with their parents, which themselves become flightless in August and September while moulting" (Royal Society for the Protection of Birds, 1979, p. 19).

Whether any population can be so reduced in numbers as a

result of oil pollution that recovery is impossible remains highly contentious. Cowell concludes that "most seabird populations are robust and have the potential to regain rapidly optimal populations in equilibrium with their food supply and their competitors . . . it seems very doubtful whether any bird species in Northern Europe is threatened with a reduction in numbers due to oil pollution, except on a local scale, where mortalities may be severe" (1976, p. 367). The recent RSPB report, however, is less reassuring and concludes that there is strong "circumstantial evidence" linking the marked decline in breeding colonies of auks along the English Channel over the past fifty years to the effects of chronic oil pollution aggravated by such incidents as the *Torrey Canyon* and the *Amoco Cadiz* (Royal Society for the Protection of Birds, 1979, p. 14). The National Academy of Sciences study also noted reductions in breeding populations of a wide variety of species in locations as far apart as the Baltic (velvet scoters, long-tailed ducks), the South Atlantic (particularly populations of jackass penguin along the African coastline), and Newfoundland (where razorbills, once numerous, are no longer present as breeders), and concluded that oil was a definite contributory factor (National Academy of Sciences, 1975, p. 93). The real difficulty in any such discussion is that it is rarely possible to attribute an observed decline in population numbers to a single cause. Rather a wide range of factors, acting synergistically, is likely to be responsible (Bourne, 1976). With respect to auk populations in the United Kingdom, for example, both the RSPB and the NAS reports acknowledge that climatic changes, affecting food supply, may have played a role in the northward retreat of this family, which in the English Channel is close to the southern limit of its breeding range. Others have noted the presence of heavy metal and organochlorine residues in seabird populations, including auks, and the threat this represents to reproductive success.

While the outcry over seabird mortality may on occasion appear to be somewhat extreme, it is not entirely misplaced. There does exist a potential long-term threat to a number of species resulting from (1) the increased exposure of local populations to oil pollution, particularly where such exposure takes the form of repeated spills, and (2) the indirect and synergistic effects of oil pollution on populations that may already be under severe stress due to loss of habitat, the presence of toxic chemicals in their food supply, or the introduction of predators. In such circumstances, oil contamination could prove to be the "final straw." And, from a conservation viewpoint, there must be particular concern for the long-term effects of oil-induced mortality on species where total numbers are small or where the potential recruitment rate into a breeding population is low (Royal Society for the Protection of Birds, 1979, p. 14). For the United Kingdom the first category would include the long-tailed duck as well as the great northern, black-throated, and red-throated divers. A recent RSPB census of the red-throated diver, probably the most vulnerable of the divers since it feeds at sea throughout the year, placed the total breeding population in Britain and Ireland at around 1,000 pairs. Of these about half are to be found in the Shetlands, many feeding in the waters of Sullom Voe in close proximity to the oil terminal. The rarer great northern diver breeds in Iceland, but a large proportion of the population overwinters in northern Scotland. Significant numbers of great northern divers were among the recorded casualties from both the *Esso Bernica* spill at the Sullom Voe terminal and the *Amoco Cadiz* grounding off Brittany. "Abnormal mortality in such species as these, where the total population is small, may be more harmful than comparable levels of mortality in more numerous species" (Royal Society for the Protection of Birds, 1979, p. 32).

Of equal concern is the effect of oil pollution on species that lack the potential to quickly replace losses due to increased adult mortality. This inability may result from low breeding success, small clutch size, delayed maturity, or high mortality rates among young birds. Auks, for example, usually lay a single egg each year. "It has been calculated that a typical 100 pairs of razorbills or guillemots will produce 70 young each year, of which maybe only 20% survive to breeding age . . . Thus the potential rate of recruitment into a breeding population is low, but under normal conditions is balanced out by a high adult survival rate of over 90% per annum" (Royal Society for the Protection of Birds, 1979, p. 19). Recovery of a population after a spill that eliminated a large proportion of breeding adults and potential first breeders would take an extremely long time, while repeated spills, even widely spaced in time, could have a devastating impact.

Moreover, the most important concentrations of vulnerable species in U.K. waters are to be found precisely in those areas where the level of oil-related activity, offshore exploration and development as well as surface shipments by tanker of crude products, is likely to increase dramatically over the next few years. These include populations that are of international significance in the sense that they represent a substantial proportion of the European or global total. According to RSPB estimates, for example, the U.K. razorbill population of around 150,000 pairs represents approximately 70 percent of the world total of this species, with over one-third of the U.K. population concentrated in the Orkneys, Shetlands, and along the northeastern coast of Scotland. "Hitherto [these populations] have been largely free of risk from oil pollution but this is no longer the case. Recently there has been a series of kills: more appear inevitable, and there is a real danger of massive bird mortality on a scale not yet seen in western Europe" (Royal Society for the Protection of Birds, 1979, p. 5).

As the oiling of seabirds raises particularly emotional issues, it has been treated in some detail. Clearly oil pollution, whether in acute or chronic form, has other major consequences for society. These include possible effects on fishery productivity, the degradation of the aesthetic and recreational value of coastlines, and the risk to health. In the first instance, there appears to be a general consensus that acute oil spills present little risk to commercial fisheries. The obvious exception to this assessment would be a spill occurring within a nursery or spawning ground or in an estuary during a salmon run. "Generally, fish do not suffer directly from sinking of oil but may acquire an aberrant flavor by feeding on benthic organisms carrying oil droplets. Even if the area were an important feeding ground for fishes of commercial importance, an appreciable mortality would not occur among such fishes, but they might become tainted, which would affect the fish industry economically" (National Academy of Sciences, 1975, p. 90). Similarly most evidence indicates that chronic pollution has not adversely affected fisheries' productivity in such areas as the North Sea or the Gulf Coast, although tainting of seafood, and particularly shellfish, is widely reported. In the North Sea, where catches of valuable species such as herring have declined significantly, it is virtually impossible to distinguish the impact of oil pollution from the effects of other pollutants and overfishing. Other marine organisms of commercial importance, including oysters, mussels, and edible seaweed, may prove more vulnerable due to their lack of mobility. By and large, however, the most serious and longest-lasting problem is tainting. Concentrations of less than 0.01 parts per million are reported to give rise to a marked oily taste in oysters, finfish, and seaweed which may persist for several months (Nelson-Smith, 1977, p. 60).

As with other aspects of North Sea development, it must be recognized that the concern of the fishing industry extends beyond oil pollution per se to the collective impact of offshore exploration and production. Additional restrictions on fishing grounds due to "safety zones" around rigs, platforms, and pipelines; possible long-term effects on fishery stocks due to disruption of migration and breeding patterns; compensation for fouled fishing gear; all acquire added significance in view of the possible opening up of British waters to EEC boats under the Common Market's fisheries policy.

In terms of direct effect on humans, there is little disagreement about the aesthetic and recreational losses caused by fouled beaches. Human health effects resulting from consumption of contaminated seafood are far more difficult to assess. In general it seems safe to assume that seafood that has become tainted by petroleum will not be consumed. However, as the National Academy of Sciences notes, taste is not necessarily a reliable way of preventing ingestion of food that represents a health risk. "Although it is the bulk hydrocarbon that gives the bad taste, it is the incorporated traces of carcinogen that represent the possible health hazards. If the carcinogens should be concentrated in the food chain relative to the total hydrocarbons, we could be faced with high carcinogen content without any bad taste" (National Academy of Sciences, 1975, p. 97). Of particular concern are carcinogens, particularly polycyclic aromatic hydrocarbons (PAH), known to be present in small amounts in crude oil. It is, of course, important to keep a sense of proportion, and, as the NAS study notes, "it must be kept in mind that these compounds are also assimilated from smoke of cigarettes and may be inhaled in the smoke from burning coal or petroleum, from burning refuse, from motor vehicle operations, and from the production of coke" (ibid.). Moreover, carcinogenic PAH compounds have been noted in a wide variety of vegetables, grains, and fruits. Thus the amount of carcinogen that could be ingested by eating contaminated seafood would appear to be no greater than that acquired from eating any other food.[20] Such an observation of course does not justify dismissing the matter. Moreover it is not exactly reassuring to learn that there exist no standards at all for oil-contaminated shellfish taken from the U.S. Gulf Coast, an area subject to chronic inputs of petroleum where reports of tainting have been frequent. It is generally assumed that, when transferred to clean water, tainted shellfish will quickly flush the oil from their tissues. Richard A. Kerr (1977, p. 1136) notes that recent research based on field rather than laboratory studies casts doubt on this assumption and observes that in any case a level of oil contamination that is considered safe for human consumption has yet to be determined.

In the final analysis, assessing the degree of risk is almost impossible when so little is known about the circumstances and processes whereby chemicals may induce cancer or about the rate of biological uptake, metabolism, and concentration of those carcinogens known to exist in petroleum.

> To maintain a reasonable degree of prudence in these matters, it seems clear that we must operate under the assumption that there is no safe value, i.e. no threshold, below which complete safety can be guaranteed. On the other hand, we should attempt to evaluate the various hazards to man from compounds in his food in such a way as to minimize the total risk. It makes no sense to stop eating fish for fear of their possible carcinogenic content and replace the fish by another food source that poses an equal or even greater danger. Thus, we believe that a special effort should be made to measure the concentration of the carcinogenic PAHs in a variety of foods on a continuing basis. Although it is clear that much more information relating to possible low-level toxic effects of contaminants in all foods would be of great importance, it does not appear that our present information provides a basis for alarm about the health effects of oil spills. (National Academy of Sciences, 1975, p. 98)

Coping with Oil Pollution

The essential features of marine oil pollution, its causes and consequences, have been discussed in some detail. Some understanding of these features is essential if the risks of offshore oil activities are to be accurately assessed and if the most appropriate ways of dealing with the problem are to be identified. Clearly this form of marine pollution will continue for as long as society depends upon hydrocarbon fuels. In these circumstances perhaps the most important questions are practical ones. What can society do to reduce the magnitude of the problem and the severity of the impact? How adequate are current pollution control programs and practices? Are there alternative approaches (technological, behavioral, institutional) that might reduce the social and environmental costs of energy development in a more effective and immediate way?

Bruce Mitchell suggests that resource management involves "actual decisions concerning policy or practice regarding how resources are allocated and under what conditions they may be developed" (1980, p. 32). Implicitly the concept of resource management assumes that society has the opportunity, the inclination, and the ability to manage nature in such a way as to maximize the benefits and minimize the costs incurred by present and future generations. There undoubtedly exist a variety of ways in which society can deal with the causes and consequences of marine oil pollution due to offshore activity (Table 20). These range from efforts to reduce the amount of oil actually entering the oceans—as, for example, the installation of blow-out preventers during drilling and the use of load-on-top (LOT) systems to prevent the chronic discharge of oil from tankers during ballasting—to programs that will enable society to bear the costs of oil pollution, as in compensation to individuals and local communities for environmental damages or reimbursement for expenses incurred in the course of clean-up operations. The measures identified in Table 19 may involve the adoption of new pollution control techniques, a modification of human behavior, managerial and institutional changes, or any combination thereof.

Despite the wide theoretical range of management policies and practices, only a very restricted number of approaches are being vigorously pursued.[21] Moreover there exist marked interagency and even international differences in both general approach to marine pollution control and particular practices of oil spill treatment, as for example in the sharply contrasted attitudes toward the use of chemical dispersants evident in the United Kingdom, Norway, and the United States. In general it seems fair to say that the prevailing approach, in the United Kingdom as elsewhere, emphasizes immediate relief rather than prevention. Given the public outcry that invariably accompanies a major spill, a preoccupation with post-spill emergency measures to reduce environmental damage and to remove the spilled oil as quickly as possible is understandable. By way of contrast, however, efforts to deal with the causes of the problem, for example through international conventions governing tanker safety and operating practices, have proceeded somewhat haphazardly. Those international conventions that are in force tend to be restricted in scope and all too often only half-heartedly enforced. As will be discussed, advisory bodies such as ACOPS as well as conservation groups (notably RSPB) have been sharply critical of U.K. pollution control policies, particularly the government's reluctance to vigorously pursue and implement conventions governing the intentional discharge of oil at sea.

Although a detailed discussion of the evolution of marine pollution control efforts in the United Kingdom is beyond the scope of this chapter, certain obvious questions require consideration. In what ways and how adequately is the United Kingdom coping with

Table 20. Strategies for Coping with Oil Pollution

Type of Adjustment	*Technological*
(1) Reducing the amount of petroleum entering the marine environment (i.e., affect the cause)	Controlling indirect sources of petroleum-derived hydrocarbons (e.g., automobile emissions; urban runoff, etc.) Controlling discharges from marine transportation sources through: (a) pollution control systems segregated ballast tanks (SBT) load on top (LOT) crude oil washing (COW) (b) effluent discharge standards (c) provision of shore facilities to treat oily wastes Controlling discharges from offshore drilling and production platforms (e.g., blow-out preventers)
(2) Minimizing the potential for significant environmental damage from spilled oil	Recovery and containment of spilled oil (e.g., booms, pumps, surface skimmers) Preventing oil from entering sensitive environments (fishing grounds, marshes, estuaries, etc.) through: (a) sinking (b) chemical dispersal
(3) Adjusting to losses	

Behavioral
Reducing consumption of petroleum hydrocarbons
Reducing risks of pollution incidents through: (a) navigation aides and designated shipping lanes (b) inspection schemes and active enforcement of international regulations and agreements to ensure high standards of seamanship and offshore oilfield operating practices (c) establishment of principle of port state jurisdiction
Multilevel contingency planning to: (a) identify particularly sensitive environments, equipment, and trained manpower needs (b) ensure clean-up methods are suited to recover and/or disperse oil with minimal environmental risk
Regional and international warning and spill coordination agreements
Land and sea-use planning to minimize siting conflicts between energy-related facilities and environmental amenities, fishing grounds, recreational resources, etc.
Public relief measures and compensation for environmental losses, etc.
Reimbursement of clean-up costs
Individual loss bearing

the threat of oil pollution? Why are certain approaches being pursued while others remain neglected? And are the measures being implemented the most appropriate ones?

THE CONTAINMENT AND REMOVAL OF SPILLED OIL

In coping with the problem of oil pollution, those agencies and institutions responsible for protecting environmental quality, as well as impacted communities, are understandably concerned with appropriate remedial measures intended to minimize the impact of major spills. Quite apart from any steps that may be required to reduce the flow of oil into the sea, an immediate issue in the aftermath of a pollution incident will be whether or not the spill should be treated and if so in what way.[22] The distance of the slick from the shoreline as well as the volume and type of oil spilled will be important considerations, but any decision is likely to prove controversial, involving as it must human judgments about the degree of risk posed by the spill and the chances of successful treatment.

Available treatment techniques include the use of mechanical recovery devices, such as skimmers, or chemicals to disperse the oil. Recovery of the oil obviously represents the most attractive form of oil spill treatment, yet for technical and economic reasons this approach has serious limitations (Johnston, 1976b, pp. 119–120). In open oceans in particular, recovery efforts are likely to be expensive and only marginally successful. Yet the use of chemical dispersants is highly controversial. As noted in the National Academy of Sciences study:

> There are those who strongly urge the use of detergents to disperse point sources of petroleum input . . . an obvious argument in favor of detergents is that the conversion of an oil spill into diffused and disseminated form will minimize the quantity of oil eventually reaching the beaches. Thus, the use of detergents is one way to eliminate the most visible evidence of petroleum spills. The difficulty with this practice is that we do not know what happens to the dispersed hydrocarbons. Are they truly degraded, or do they simply spread the toxic effects of oil over a larger area? (National Academy of Sciences, 1975, p. 106)

(a) Mechanical Clean-Up Methods. Mechanical clean-up methods, involving containment of the oil slick through the deployment of specially designed booms, recovery of the oil using surface skimmers, and collection of the watery oil in container barges, obviously represent an environmentally acceptable way of dealing with oil

spills (Johnston, 1976b, p. 120). Where such systems can be successfully deployed it may be possible to avoid more controversial measures to sink or disperse the oil that involve substantial environmental risks. Moreover there is the added economic advantage that any recovered oil can be used or sold to recoup part of the costs of cleaning up the spill.

Mechanical methods of containment and recovery are admirably suited to treatment of crude oil spills in sheltered environments.[23] R. Johnston (1976b, p. 120) suggests that under favorable conditions a well-designed system operated by skilled personnel is capable of retrieving more than 75 percent of the spilled oil. Even for nearshore spills, however, successful recovery is dependent upon the ready availability and rapid deployment of equipment and vessels. As Jerome Milgram (1977, p. 91) has emphasized, if any one of the five critical items is missing—booms, skimmers, towing vessels, storage barges, or trained personnel—recovery will be ineffective. In these circumstances policy-makers must decide whether the high capital cost of outfitting and maintaining recovery fleets in major ports and oil centers represents the most satisfactory form of oil spill control investment. And despite the attractions of recovery, there remain very serious doubts as to the ability of current recovery devices to operate effectively in exposed offshore environments. Booms, for example, are incapable of containing oil in the presence of strong currents or under adverse wind or wave conditions. Similarly, few skimmers have been designed that function well in heavy seas: "The problem of achieving an optimum encounter rate, so that skimmers are worked productively for as long as is practicable, is a major constraint. Unit costs of operation may also be very high for those used in concert with separate boom systems which have to be handled by two extra vessels. Furthermore, a skimmer which does not ride the waves easily will one minute be pumping oil and the next air" (Royal Society for the Protection of Birds, 1979, p. 103). A recent Massachusetts Institute of Technology study concluded that any effort to recover oil spilled in the course of offshore drilling on the Georges Bank (off the New England coastline) "would seem at best marginally feasible, extremely expensive and probably a waste of resources" (Johnston, 1976b, p. 120). In particular the MIT study cited the adverse weather conditions that would prevail in winter (when containment and recovery equipment could be used only intermittently) and the high risk to operating personnel.

Although the MIT study dealt with a purely hypothetical situation, its conclusions raise serious doubts about the wisdom of attempting to recover oil spilled in the exposed waters of the North

Sea or the Atlantic approaches to the English Channel. The Ekofisk platform *Bravo* blow-out offered an opportunity to test a variety of recovery techniques under actual oil spill conditions. And while several types of skimmer apparently functioned successfully in moderate seas, others are reported to have picked up large amounts of seawater which quickly filled the available storage capacity without making any great contribution to the clean-up operations (Norges Offentlige Utredninger, 1977, pp. 56–57; Royal Society for the Protection of Birds, 1979, p. 103).

Clearly there is scope for major improvements in the technology of oil spill containment and recovery. In the United Kingdom, for example, the Warren Spring Laboratory has developed a V-shaped boom (known as Springsweep) that is towed alongside a recovery ship. Although originally designed to retrieve crude oil, the Springsweep system is reported to have been extensively modified as a result of the *Eleni V* experience and to show promise in dealing with fuel oil and crude oil mousses (Select Committee on Science and Technology, 1978, p. 123). "This device, expected to be available commercially in 1980, is used to actively hunt down patches of oil, needs only one vessel to operate it, is demountable, can be used on small tankers or offshore supply boats with tank space to hold recovered oil, and is of low inertial mass so that it rides the waves with little trouble under moderate sea conditions" (Royal Society for the Protection of Birds, 1979, p. 103).

In general, however, the development of enhanced recovery techniques has not been a high government priority in the United Kingdom. The Advisory Committee on Oil Pollution of the Sea has been strongly critical of British policy in this respect, noting for example that "much money has been poured into research, not least by oil companies, into dispersants, the manufacture and export of which is a profitable business . . . while there is little encouragement for manufacturers of mechanical clean-up devices" (Advisory Committee on Oil Pollution of the Sea, 1977, p. 12). The ready availability of large stocks of chemical dispersants and the limited effectiveness of mechanical recovery devices clearly restricts the options available to officials in charge of clean-up operations. The committee concluded that much better recovery techniques could be developed within a reasonable period of time if the British government gave its wholehearted support to research and development efforts, although it acknowledged that mechanical devices would always prove more effective in the Baltic or the Mediterranean than in the harsher environment of the North Sea.

(b) Chemical Dispersants. The primary objective of any clean-up operation must be to prevent spilled oil from reaching the shoreline. Since complete recovery at sea is unlikely even under favorable circumstances, it may well become necessary to consider sinking or dispersing the slick. Unfortunately such methods may themselves have undesirable consequences for the marine environment.

Sinking the oil is usually accomplished through the addition of specially treated sand or pulverized fuel ash. While the oil is carried to the bottom sediments, where bacteria capable of degrading it are concentrated, there is the very real danger of chronic contamination to feeding and breeding grounds of fish and invertebrates (Nelson-Smith, 1977, p. 68). In concluding that sinking represented the least preferable way of treating oil spills, the National Academy of Sciences study (1975, p. 106) noted that following experiments in the North Sea catches of finfish and shellfish remained tainted for several months. In view of the risk to important commercial fisheries, the sinking of oil is not authorized in U.K. waters at the present time.

Instead, British authorities rely heavily on chemical dispersants. Since the use of dispersants is so highly controversial, it is unfortunate that the term is often used incorrectly to describe *any* type of chemical agent used to combat oil spills. In fact three different agents may be used in oil spill control. These include (1) *detergents* to remove oil or prevent it from adhering to solid surfaces such as beaches or rocky shorelines; (2) *emulsifiers* that are used to promote a stable distribution of one liquid (oil) in another liquid (water); and (3) *dispersants* to facilitate the break-up of a slick into a mass of tiny droplets (Lindblom, 1978, p. 129). Dispersants, the chemical agent most frequently used in treating oil slicks, come in various formulations and are usually applied by spray booms mounted on a vessel towing agitators ("breaker boards") to ensure that the dispersant is thoroughly mixed with the oil and seawater. Water-based dispersants require more time and mixing energy to effectively disperse the oil than solvent-based dispersants. The latter work fairly quickly but require high dosages, since the formulation becomes less efficient when diluted. More recently the aerial application of concentrated dispersants that require minimal mixing energy has been proposed as a way of responding quickly and effectively to spills that occur far from land (Lindblom, 1978, p. 130; Exxon Chemical Company, 1980, p. 4). In theory at least, dispersal of a slick into fine droplets allows for more rapid degradation of the oil by natural processes. However as Anthony Nelson-Smith points out, chemical agents are themselves bioactive. "Increasing the surface area also enhances the extraction of water-soluble toxicants, and some droplets are bound to fall within the size-range favoured by filter-feeders" (Nelson-Smith, 1977, p. 68). The early "first generation" dispersants contained aromatic hydrocarbon solvents that were highly toxic to marine fauna and undoubtedly did far greater damage to the environment than the oil itself. Although current dispersants are both more efficient and less toxic, virtually nothing is known about their possible long-term, sublethal effects on marine flora and fauna. Moreover there is still no effective way of dispersing heavy fuel oils or many types of mousse. In the *Eleni V* incident, for example, dispersants were relatively ineffective once the fuel had cooled and congealed (Select Committee on Science and Technology, 1978, p. xix). Thus, quite apart from uncertainty over their long-term effects, chemical dispersants are really effective only when used to treat relatively thin patches of freshly spilled crude oil.

Much of the public distrust of chemical agents stems from the biological damage inflicted on the Cornish coastline following the stranding of oil from the *Torrey Canyon*. In this highly publicized incident, overzealous and inexperienced clean-up workers "motivated by a desire to placate public feeling" enthusiastically applied concentrated, highly toxic detergents to the beached oil (Cowell, 1976, p. 370). The resulting public debate focused chiefly on the chemical agents themselves rather than on their improper application. Eric Cowell (1976, p. 376) notes that the biological consequences of using detergents were well known before the *Torrey Canyon* disaster, but that the warnings of marine ecologists went unheeded.

As already noted, there exist marked differences of opinion within the international community with respect to the use of chemical agents. In the United Kingdom the government's position is that dispersants represent "the only method of dealing with spills proven to be really effective in the seas surrounding the U.K." (Select Committee on Science and Technology, 1978, p. 10). As noted by ACOPS, this view is a minority one:

> . . . in countries where the environmental lobby is strong, such as in most Nordic countries and the United States, the presumption is that until proven "innocent" dispersants should be considered as being sufficiently toxic to be capable of inflicting damage to the marine environment, while in other countries, including Britain, the opposite opinion rules. Opponents of dispersants which now include the Soviet Union, have imposed a

> total ban on their use in clean-up operations, save in exceptional circumstances as when a fire risk exists. Britain has isolated herself on this question, but argued . . . that dependence on chemical means was temporary, pending development of fully effective recovery equipment.[24] (Advisory Committee on Oil Pollution of the Sea, 1976, p. 28)

Even in the United Kingdom, of course, certain conditions must be satisfied before the use of a dispersant is authorized. Under the 1974 Dumping at Sea Act, the Ministry of Agriculture, Fisheries, and Food may issue a license only if it is satisfied that use of the dispersant will not result in significantly greater damage to biota than would occur if the slick remained untreated. The use of an approved dispersant may also be controlled where there is a risk (1) to marine communities of ecological importance, (2) to fish stocks (for example in enclosed or shallow waters where dilution of the dispersant may be restricted), and (3) to the quality of seafood. In one incident in the Firth of Forth, for example, it is reported that the use of dispersants was not authorized in order to prevent possible tainting of a sprat catch, even though this decision may have contributed to increased bird mortality as a result of exposure to untreated oil (Royal Society for the Protection of Birds, 1979, p. 105).

The United States and Norway are among those nations that have pursued a far more restrictive policy with respect to chemical dispersants. At the time of the Ekofisk platform *Bravo* blow-out, for example, the use of chemical agents in Norwegian waters was permitted only for fire suppression in the immediate vicinity of a well or platform (Norges Offentlige Utredninger, 1977, pp. 57–60). Although there have been indications that the Norwegian government is currently re-evaluating its policy, continuing differences in attitude and in criteria for the use of dispersants make international cooperation on pollution control in the North Sea extremely difficult (Advisory Committee on Oil Pollution of the Sea, 1977, p. 12).

In the United States the use of chemical dispersants has generally been discouraged. As Milgram points out, the position of the Environmental Protection Agency (EPA) is that "dispersants should almost never be used. The oil industry, on the other hand, takes the position that dispersants should almost always be used. Neither position is particularly contructive" (1977, p. 93). Robert W. Holcomb (1969, p. 206) suggests that official distrust of chemicals is at least partly attributable to perceptions derived from the *Torrey Canyon* and Santa Barbara incidents. Even between agencies, however, there exist marked differences in attitude. In the clean-up operations that followed the massive spread of oil into U.S. waters from the *Ixtoc I* blow-out, the U.S. Coast Guard and the National Oceanic and Atmospheric Administration (NOAA) both favored the use of chemical dispersants. Such measures were successfully opposed by EPA on the grounds that the use of dispersants would be ineffective under the prevailing conditions. However EPA does maintain an "accepted" list of chemical agents for oil spill treatment. Currently this list includes some fourteen dispersants added on an ad hoc basis since 1977, relying on test data submitted by the manufacturers. Dispersants from this "accepted" list are reported to have been used in Mexican waters with apparent success (Gordon P. Lindblom, personal communication). Moreover EPA has on several occasions approved the use of a dispersant, for example in the clean-up of an oil spill in Rockaway Harbor, New York, in 1978.

No matter how successful efforts to treat an oil slick at sea may prove to be, some oil is likely to reach the shore. And regardless of the origins of the oil, a major spill from an offshore platform or the release of oil during routine shipping operations, coastal authorities must decide how best to deal with the stranded oil. The overwhelming preference in the scientific community would appear to be to leave the oil untreated and allow processes of weathering and degradation to run their natural course (Cowell, 1976; Department of Trade and Industry, 1972; Nelson-Smith, 1977).[25] Clearly such a course of action is unacceptable when areas of high amenity or recreational value are affected. Least damage is likely to result when clean-up efforts are restricted to simple mechanical methods (pumps, shovels) and the use of absorbent materials (straw, peat, chalk, or sawdust). The temptation, however, is to mount an all-out effort using all available equipment and manpower. According to Cowell, Cox, and Dunnet (1979, p. 8) even bulldozers were seen in coastal salt marshes after the *Amoco Cadiz* spill. In the case of salt marshes it is doubtful whether any form of treatment can offset the damage already inflicted by spilled oil; certainly recovery from the effects of the *Amoco Cadiz* spill is likely to be significantly faster in those oiled salt marshes not physically destroyed by bulldozers.

As demonstrated in the aftermath of the *Torrey Canyon* disaster, even greater harm is likely to result from the indiscriminate use of detergents. Current detergents are a good deal less toxic than those used on Cornish beaches more than a decade ago, yet, not surprisingly, in the hands of overenthusiastic, inexperienced clean-up crews, improper application (in terms of habitat as well as dosage) is all too frequent, with serious consequences for the coastal fauna and flora. Cowell, Cox, and Dunnet go to the root of the problem when they suggest that "In the emotional heat of these disasters, we rarely stop to evaluate the effect of our clean-up techniques. In fact, occasionally we find that the pressure of public opinion is so strong that *any* clean-up technique may be used just to show the enraged public that some progress is happening" (1979, p. 1).

Clearly conflicting assessments over the most appropriate form of oil spill treatment will continue to hamper clean-up efforts. There is particular need for agreement at the international level on procedures for evaluating the relative effectiveness of alternative measures under a variety of spill conditions. Much more could be accomplished too in terms of exchanging information between different national agencies. Delegates to a recent Intergovernmental Maritime Consultative Organization (IMCO) meeting, for example, expressed frustration with the lack of comprehensive information on the toxicity and effectiveness of dispersants that prevented them from making informed decisions regarding pollution control policies and equipment. Essentially similar dilemmas confront local authorities in the United Kingdom. Despite the relatively permissive attitude toward chemical agents in the United Kingdom, any decision to use dispersants or detergents will be highly controversial. Yet little has been accomplished in the way of advance planning to identify ecologically sensitive areas requiring special protection and treatment or to establish priorities with respect to the most suitable form of clean-up treatment in case of a spill. In the final analysis, however, the most appropriate measures are likely to be dependent upon the particular circumstances of the spill: "The secret of success is to suit the clean-up method to the conditions prevailing at the time, while remembering that the ultimate aim should be to reduce damage to the marine environment as a whole, rather than merely to shift an undesirable problem from your beaches into some other authority's area as quickly and cheaply as possible" (Nelson-Smith, 1977, p. 68).

CONTINGENCY PLANNING

Minimizing the impact of an oil spill (or, what is perhaps more important in view of the record to date, ensuring that environmental damage is not exacerbated as a result of ill-advised clean-up measures) requires advance planning. Clearly, advance planning cannot anticipate every situation, but much can be accomplished to ensure a rapid response, using measures that have already been agreed upon as appropriate given the circumstances and location of the spill:

> . . . the essence of any form of treatment is . . . speed of action, which can only come from advance planning. It is important to

> arrange both a working structure (which usually involves the emergency services such as firefighters and coastguard, together with organizations possessing the necessary plant and supplies) and a plan of action related to the area in question. A heavily used and already much-polluted dock will present problems or offer possibilities which differ greatly from those of an inaccessible or rugged rocky coastline of particular biological value, between these extremes there may be beaches of some interest to biologists but which also support a valuable tourist trade. The best mode of treatment for each stretch of coast or offshore region, taking into consideration its topography, prevailing currents and so forth , as well as its particular use or value, must be firmly agreed amongst all interested parties well before any need arises to use the plan, or emergency operations will bog down in arguments, recriminations or even legal action. (Nelson-Smith, 1977, p. 66)

The importance of contingency planning is invariably emphasized in oil spill pamphlets and guidelines issued by government departments. In practice, however, it seems fair to say that little has been accomplished in the United Kingdom, at least at the national level.[26] Only in 1979 was a Marine Pollution Control Unit finally established within the Department of Trade with overall responsibility for coordinating clean-up efforts in the event of a major spill and for developing a national plan to deal with oil pollution. Yet in 1980 work on a national contingency plan was still incomplete due to disagreements between the government and local authorities over (1) the division of responsibility and the allocation of resources in the event of a major spill, and (2) the question of who should bear the cost of clean-up operations, including preplanning and stockpiling of equipment and materials (Advisory Committee on Oil Pollution of the Sea, 1980, pp. 22–23). Thus, despite the laudatory exhortations of government departments, clean-up operations have by and large been conducted on an ad hoc, spill-by-spill basis, with trial and error and improvisation being the rule rather than the exception.

At the present time primary responsibility for dealing with any spill resulting from offshore oil production in the U.K. sector of the North Sea rests with the field operator. Contingency plans must be approved by the Department of Energy before production can begin. These plans specify proposed spill control and clean-up techniques and must cover every stage in the production, transportation, and storage of oil. British Petroleum's contingency plan for the Forties field, for example, covers marine operations (platform, subsea pipeline), terrestrial operations (the Cruden Bay to Grangemouth refinery land pipeline), and terminal operations (the Hound Point tanker loading facility on the Firth of Forth) (Fulleylove and Lester, 1977). In general therefore the government's role is restricted to monitoring clean-up activities and offering advice although "if . . . the operator's reponse was inadequate the Government might give positive directions to the operator or, in an extreme situation, take over control of the clean-up operation" (Department of Energy, 1979, p. 15). In 1978 the government advised all offshore operators that "where an oil spill is not causing or likely to cause damage, it is preferable, in the absence of reliable recovery equipment, for it to be allowed to evaporate and degrade naturally rather than for it to be dispersed chemically" (Department of Energy, 1979, p. 15). Since the operator's own contingency plans are based in large part on the use of jointly owned stocks of dispersants maintained by the United Kingdom Offshore Operators Association at five major ports (Aberdeen, Lerwick, Lowestoft, Pembroke Dock, and Plymouth), adhering to the government's revised guidelines will require a change in both clean-up philosophy and stocks of equipment on the part of offshore operators.

Spills that threaten the coastline and particularly those from sources that cannot be identified raise additional issues of both an ecological and a jurisdictional nature. At present the Department of Trade is responsible for controlling oil pollution in coastal waters and maintains spraying equipment and dispersant stocks at a number of ports around the United Kingdom. "Theoretically, a task force capable of cleaning up to 6,000 tons per day is deployable within 48 hours of mobilization. In practice this and lesser targets have never been remotely approached, not necessarily because of failure to provide adequately equipped vessels and stocks of dispersant but because of failure to attain the planned clearance rate expected per ship" (Royal Society for the Protection of Birds, 1979, p. 110). Local authorities also have powers to deal with oil pollution along the coastline and in waters up to one mile offshore. As the *Eleni V* incident demonstrated, in the absence of an agreed-upon plan, there are likely to be sharp differences of opinion and recriminations between central and local government authorities. However local authorities have by and large been given little encouragement or support to acquire the equipment and the expertise with which to respond effectively to oil spills. "Their ability to protect ecologically sensitive areas is limited. In the event of massive coastal pollution, their resources will rapidly become swamped. Local authority oil pollution officers are often directors in charge of manpower departments such as Highways or Engineering Services. Not surprisingly, because of other duties, some of these officials have little time to develop expertise in the subject or keep fully up to date on developments and experience elsewhere" (Royal Society for the Protection of Birds, 1979, p. 111). Where authorities have acquired some oil spill cleaning capability, as in the case of the Suffolk County Council and the Shetland Islands Authority, it has been at their own initiative and expense. Only recently has the Department of the Environment begun to stockpile equipment for use by local authorities in the event of an emergency, and it is not entirely reassuring to learn that half of the available funds have been expended on equipment for treating beaches with chemical dispersants and detergents.

Local authorities are at a particular disadvantage in dealing with smaller spills, usually from coastal shipping, where the source cannot be identified. Throughout the 1970s the U.K. government has adhered to the principle that "the polluter shall pay," an admirable principle except that in the case of unattributable spills there is no mechanism for local communities to recover the costs of clean-up operations or obtain compensation for environmental damages.[27] "Broadly there will be no grant aid unless the pollution becomes a national emergency but compensation can be claimed from the ship owners by the government. The cost of oil pollution clearance should be recorded for future claims on tanker owners" (Select Committee on Science and Technology, 1978, p. 34). The 1977 ACOPS survey found that nearly three-quarters of the oil pollution incidents occurring outside ports could not be attributed to a specific source, and the bulk of the costs of clean-up operations are being borne by those authorities on whose shoreline the oil lands. "The government has stated that no undue burden should be imposed on the shipping industry. However, any costs that shipowners and cargo owners would be required to bear must be compared with the potential costs falling on local authorities and also on the fishing and tourist industries . . . The sums involved are not negligible and should not be ignored by the Government" (Advisory Committee on Oil Pollution of the Sea, 1978, p. 8).

Even where polluters can be identified, local authorities are by no means assured of adequate compensation. In seeking a clarification of government policy in the aftermath of the *Eleni V* spill, the clerk to the Suffolk District Council emphasized the need for "adequate arrangements for compensation to coastal authorities who are consistently faced with the threat of oil pollution . . . it seems only fair that the whole country should pay the costs of dealing with any pollution that arises and coastal district authorities should not be put in the position of having to fight to obtain compensation" (Select Committee on Science and Technology, 1978, pp. 65–67). In responding, the Secretary of State for the Environment implied that the government would consider a change in its policy:

> We shall not allow local authorities to suffer losses as a result of the work they are having to do to clear up the mess left by the *Eleni V* incident . . . We shall first of all try to recover the money from the insurers of the vessel itself. I can't of course anticipate the outcome of the insurance claim action and give you an absolute assurance that the claims will be met, but at this stage we are pretty confident that we shall be able to recover in full all reasonable costs. . . . [If the claim] exceeds the limit of liability authorised by the Civil Liability Convention of 1969, our next recourse would be to the supplementary fund set up and financed by the oil companies to deal with just such a situation . . . and we stand ready if necessary to make Exchequer assistance available. Whatever happens the local authorities are protected.[28] (Select Committee on Science and Technology, 1978, p. 139)

However, for all the reassurances given at the time of the *Eleni V* incident, the government appears committed to the view that local authorities involved in major oil pollution incidents must meet all their own clean-up costs and carry them until reimbursed by the polluter, regardless of the delays and uncertainties involved in efforts to recover such costs. Moreover, an interdepartmental group charged with preparing a comprehensive review of national policy on compensation recently recommended against the establishment of any compensation fund to cover the costs of pollution from unidentified sources on the grounds that the extent and amount of such damage had not been fully documented (Advisory Committee on Oil Pollution of the Sea, 1979, p. 29).[29] Thus, at the beginning of the 1980s, the United Kingdom still lacked any national fund to cover losses, damages, or costs incurred by local communities as a result of unattributable spills.

Efforts to ensure that oil spills occurring in U.K. waters are dealt with in a coordinated, effective, and equitable manner clearly leave much to be desired. While offshore operators have by and large taken their responsibilities seriously, preparing contingency plans and establishing compensation funds, far too little attention has been given to advance planning measures for the shoreline. Yet it might well be argued that it is precisely in this area that there is the greatest need for a carefully considered plan of action to resolve issues of jurisdiction and method of treatment before any oil is washed ashore. The presumption (at least in London) all too frequently appears to be that little crude oil from offshore production is likely to come ashore and that smaller discharges from routine shipping operations are insignificant. Neither presumption appears particularly well founded. In the absence of clearly formulated, carefully rehearsed plans for the coastline, it is not surprising that personnel, equipment, and, perhaps most important of all, information are rarely used to the best advantage.

Investigations of recent accidents around the British Isles have been sharply critical of (1) the lack of advance planning (resulting in hasty improvisation, improper use of equipment, and reliance upon inexperienced work crews), and (2) the failure of operators and authorities to ensure compliance with approved control procedures and guidelines. The U.K. Select Committee investigating the *Eleni V* incident, for example, noted that "virtually no preparations have been made to deal with problems posed specifically by spills of heavy fuel oil" (Select Committee on Science and Technology, 1978, p. xv). In general the committee concluded that existing contingency plans were inadequate and quite incapable of dealing with a major spill. Their report sharply criticized the Department of Trade for its failure to cooperate more closely with local authorities. The deficiencies in contingency planning revealed by the Select Committee certainly contributed to the decision to set up a new Marine Pollution Control Unit, its principal duties being (1) to ensure that the resources available for dealing with oil pollution are used effectively; (2) to develop national contingency plans for dealing with marine pollution; (3) to relate these plans to those of neighboring countries; and (4) to take charge of operations at sea in the event of a marine pollution emergency in British waters. Perhaps the existence of this unit will ensure that advance warnings of the likelihood of a spill and the inadequacy of contingency plans, as at Sullom Voe, will be taken seriously and not simply dismissed as alarmist.[30] Undoubtedly there has been progress since the Ekofisk blow-out, when the Norwegian Action Committee investigating the incident confirmed that "neither Phillips nor the official authorities had ready worked-out preparedness plans, sufficiently qualified and trained personnel or practical experience from collection operations under the kind of conditions which obtained during this operation" (Norges Offentlige Utredninger, 1977, p. 57). However, there is considerable room for improvement. Cowell, Cox, and Dunnet (1979, p. 7), in a strong indictment of existing planning efforts in the United Kingdom, argue that the problem can be effectively resolved only through (1) preplanning, including identification of necessary equipment, training of personnel, mapping of ecologically sensitive areas, and identification of experts to be called in after a spill, and (2) on-site spill coordination involving selection of appropriate clean-up techniques for different habitats and inventorying ecological changes. The establishment of the Marine Pollution Control Unit offers an opportunity to move in this direction, although mapping of sensitive ecological zones can best be accomplished at the local level. In the final analysis, of course, even the most satisfactory contingency plan cannot account for poor judgment or failure to comply with regulations. At the time of the Ekofisk blow-out, for example, the Norwegian State Pollution Control Authority's regulations required that license holders have sufficient mechanical recovery and collection equipment on hand to be able to deal with an uncontrolled blow-out of eight thousand tons of oil per day under any condition likely to be encountered in the North Sea. The Action Committee found that Phillips had not taken any steps toward fulfilling this requirement (Norges Offentlige Utredninger, 1977, p. 53).[31]

REDUCING OIL POLLUTION

Ixtoc I, Amoco Cadiz, Ekofisk platform *Bravo,* Santa Barbara are reminders of the importance of remedial measures to minimize the damage inflicted after oil has been spilled. Yet it might well be argued that by itself such an approach is inherently limited, treating the symptoms of the disease without attempting a cure. It is not simply that remedial measures will almost always prove to be inadequate, although the *Eleni V* incident was a clear demonstration of the limitations of current clean-up techniques. Nor is it solely that a reactive, crisis-oriented approach tends to distract attention from less dramatic aspects of the problem, such as accidental and deliberate spillages during routine shipping operations. The argument is rather that the time and energy required to develop an "oil spill treatment capability" leave little opportunity (or perhaps inclination) to pursue and coordinate alternative, more imaginative solutions that might in the fullness of time help to reduce the number and frequency of oil pollution incidents.

The most serious criticisms in this regard relate to the limited success of international efforts to control oil pollution from shipping; in the words of the RSPB report, the history of these efforts "is an instructive saga of inadequacy and delay" (Royal Society for the Protection of Birds, 1979, p. 1). Clearly this is only indirectly related to offshore energy development in the North Sea. Yet, as has been repeatedly emphasized, it is impossible to look at offshore activities in isolation. The effects are felt in a multitude of functionally related areas and activities. Given the increasing volume of tanker traffic through Scottish coastal waters (and particularly in the hazardous and difficult approaches to the oil terminals at Flotta and Sullom Voe), the failure of the U.K. government to act more decisively in this area of marine pollution control is disturbing.

The immediate need is for speedy ratification and active enforcement of existing conventions. Thus, although the 1970s might aptly be described as a period of increasing environmental awareness, the only international agreement on oil pollution at sea actually in force at the end of the decade was the 1954 Convention for the Prevention of Pollution of the Sea by Oil.[32] More recent agreements, notably the 1973 International Convention for the Prevention of Pollution from Ships (MARPOL) which sought to achieve "the complete elimination of international marine pollution by oil and the minimization of accidental spills" within ten years, have yet to be ratified by a sufficient number of nations to take effect.[33] Victor Sebek in 1978 referred to the "snail-pace at which the signatory states had been proceeding in giving dormant conventions the kiss of life" (Sebek, 1978, p. 85). Clearly the adoption of a convention, while a diplomatic success, does not bring immediate benefits. And by the time the ratification process has been completed, there are likely to have been advances in both scientific knowledge and marine pollution control technology that require amendments to the umbrella convention (Baroness White, 1977, p. 55).

International agreements are only arrived at as a result of hard bargaining and major concessions by those most anxious to obtain agreement. In the process there are likely to be less than desirable compromises that severely weaken the final protocol. The difficulties involved in reconciling differing national attitudes and priorities with respect to marine pollution control are obvious:

> . . . political considerations often dictate in advance negotiating positions to be taken up, even before expert evidence has been considered. The highly technical provisions in international conventions on marine pollution may cover a number of scientific non-sequiturs as a result of hard bargaining round the conference table. There is regular criticism that provisions are either too lenient or unnecessarily strict, but it would be naïve to suppose that they rested entirely on sound scientific data, even when these are available. A notorious example is to be found in one of the standards adopted in the 1973 International Convention for the Prevention of Pollution from Ships which specifies that the total quantity of oil discharged into

> the sea must not exceed for existing tankers 1/15,000 of the total quantity of the particular cargo of which the residue formed a part, while the 1/30,000 ratio is provided for new tankers. I doubt if any scientist would find himself able to justify either of these two figures on purely technical grounds. One might almost suppose that the negotiators thought of a number and doubled it. (Baroness White, 1977, p. 55)

The U.K. government has been sharply criticized for its "hard line" with respect to international marine law (Royal Society for the Protection of Birds, 1979; Advisory Committee on Oil Pollution of the Sea, 1977, 1978). At the 1978 International Conference on Tanker Safety and Pollution Prevention, for example, the U.K. delegation was particularly active in opposing the United States proposal that all existing tankers over twenty thousand tons should be fitted with segregated ballast tanks (SBT), a system designed to retain the oil and the water required for ballast in separate compartments. An editorial in the *Marine Pollution Control Bulletin* suggests that "the SBT modification would eliminate the problem caused by vessels discharging ballast tank water contaminated by oily residues prior to entering a port for loading. This is one of the main sources of oil pollution around Britain's shores yet the British government while recognizing that the SBT system is superior, appear to favour the Crude Oil Washing (COW) system faced with estimated tanker conversion costs of £150 million" (*Marine Pollution Control Bulletin*, vol. 9, no. 3, 1978, p. 62).[34] According to ACOPS the final agreement, permitting the use of crude oil washing as an *alternative* to segregated ballast tanks on existing crude carriers of forty thousand tons or more, was not simply a missed opportunity but downright illogical: "COW is a useful operational device for cleaning cargo tanks without the use of water. But it can never be regarded as a substitute for SBT which is a structural device, preventing contact of oil and ballast water in cargo tanks" (Advisory Committee on Oil Pollution of the Sea, 1978, p. 32). Its recognition as an alternative appeared to reflect the lobbying efforts of oil and tanker companies. Since segregated ballast tanks reduce carrying capacity by between 15 and 20 percent, few companies are likely to pursue this option.

In general, international maritime law has sought to control oil pollution through (1) prohibiting discharge of oils and oily wastes in coastal waters or other special areas; (2) establishing discharge standards for all vessels operating on the open seas; and (3) requiring the installation of specified pollution control systems on crude oil carriers. Even the most comprehensive agreement of course cannot prevent accidents caused by human misjudgment or noncompliance, and consequently expectations that international regulations (when in force) will significantly reduce oil pollution appear overoptimistic. Based on past experience, compliance, monitoring, and legal enforcement are likely to prove far more difficult than anticipated. Too much faith is often placed in the pollution control system per se, with little consideration of the human element that will determine how well such systems are operated and maintained. An example would be the widely acclaimed load-on-top (LOT) system of controlling tanker waste discharge, required under the 1969 amendments to the Convention for the Prevention of Pollution of the Sea by Oil. This procedure relies on the shipboard retention and separation of the oil wastes resulting from ballast and tank cleaning; relatively clean water is discharged, while the oil residue is retained in the settling tank and the next cargo loaded on top. As S. Z. Pritchard has noted, "LOT is completely reliable only for journeys allowing time for the oil-water separation onboard, which may not meet the case of North Sea routes. At least on paper, over 80% of tankers are said to be committed to the LOT system, but industry checks at the largest Middle Eastern terminals revealed that two thirds of tankers capable of operating LOT did not do so (half made no attempt to retain slops), and while the selected inspection schemes have brought many operators into line, the maritime industry as a whole seems reluctant to install automatic monitoring systems on tankers" (1978, p. 66). Moreover *total* separation of oil and water in the slops is never possible, and in the absence of automatic monitoring systems great care must be taken in the discharge of slop water to ensure that the oil-water interface is not passed.

Additional problems arise in complying with international regulations where onshore infrastructure (for example reception facilities for oily wastes and ballast water at ports and terminals) is inadequate. This is particularly critical in areas such as the North Sea, for "as long as shore reception facilities are lacking, ships on shore voyages, which preclude the utilisation of load on top, will have problems and, as long as night remains dark, there will be a temptation, perhaps even a need, to discharge oily waters" (Portmann, 1977, p. 127). The Marine Environment Protection Committee has expressed particular concern over the lack of reception facilities in such areas as the Mediterranean, the Red Sea, and the Arab/Persian Gulf. As an indication of U.K. government attitudes and priorities it is perhaps sufficient to note that the Sullom Voe terminal was allowed to operate for nearly a year without proper facilities which "undoubtedly contributed in part to the severe oil pollution problems experienced off Shetland during winter 1978/79" (Royal Society for the Protection of Birds, 1979, p. 69).

Nor has the U.K. government been particularly active in creating the conditions or circumstances that would allow for effective enforcement of those agreements that are in effect. A major problem is the lack of jurisdiction over nonflag vessels operating outside territorial waters, even where such vessels are suspected of discharging oil and imposing environmental losses on the coastal nation. In extraterritorial waters, and it should perhaps be emphasized that the United Kingdom still claims only a three-mile jurisdiction, responsibility for enforcement lies with that country in which the ship is registered. Clearly there will be significant differences in enforcement standards between nations, and wherever states have demonstrated an inability or unwillingness to prosecute, vessels carrying that flag may in effect pollute the open seas with immunity. The Advisory Committee on Oil Pollution of the Sea has "consistently maintained . . . that operative provisions of all international and national legislative acts on the prevention of marine pollution will remain a dead letter until the port State is allowed to investigate and prosecute offending vessels which call at its ports" (1977, p. 6). Yet at the Third United Nations Conference on the Law of the Sea (UNCLOS III) the U.K. government was one of the last nations to accept the principle of port state jurisdiction.[35] Even within British territorial waters, however, successful prosecution of offenses occurring outside ports under the 1971 Prevention of Oil Pollution Act has been rare, averaging less than three a year over the past decade, with only one successful prosecution in 1978.[36] Even where a case is successfully brought, fines (which under U.K. law must be related to the income of the master of the offending vessel) provide little deterrent. "Governments are reluctant to apply and enforce strict requirements on their own vessels and ports because of the cost penalties involved, and the consequent trading overheads and competitive disadvantage in international markets which result. Governments are also reluctant to interfere with foreign flag ships either in port or on the high seas, whether for oil pollution or other reasons, because of fear of 'reprisals' on their own vessels trading overseas" (Royal Society for the Protection of Birds, 1979, p. 75). The end result is a further unfortunate illustration of Garret Hardin's "Tragedy of the Commons" syndrome (Hardin, 1968).

Clearly much more could be done to prevent oil pollution in British coastal waters. In general the U.K. government has not aggressively sought nor enforced marine pollution agreements and has been reluctant to take advantage of those options that are available.

Pritchard, for example, has pointed to the ironic fact that the North Sea, despite the expansion of offshore oil production and an increase in tanker traffic, now enjoys less protection than previously. The 1973 MARPOL convention created a number of "special areas" where more stringent regulations, including a prohibition of the discharge of any oil, were deemed appropriate. Although the North Sea was designated as a special area for environmental protection under the 1962 amendments to the Convention on Oil Pollution, it was conspicuously omitted from the MARPOL convention. "It is, in fact, the only major maritime regime which fares less favourably in the 1973 treaty vis-à-vis previous environmental conventions. Other areas—the Baltic, the Black Sea, Mediterranean, Red Sea and the Arabian/Persian Gulf—have either retained or upgraded their claims to special protection due to geographical, ecological or shipping conditions" (Pritchard, 1978, p. 66). Similarly the RSPB report drew attention to the fact that under the operating provisions of IMCO, the United Nations agency responsible for marine safety, high risk areas can be designated prohibited zones that must be avoided by particular categories of vessels. A number of such areas have been established—portions of the Bay of Biscay (France), off Cape Agulhas (South Africa), and Cape Terpeniya (U.S.S.R.)—to minimize the risk of stranding in hazardous waters that are adjacent to ecologically sensitive habitats. "A number of cases can be made for the urgent designation of such areas off our shores—around Shetland for instance, where the scheme could be operated in conjunction with some form of controlled approach for tankers using the Sullom Voe oil terminal" (Royal Society for the Protection of Birds, 1979, p. 72). Only in the aftermath of the public outcry over the *Esso Bernica* spill were measures of this kind eventually implemented at the initiative of the local authority. The measures, as announced by the Shetland Islands Council in 1979, included agreements with the oil and shipping industries on a prohibited zone for tankers, on reporting-in procedures for tankers approaching and leaving Sullom Voe, and on minimum ballast loads for incoming tankers. "Enforcement measures include an air-patrol rendezvous with tankers incoming and outgoing; close attention to ship safety standards (assisted by the new resident Department of Trade surveyor); and strong port sanctions against offenders, including time penalties and the banning of certain tankers from Sullom Voe" (Advisory Committee on Oil Pollution of the Sea, 1980, p. 46).

As noted at the outset of this chapter, all applications of advanced technologies involve risk; incidents such as the Ekofisk blow-out, the *Esso Bernica* spill, the collision of the *Eleni V* and the

Roseline, and the collapse of the *Alexander L. Kielland* semisubmersible rig, pose the question of just how much of a risk society is willing to take in order to obtain the benefits of petroleum (Sills, 1977). Society of course does not undertake any formal assessment of such risks, and policy-makers are confronted with a range of conflicting, ambiguous, shifting judgments; judgments moreover that are likely to be greatly influenced by the manner in which the risk has been articulated (for example, occasional high-profile "events" as opposed to more frequent but less dramatic events), by the way in which the losses have been distributed, and by the channels through which (and the extent to which) information has been presented. In such circumstances it is perhaps not surprising that policy-makers may well prefer not to address the difficult but basic question of whether or not the development of offshore oil and gas resources and the use of the marine environment are taking place in such a way as to maximize the benefits and minimize the costs incurred by society.

The increase in the volume of oil discharged into U.K. waters represents one of the externalities of North Sea oil development. Hitherto the United Kingdom has been fortunate in that no major spill has occurred in the course of offshore exploration or production. At the same time, the number of incidents involving tankers carrying crude or refined products is disquieting. Moreover the risk of chronic pollution along previously unaffected coastlines that are of immense value for their amenity and biological resources appears certain to increase as the volume of North Sea oil transshipped at onshore terminals increases and as a newer generation of smaller discoveries are brought into production.

Recent expressions of concern over the frequency of oil pollution incidents in U.K. waters undoubtedly reflect a feeling that the environmental costs and risks of offshore energy development have been underestimated and underemphasized. This view is not accepted by those representing the offshore industry, who argue that public perceptions have been distorted by the occasional spectacular spill and by unwarranted media descriptions of such incidents as "ecological catastrophes." Such disagreements reflect very different judgments with respect to the values and priorities that should be assigned to alternative uses of the oceans and coastal waters. And, as this chapter has shown, the opportunity for and the likelihood of such disagreement are greatly increased when there exist so many unknowns with respect to the sources, fate, and effects of petroleum hydrocarbons in the marine environment. Thus, as noted in the National Academy of Sciences study, there remains tremendous uncertainty within the scientific community over such basic issues as the threshold at which hydrocarbon inputs begin to produce undesirable and perhaps even irreversible changes in marine systems. Max Blumer once introduced a review of the limits of our knowledge in the field of organic geochemistry (Blumer, 1975) by quoting the words of the Tamil poet Avvaiyar:

> What we have learnt, is like
> a handful of Earth.
> What we have yet to learn
> is like the whole World.[37]

And while there is now a general understanding of the movement of oil through the marine environment and of the processes that contribute to the dispersal and degradation of oil, it seems fair to say that nowhere is the incompleteness of our knowledge of the world around us better revealed than in efforts to understand, control, and minimize the effects of oil pollution.

The task of those who must actually formulate and implement pollution control policies is undoubtedly complicated by the absence of clear-cut answers to basic questions. Yet it is impossible to avoid the conclusion that much more could have been accomplished in the way of (1) reducing the amount of oil that actually enters the marine environment, notably from tankers, and (2) preparing and coordinating contingency plans to protect the coastline from the effects of those spills that do occur. Quite apart from the environmental losses incurred by those communities most affected by oil pollution (amenities, biota, recreation), such "preventive" measures must appear increasingly attractive in light of the sheer cost of clean-up operations. Yet, by and large, oil pollution control planning in the United Kingdom has been marked by a lack of anticipation and little sense of urgency. Rather, in the words of Lord Ritchie-Calder, there has been "Government by catastrophe, because policies are formulated only when disasters occur" (Advisory Committee on Oil Pollution of the Sea, 1979, p. 4). Thus the creation of the Marine Pollution Control Unit followed the *Eleni V* incident. Only in the aftermath of the *Esso Bernica* spill did the Shetland Island authorities introduce stringent pollution control measures. Moreover, for all the talk of new priorities and initiatives, the refusal to establish a national compensation fund to reduce some of the present inequities in bearing the cost of oil pollution, the proposed reduction of personnel in the Warren Spring Laboratory, and the reluctance to involve other interested parties (notably the Department of the Environment and local authorities) in national policy discussions concerned with oil pollution and compensation, all suggest that there is much still to be accomplished in the way of reducing the risks and costs of oil pollution.

Perhaps what is revealed most strikingly is a fundamental weakness in the institutional arrangements for formulating and implementing maritime policies. The Advisory Committee on Oil Pollution of the Sea has been particularly critical of fragmented ministerial responsibilities and policy-making in the United Kingdom, arguing that an integrated approach to sea use planning and management would allow for a more careful balancing of national interests with respect to alternative uses of the marine environment (Advisory Committee on Oil Pollution of the Sea, 1978, p. 11; 1979, pp. 24–25; 1980, pp. 50–51). Unfortunately there has been little progress in this direction, and in 1979 it was announced that the Lord Privy Seal's role in coordinating policy on maritime matters had been abolished. This role has now been assumed by the Secretary of State for Trade, with other departments, including Defence, Energy, Environment, and Fisheries having a primary role in specific fields (*Hansard*, vol. 405, no. 84, February 20, 1980, p. 800). As noted by the Advisory Committee on Oil Pollution of the Sea, "with its great dependence on the sea, Britain can ill afford to develop policy without balancing and reflecting the concerns of a large number of interested parties" (1980, p. 50).

Quite apart from the issue of comprehensive sea use planning, there exist unexplored opportunities to coordinate pollution control policies and programs for the marine environment with land use and contingency planning in the coastal zone. The difficulties that local authorities have encountered in assessing the need for and the impact of such onshore support facilities as service bases, platform fabrication yards, storage terminals, and processing facilities will be analyzed in the next chapter, but it should perhaps be emphasized here that, in marked contrast to the situation in the United States, current U.K. licensing procedures do not allow for any formal assessment or public discussion of the potential environmental or social impacts of offshore exploration and development.[38] Moreover, there is provision in the United States under the Outer Continental Shelf Lands Act (as amended in 1978) for state and local government participation in policy-making and managerial decisions relating to the development of offshore energy resources (Manners, 1980b). The act further establishes a State Participation Grant Program to provide grants to affected communities in order to cover the administrative costs of participating.[39] An alternative approach is being pursued in New Zealand, where the Town and Country Planning Act of 1977 encourages regional and local authorities to engage in maritime planning for coastal waters to the limits of the territorial sea. Why are such opportunities not more aggressively pursued in the United Kingdom? Why are existing regulations not actively enforced? What are the alternatives, and who should determine what course of action is in the best interest of society? As David L. Sills has commented, "there are no easy answers to these questions—nor to most questions of technology management. But they can never be answered unless they are asked" (1977, p. 636).

4. Onshore Development: Land Use Planning in the United Kingdom

Introduction

The offshore search for oil and gas is inevitably accompanied by onshore development. It is, of course, impossible to foresee the *precise* nature, magnitude, or location of onshore development until the exploration phase of offshore activity has been completed and the existence of commercially viable fields has been established. Much of the uncertainty that surrounds onshore development, at least in a planning context, is removed only as companies begin to formulate production plans both for individual fields and for the offshore province as a whole. Nevertheless, "unlike large oil spills, which *may* occur once offshore oil activity begins, onshore support development will certainly occur. Furthermore, when offshore fields are depleted—as they must be eventually . . . —the social, economic, and environmental changes will remain" (Baldwin and Baldwin, 1975, p. 163).

The onshore facilities required to support offshore exploration and production include supply bases, platform fabrication yards, pipeline terminals, pumping stations, and storage tanks; these facilities in turn create the opportunity for further downstream activities including refining, gas separation, and petrochemical production. Each of these activities in turn brings social, economic, and environmental changes of both an immediate and a long-term character. Local labor markets will be affected by the creation of new jobs and an influx of construction workers; property values will be sharply increased; and housing, education, and social welfare programs will need to be expanded. Such alterations, resulting from rapid economic and demographic growth, combine to produce profound changes in both individual lifestyles and community structure while simultaneously diminishing the amenity resources of particular localities (Figure 8).

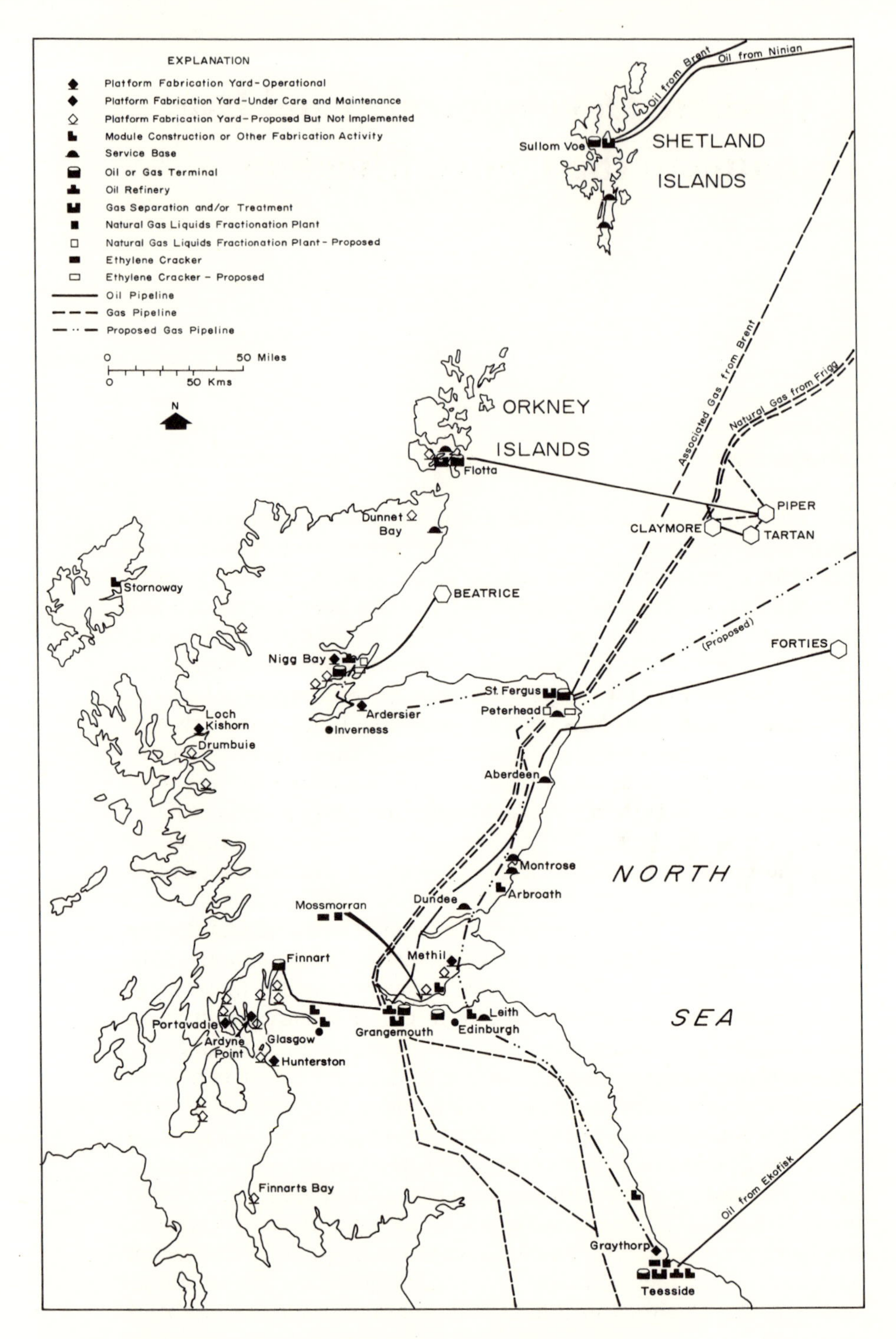

Figure 8. Onshore Development for North Sea Oil, 1980

The major onshore impacts, both environmental and social, are summarized in Table 21. Not all of these impacts will be felt in full measure by all communities. Clearly, larger communities and those with existing infrastructure are in a better position than others to cope with rapid offshore development. Smaller coastal communities with little previous experience of planning for rapid growth will be hardest hit by an energy-based boom. Such communities often lack the full-time public officials and staff or the budgeting and planning programs that their situations may require. With a rapid population influx, social services soon fall far short of community needs. In such a situation there is likely to be growing antagonism toward "outsiders" and anxiety over the cost and risk of public infrastructure investment. Unless effectively addressed, these social impacts may have significant policy repercussions, feeding back to influence the attitudes of those in the local community who initially favored offshore development in anticipation of the tangible economic benefits.

An indication of the scale of the changes that may be initiated as a result of offshore activity is to be found in Tables 22 and 23: the former provides an estimate of the level of direct employment that is likely to be generated by different offshore activities, and the latter provides data on actual changes in oil-related employment in Scotland.[1] As noted in a previous section, not all of these impacts will be experienced at once—rather there is an overlapping sequence of activities that differ in the form and duration of their impact. Some onshore development will take place even in the earliest stages of offshore activity. Maintenance of drilling rigs and service vessels, drilling platform construction, and pipeline installation are activities that have created widespread and significant impact long before a barrel of oil has been produced.

Other support activities, notably the construction of production platforms, are subject to sudden and unpredictable fluctuations in demand. During 1978, for example, several platform fabrication yards in Scotland were forced to shut down or sharply curtail their activities due to a lack of orders. In remote, isolated communities with a poorly developed service infrastructure, a limited pool of skilled labor, and inadequate housing stock, the establishment and subsequent closure of such yards has been profoundly disruptive. At the present time, the major onshore planning conflicts revolve around the siting of such downstream facilities as oil refineries, gas separation plants, ethane crackers, and other petrochemical processing plants, reflecting the fact that the United Kingdom has now moved from the development phase to the production phase of offshore ac-

Table 21. Onshore Support Facilities and Associated Impacts

Phase	Major Support Facilities	Major Onshore Impacts/Issues	
		Environmental	*Social*
I. Exploration	Supply bases	Use of existing allweather harbors and facilities; potential for conflict with existing uses (e.g., fishing)	Minimal impact on labor/housing markets but potential for diversion of labor from traditional maritime occupations
		May prove temporary if no commercial discoveries are made in immediate offshore area	
II. Development	Supply and service bases for offshore rigs	Strain of existing harbor space and facilities; probable need for harbor expansion including land reclamation and dredging (e.g., Aberdeen, Peterhead)	Expansion of engineering, transportation, communications, supply handling and storage, and administrative services; severe impact on local housing and labor markets
	Production platform construction	Locational restrictions (sheltered, deepwater site) may restrict range of choice; potential for conflict in areas of high amenity value due to noise, visual intrusion, etc.	Severe impact particularly in remote, rural areas due to high but fluctuating labor requirements; work camps and other forms of temporary accommodation as well as character of work force may create social strains
	Pipe-laying	Minimized through careful route selection to avoid sensitive ecosystems; multiple landfalls may be avoided by means of coordination and joint-user schemes	Minimal and temporary

	Terminal and storage facilities (including tank farms, oil/gas separation, etc.)	Varies according to product characteristics and circumstances (e.g., oil requires no immediate treatment as gas does); potential for conflict if located in area of high amenity value or where land is already designated for alternative use (agriculture, forestry); risk of marine pollution greatest when surface transport and transshipment of oil occur	Dependent upon scale of facilities; severest impact during construction
III. Operation	Platform maintenance	Threat of marine pollution associated with day-to-day rig operation and surface transport of oil; "blow-outs"; tanker collisions represent lower probability but higher stress incidents	Minimal
	Processing (refinery; LNG, ethylene processing and cracker plants; petrochemicals)	Dependent upon scale but may include visual intrusion, air and water pollution, noise; safety issues; conflict with existing land use, etc.; constraints on siting may arise due to preexisting location of landfalls, storage, and other primary facilities	Dependent upon scale of facilities but severest during construction

Source: Manners, 1978, p. 87.

Table 22. Direct Employment Generated by Offshore Activities

		Direct Employment	
Phase	*Activity/Facility*	*Short-Term*	*Long-Term*
I. Exploration	Exploratory drilling/service base	45 onshore jobs per rig	
II. Development	Development drilling/platform	200 workers per rig during drilling	16 workers per rig during operation
	Permanent service base		50–60 workers per rig serviced
	Steel platform fabrication yard	250–550 workers per platform[a]	
	Concrete platform fabrication yard	350–450 workers per platform on average (peak figure of 600–1,200)	
	Marine terminals	560 workers per land terminal[b]	10–90 workers for land terminal
III. Operation	Partial processing (separation of gas, water, mineral impurities)	150 construction jobs (during approximately 15 months)	10 operation/maintenance jobs
	Gas processing and treatment	500 construction jobs (during approximately 18 months)	45–55 operation/maintenance jobs

Note: Employment generated locally will depend upon the nature of the work and the circumstances of individual communities.
[a]Peak employment at the Brown and Root platform fabrication yard at Nigg Point, Scotland, exceeded 3,000 persons.
[b]By 1979 the construction workforce employed at the Sullom Voe site exceeded 5,000 persons.
Source: Conservation Foundation, 1977.

Table 23. Scotland: North Sea Oil–Related Employment, 1978

Regions	Oil Employment			Oil Jobs as % of Total Employment	Changes in Employment			
					1974–1976		1976–1978	
	In Wholly Involved Companies[a]	In Partially Involved Companies	Total		No. of Jobs	%	No. of Jobs	%
Grampian	22,900	2,700	25,600	14.7	+4,000	+47	+11,200	+78
Strathclyde	1,200	5,800	7,000	0.7	+ 150	+ 1	− 4,400	−39
Highland and Islands	6,900	350	7,250[b]	8.1	+3,900	+96	− 1,050	−13
Lothian, Central, Borders, Dumfries, Galloway	1,200	1,050	2,250	0.4	− 650	−22	− 100	− 4
Fife	1,500	350	1,850	1.4	+ 600	+37	− 400	−18
Tayside	2,150	300	2,450	1.5	+ 400	+39	+ 1,000	+69
Total	35,850	10,550	46,450[c]	2.2	+8,400	+28	+ 6,250	+16

Note: The Scottish Economic Planning Department undertook a survey of all oil-related employment in 1978. Data since that survey are only for employment in companies wholly involved in North Sea oil development. Employment in these companies stood at 41,760 in June 1979.

[a]As of June 1979, employment in wholly involved companies was Grampian 28,060; Strathclyde 770; Highland, Western Isles, Shetland, and Orkney 7,730; Tayside 2,320; other regions 2,880.

[b]This excludes certain temporary construction workers. This employment significantly affects the Highland and Islands. If included it would raise the percentage of the regional workforce employed in oil-related jobs to 11.9 percent and the Scottish total to 2.4 percent.

[c]After allowances for sectors not covered and for the consumption multiplier effects, overall employment was estimated to be in the range 60,000–70,000 in mid-1978.

Source: Scottish Office, 1979a, pp. 30–33.

tivity. At Peterhead, for example, Shell withdrew its proposal for a natural gas liquids separation plant in the middle of a public inquiry after it became apparent that the risks involved in shipping hazardous materials through the already-congested harbor had been underestimated.[2] A subsequent proposal by Shell/Esso to construct a natural gas liquids separation plant and an ethane cracker at Mossmorran in Fife (with an associated jetty at Braefoot Bay on the Firth of Forth) received outline planning permission in August 1979, although efforts were still being made by a local group, the Dalgetty Bay Action Group, to block the project in the courts.[3] The recently approved gas gathering system for the North Sea will inevitably bring further developments of this kind to Northeast Scotland, and similar siting controversies must be anticipated.

Even from this brief description of onshore activities and impacts, it should be apparent that the development of North Sea oil has focused attention on the difficult decisions that must be made when one particular use or development of the environment appears to be in conflict with another. It is, as Peter Smith has pointed out, precisely this type of conflict that brings resource management issues home to the public:

> Confronted with argument and discussion of the *Limits to Growth* variety, the average person could be forgiven for supposing that the whole resource debate is simply a philosophical, or perhaps even an ideological, exercise, with little direct relevance to day-to-day living . . . By contrast, the one aspect of resource exploitation which really brings the matter closest to home—and that in a most literal way—. . . is the conflict which arises when the "need" (however defined) for more resources impinges on the less material aspirations which have come to be known collectively as the "quality of life." (1975, p. vii)

Such issues have been debated with particular intensity in Scotland, in part as an inevitable result of the location of the offshore fields. Thus, Scotland, with stretches of coastline of unparalleled natural beauty and scientific interest, has borne the brunt of the social, economic, and environmental "costs" of providing essential onshore support facilities. At the same time, however, the amenity/development issue has become inextricably linked with the broader political debate over devolution. In these circumstances, the "national interest" in developing offshore oil as rapidly as possible has inevitably confronted the growing Scottish sentiment for a fuller expression of cultural identity through greater political autonomy. From this perspective, the oil in Scottish waters offers a once-in-a-lifetime opportunity to achieve economic independence from London and to rectify the years of neglect that have contributed to Scotland's social and economic deprivation (at least when compared with other, more affluent, regions of the United Kingdom). It has therefore been argued by such groups as the Scottish Nationalist Party that the benefits of North Sea oil should be directed initially (and exclusively) to Scotland and their flow regulated in such a way as to ensure both long-term economic stability and preservation of other valuable attributes of Scottish culture and landscape.

It may be argued that one of the primary tasks of the planner is to reduce the potential for amenity/development conflicts. In this respect, forward or advance planning seeks to facilitate development while simultaneously conserving those social and environmental attributes of a community or locality that have been identified as being of value. It is, in some ways, an attempt at mediation. However, given that the arguments are often about priorities, what Roy Gregory in his book *The Price of Amenity* (1971) calls the "allocation of values—and of costs," the planner operates within a well-defined legal and political framework. The procedures for resolving any conflict over development are established in law, while in the final analysis the decision as to whether or not to proceed with a particular project will be a political one.

> The final decision for or against a proposal will be taken on the basis of a value judgement by one or more politicians, generally by the application of policy to the facts of a particular case. In exceptional cases . . . a proposal may raise issues of wide public importance which put it dramatically in the public, and thus the political domain. At such times the behavior of the participants in the controversy will be influenced by the need to be politically effective, for [in the United Kingdom] the Ministers with ultimate powers to say yes or no to most developments are responsible to Parliament. (Grove-White, 1975, p. 1)

In the context of offshore oil development, a number of American authors have concluded that the more comprehensive British approach to planning, with its emphasis on consultation rather than confrontation, has expedited the development of North Sea oil resources while simultaneously anticipating and absorbing the major onshore impacts without serious social or environmental costs. Irvin L. White et al. for example suggest that in dealing with such onshore impacts as the location of support facilities and pipeline landfalls, refinery siting, the provision of housing and social services, or the "bolstering" of local economies, "the British have demonstrated one of the strong points of their system, the way in which they formulate and implement national and local planning" (1973, pp. 40–41). The authors of this report, which is specifically directed toward looking at the implications of the North Sea experience for future United States development, are particularly laudatory of the way in which planning policies and actions at the various levels of government are coordinated.

White et al. further suggest that the British system of planning "provides open access to all parties" (1973, p. 41), a feature of the planning process that they argue elsewhere is of critical importance in reducing the possibility of friction and of interested groups "being caught by surprise." "The British system demonstrates the benefits of making more information publicly available early enough for all interested parties to have time to evaluate and respond to the proposed action" (ibid., p. 43).

Pamela and Malcolm Baldwin similarly conclude that "the British planning system . . . gives that country a framework of land-use control to cope with onshore development induced by oil" and that despite some strain, planning procedures have been particularly effective in minimizing social disruption and preserving scenic amenities. They suggest that the effectiveness of the British approach contrasts with that of the United States, where the "practice of environmental impact assessment . . . enables us to anticipate and analyze the effects of federal oil development on the natural environment, but has been markedly deficient in social assessment" (Baldwin and Baldwin, 1975, p. 16).

There can be little disagreement about the importance of coordinating national and local planning, about the need to assess and mitigate social and environmental impacts, or about the desirability of public involvement; but it is difficult in view of the realities of planning in the United Kingdom to fully share these authors' conclusions. Indeed, in some respects their enthusiasm seems misplaced. While planning in the United Kingdom may well offer certain advantages over existing procedures in the United States, it is not without serious limitations. The following sections therefore examine in some detail the planning framework for oil-related onshore development in the United Kingdom, both as it exists in theory and as it operates in practice.

Land Use Planning

Land use planning in the United Kingdom has its origins in the grim circumstances of rapid urban growth during the industrial revolution of the early nineteenth century. The squalid rows of working-class tenements and the congestion and unhygienic conditions of the industrial towns that sprang up around the sources of power and raw materials made some form of public regulation of development almost inevitable. Initially statutory planning focused not on land use but on sanitation. A series of sanitation laws, providing minimum standards of hygiene and reducing permissible housing densities, was enacted in the mid-nineteenth century. These laws culminated in the Public Health Act of 1875 that created urban and rural sanitary districts and allowed local authorities to pass by-laws regulating street and building construction.

Although powers to undertake slum clearance and to construct public housing were granted to local authorities in the Working Classes Dwellings Act of 1890, most planning initiatives in the nineteenth century were essentially *reactive*—addressing and seeking to remedy specific facets of urban blight. The more progressive concept of "town planning" first appears in the Housing, Town Planning, Etc., Act of 1909 wherein authorities were encouraged to undertake land use planning "with the general object of securing proper sanitary conditions, amenity and convenience in connection with the layout and use of the land and of any neighbouring lands." This act, the first to acknowledge the significance of amenity in land use planning, was not mandatory and few local authorities appear to have made use of its provisions.

Since 1909, however, a succession of planning acts has significantly expanded the role of the central government (exercised through the relevant ministry) and of local authorities in land use planning (Table 24). As defined in a government pamphlet on town and country planning, "from the passing of the Housing, Town and Country Planning Act 1909 until the present day, the basic purpose of land use planning has been to ensure, as far as possible, that land is used in the best interests of the nation as a whole, rather than being simply subject to market forces" (Central Office of Information, 1975, p. 1). The overall objective of such planning, essential in such a densely populated country, is "to preserve a balance between the competing claims made on the environment by homes, industry, transport and leisure. This objective is pursued through a general framework of planning legislation designed to ensure that in both

Table 24. United Kingdom: Land Use Planning Legislation

Date	*Act*	*Major Provisions*
1875	Public Health Act	Created sanitary districts; authorized local authorities to enact building codes
1890	Working Classes Dwellings Act	Authorized slum clearance and construction of public housing
1909	Housing, Town Planning, Etc., Act	Empowered local authorities to prepare town planning schemes intended to secure proper sanitary conditions, amenity, and convenience in use of land
1919	Housing and Town Planning Act	Made preparations of town planning schemes obligatory
1932	Town and Country Planning Act	Scope of planning extended to rural areas and all types of land use
1940	Barlow Report	Report on the causes of the distribution of the industrial population, including consideration of the social, economic, and strategic disadvantages of industrial concentration, and appropriate remedial measures; formed basis for postwar planning policy
1947	Town and Country Planning Act	Brought all development under control by making it subject to planning permission; required preparation of development plans
1962	Town and Country Planning Act	Consolidated planning legislation, including provisions of a series of acts during 1950s restoring owners' rights to development value and market compensation for compulsory purchase
1968	Town and Country Planning Act	Development plans replaced by structure plans and local plans: the former, dealing with broad policy and strategic issues, subject to ministerial approval; the latter, dealing with day-to-day tactical issues, not requiring ministerial approval
1969	Town and Country Planning (Scotland) Act	Appropriate legislation for Scotland
1971	Town and Country Planning Act	Consolidated provisions of previous act

1972	Town and Country Planning (Scotland) Act	Consolidated provisions of previous act
1972	Local Government Act	Reorganization of local government establishing two-tier system of counties and districts throughout England and Wales
1973	Local Government (Scotland) Act	Reorganization of local government in Scotland providing for a two-tier system of regions and districts except in the Western Isles, Orkney, and Shetland; modification of planning procedures to allow regional reports rather than structure plans

towns and countryside all development harmonises as far as possible with the landscape and the general environment, and through individual Acts under which particular aspects of the protection of the environment—such as the enhancement of the countryside or the reduction of the amount of noise and smoke in the air—are supervised" (ibid.).

The statutory basis for land use planning in the United Kingdom today is the Town and Country Planning Act of 1947 as subsequently amended in 1962 and 1968 and consolidated in 1971. These acts make virtually all "development" (which in planning law includes most forms of construction, engineering, and mining as well as any change in the use of land or existing buildings) subject to prior approval and impose two main duties on local planning authorities. The first of these may be described as *forward planning* whereby each local authority is required to define and describe the context within which future developments may take place. The second responsibility is usually described as *development control* whereby local authorities must examine particular development proposals and either approve or deny planning permission. These considerable powers and responsibilities are, of course, subject to the guidance and control of the central government in London.

While the general framework for town and country planning remains the same, it should be noted that Scotland has been the subject of separate planning acts. Thus the relevant planning legislation for Scotland is the Town and Country Planning (Scotland) Act of

1972. As a result of this act and of subsequent local government reorganization, a rather different planning hierarchy exists in Scotland, with the local planning authorities (regional and district councils) responsible to the Secretary of State for Scotland rather than to the Secretary of State for the Environment as is the case of England (Figure 9).

DEVELOPMENT PLANS

Under the original Town and Country Planning Act of 1947, the cornerstone of the planning process was the development plan wherein a local authority designated appropriate areas for particular types of development.

> The plan itself does not directly control the development or acquisition of land, but it sets out policies guiding future land use. It allocates land for various uses—for example, residential, industrial, shopping or business; it shows the planning authority's proposals for roads, public buildings, parks, open spaces and other public uses, as well as existing uses which it is proposed to retain; it may define areas for comprehensive redevelopment; and it shows green belts surrounding urban areas, and indicates areas where existing uses of land are intended to remain generally unchanged. (Central Office of Information, 1975, p. 5)

A key feature of the development plan was a map on which was indicated the potential use of all land within the authority's jurisdiction. Those areas for which no particular use was specified in the plan were termed "white areas" and were essentially regarded as areas within which existing land uses (agriculture, recreation, conservation) should remain undisturbed. The plan thus provided a base against which all subsequent development proposals could be assessed and locational decisions evaluated. Since planning permission was required from the local authority even where a proposed development conformed to the provisions of the plan, the emphasis was clearly on controlling development through the monitoring and regulation of all proposed changes in land use. In this way, local authorities in the United Kingdom have for some time enjoyed a greater measure of control over land use than exists in the United States.

In order to ensure that land use decisions by local authorities conformed to national objectives and policies, all development plans had to be formally approved—in Scotland, this responsibility lay with the Secretary of State for Scotland. In this respect Scotland,

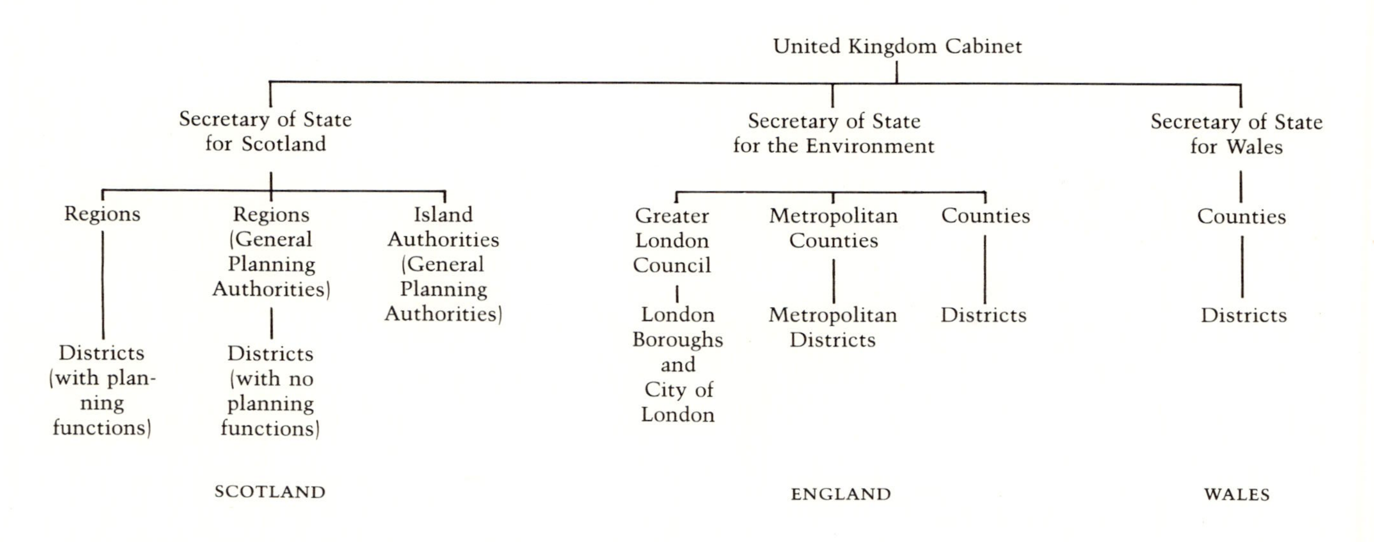

Figure 9. United Kingdom: Local Government Structure

through the Scottish Office, has for some time enjoyed a measure of administrative devolution.[4] Baldwin and Baldwin describe the Scottish Secretary as "something of a land-use czar" (1975, p. 130) whose planning authority combines and exceeds those of local, state, and federal authorities in the United States. While this may seem to exaggerate the secretary's powers somewhat, the closest analogy might be to a state governor who simultaneously holds a Cabinet post, responsible to the federal rather than to a state legislature and involved in formulating national policies while overseeing federal actions and agencies within the state.

Yet with respect to North Sea oil development, the Scottish Secretary has no individual responsibility for determining such aspects of national energy policy as the rate at which oil reserves should be developed or the way in which the financial benefits should be distributed. These policies bear the imprint of the Department of Energy. The Secretary of State for Scotland is merely one member of the British government, although one with a Cabinet portfolio (Figure 10).

The difficult relationship between the Scottish Office and the Department of Energy is perhaps best illustrated in what came to be known as the "Drumbuie legislation." Frustrated by the delays encountered in obtaining local planning approval for a platform fabrication yard in the Scottish Highlands, the Conservative government of Edward Heath introduced a bill to allow compulsory purchase, without the usual public inquiry process, of any land needed to facilitate North Sea oil development. This bill, dealing with an onshore planning issue of particular concern to the Scottish community, was introduced not by the Scottish Secretary but by Lord Carrington, then Minister of Energy. As noted by Baldwin and Baldwin, "to an important extent, the Scottish Office has been obliged to respond to and accommodate these nationwide energy policies—a condition that has in some instances rendered Scottish oil planning impotent and has helped to generate the increasing nationalist sentiment in Scotland" (1975, p. 130). Nevertheless, the broad-ranging responsibilities of the Secretary of State for Scotland include the provision of all health, education, and welfare services, the formulation and implementation of economic development plans, as well as environmental protection plans. Hence, the Scottish Office is in the forefront of onshore planning.[5]

Under the terms of the original Town and Country Planning Act, development plans were to be submitted for approval within three years of the act's coming into effect (i.e., by 1950) and thereafter reviewed and updated every five years. In practice, the process of

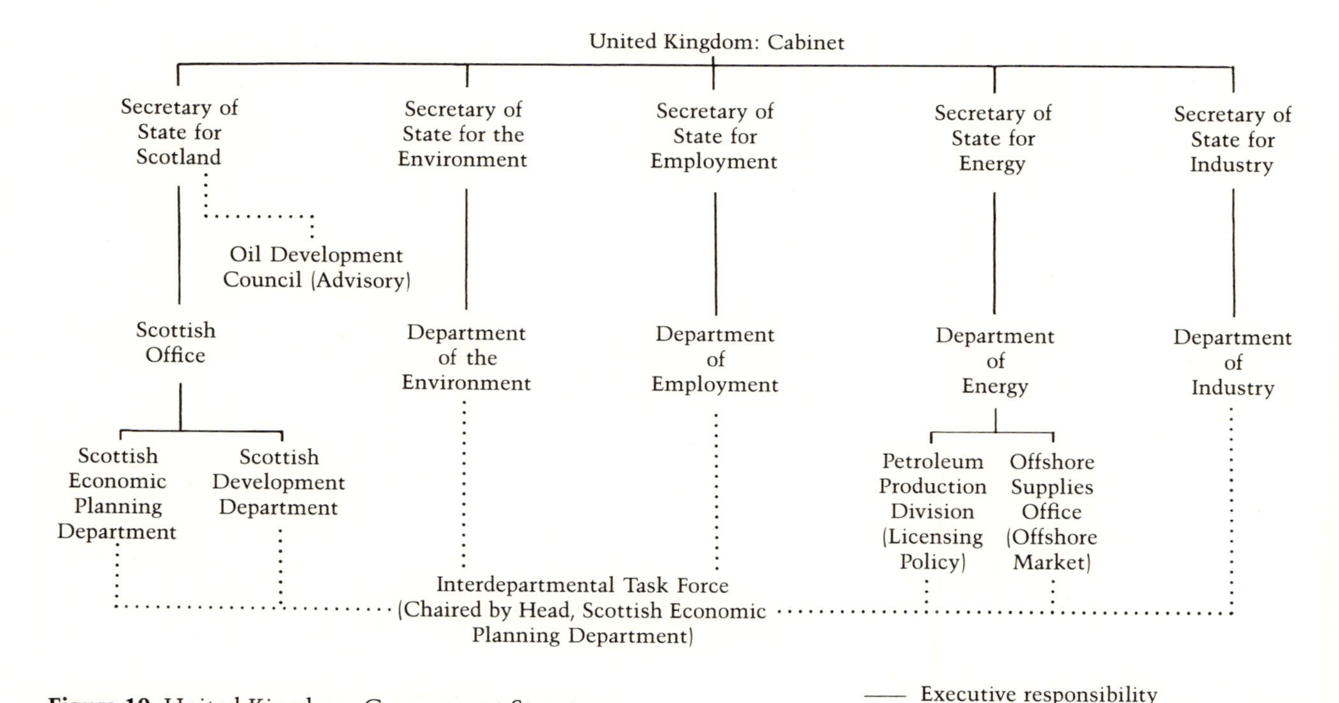

Figure 10. United Kingdom: Government Structure (with Particular Reference to North Sea Oil Development) (modified from Scottish Information Office, 1977)

formulating and approving (let alone reviewing and amending) development plans proved to be extremely time-consuming. By the end of the initial three-year period, only twenty-two plans had been submitted for approval (Ardill, 1974, p. 34). Two decades later a number of authorities still lacked development plans, while other plans were to varying degrees inadequate or incomplete.

In Scotland neither Sutherland nor Shetland had formulated a development plan in the late 1960s when the full impact of onshore oil development began to be experienced (Baldwin and Baldwin, 1975, p. 135). Even where development plans had been approved, the pace of social and economic change in the postwar period made it almost impossible to keep plans up to date, much less forward-looking and responsive to the demands of change. The inflexibility of development plans, once formulated, and the increasing preoccupation with detail and procedure further undermined the quality of planning. "The plans have . . . acquired the appearance of certainty and stability which is misleading since the primary use zonings may themselves permit a wide variety of use within a particular location, and it is impossible to forecast every land requirement over many years ahead" (Ministry of Housing and Local Government, 1965, p. 20). By the mid-1960s, many local authorities were beginning to evaluate planning applications not by reference to the approved development plan but by reference to what were described as "other material considerations."[6]

It would perhaps have been surprising if a system designed with the experience of the forties in mind had *not* revealed weaknesses when confronted with the tempo and nature of social and economic changes during the sixties. In response to the growing sense of disenchantment, the Ministry of Housing and Local Government appointed a Planning Advisory Group in 1964 to review the entire structure of the planning system. Their report, *The Future of Development Plans*, not only repeated the standard criticisms but questioned whether the development plan as then conceived could ever provide a satisfactory instrument for forward planning purpose. "In terms of technique . . . they deal inadequately with transport and the interrelationship of traffic and land use; in factual terms . . . they fail to take account quickly enough of changes in population forecasts, traffic growth and other economic and social trends; and in terms of policy . . . they do not reflect more recent developments in the field of regional and urban planning. Over the years the plans have become more and more out of touch with emergent planning problems and policies, and have in many cases become no more than local land-use maps" (Ministry of Housing and Local Government,

1965, p. 15). The emphasis, in their judgment, was on negative control of undesirable development rather than positive forward planning for the creation of a quality environment. As is noted in a subsequent section, the growing concern over the quality and concept of development planning was paralleled by frustration (particularly in the private business sector) arising from delays in processing planning applications and by complaints about lack of public involvement in the development control process.

STRUCTURE PLANS AND LOCAL PLANS

In these circumstances, the major recommendations of the Planning Advisory Group were favorably received and ultimately incorporated into the Town and Country Planning Acts of 1968 (for England and Wales) and 1969 (for Scotland). These acts have considerably modified the planning structure at the regional and local level by separating what are usually referred to as policy or strategic issues from the more mundane, day-to-day tactical issues of development control. To achieve this, the old-style development plan has been replaced by a *structure plan* and a *local plan*.

As presently conceived, a structure plan consists (primarily) of a written statement of "the local authority's policy and general proposals in respect of the development and other use of land." The intent is to formulate broad policies or strategies for large regions over a twenty- to thirty-year period. In contrast to the old-style development plan with its narrow interpretation of land use, a structure plan must take into account such related issues as employment, transportation, recreation, and population.

This change in planning philosophy is also reflected in the scope and character of the surveys that must accompany the preparation of a structure plan (Cullingworth, 1976b, pp. 81–82). Whereas earlier surveys were required to consider only the physical characteristics of the land, the new-style surveys must include the principal physical and economic characteristics of the region and must be continuously updated in order to indicate when a structure plan is in need of revision.[7] The final plan, which must be submitted to the Secretary of State for approval, must reflect national and regional policies (vis-à-vis social and economic planning), identify the potential impact of the authority's proposals on neighboring regions, and provide guidelines and bases for local planning.

In some respects, the term *plan* is a misnomer, since to avoid the excessive attention to detail that characterized the preparation of development plans, the structure plan may not include any map delineating specific boundaries. Only diagrammatic illustrations are permitted showing the general location to which specific policies refer. In sum, therefore, structure plans are intended to be flexible documents, stating broad principles and policies outlining general proposals that are not depicted or delineated spatially. "Structure planning gives the planners, the planning authority and the public the opportunity to examine in a rational framework the physical problems which beset our lives and to decide, within the limits of what is attainable, how our physical environment should be ordered and what priorities should be selected" (Ardill, 1974, p. 36).

In contrast to structure plans, local plans are intended to be more "action-oriented," providing more precise information for development control in specific areas.[8] A local plan consists of a written statement and a map showing the nature and location of future developments. In contrast to the old-style development plan, however, the local plans do not presume to zone every piece of land. The intention is to realistically allocate sites for known and potential developments, thereby avoiding the situation whereby large areas were set aside (and thereby effectively sterilized) for a particular use, such as an industrial estate, even where there was no indication or chance that such development would occur.

An even more radical change is that local plans are not subject to ministerial approval, although the appropriate Secretary of State does retain reserve power to "call-in" a local plan for approval in exceptional cases. "The rationale for this (originally) was that a local plan would be prepared within the framework of a structure plan; and since structure plans would be approved by the Secretary of State, local authorities could safely be left to the detailed elaboration of local plans. This went to the very kernel of the philosophy underlying the new legislation—that the department should be concerned only with strategic issues, and that local responsibility in local matters should become a reality" (Cullingworth, 1976b, p. 87).

The new-style planning structure was therefore intended to accomplish a variety of objectives. These included a reduction in administrative delays (both in preparation of local plans and in processing individual planning applications); the transfer of final responsibility for local development decisions to local authorities; greater emphasis on positive forward planning; and an enhanced role for the public in the planning process. Increased public participation was to be achieved through hearings and workshops during the preparation of both structure and local plans. In a number of important respects, the revised planning structure reflected the growth of regional sentiments during the 1960s and the central government's subsequent decision to undertake a measure of local government reform that would include both a reorganization of local administrative units and a redistribution of planning powers and functions. (It may be argued, of course, that this decentralization of planning powers was somewhat illusory, since the local authorities were constrained both by regional considerations and by the Secretary of State's statutory and reserve powers to approve structure plans and to "call-in" any local plan or particular planning application—see below.)

In order to understand many of the problems that have subsequently confronted local authorities in their attempts to implement the new procedures, it is important to appreciate that the reforms were based on the assumption that the sometimes bewildering array of local government entities would be regrouped into unitary authorities. "In essence the concept was one of a single authority responsible for preparing a broad strategic structure plan, within the framework of which detailed local plans would be elaborated, and development control would be administered" (Cullingworth, 1976b, p. 82).

In practice this did not happen—either in England or in Scotland. In England the Local Government Act of 1972, which came into effect on April 1, 1974, established a two-tier system of counties and districts. These new units were a far cry from the more radical reorganization into larger unitary authorities previously recommended by the Redcliffe-Maud Commission (Royal Commission on Local Government in England, 1969). J. W. House suggests that "The unitary principle had much to recommend it, in its cleaner break with the past, the designation of units large enough for effective planning, the coherence of functions intended and the possibility of capitalizing upon the rising mobility and enlarging space perceptions of the population. Such a fresh system of units would have required new concepts of community, but offered the prospect of abandoning the parochialisms, which for so long have bedevilled planning in the UK, even as high as the level of large towns or counties" (1977, p. 38). In practice these narrower "parochialisms" prevailed, and the reorganization of local government units in England involved little more than a modification of existing administrative boundaries to create the first-order units (which are termed counties and remain exceedingly disparate in size, population, ratable value, etc.) while the second-tier units (now termed districts) correspond closely with the pre-existing county districts.

In the aftermath of local government reorganization, it was perhaps inevitable that counties would be allocated responsibility for the preparation of structure plans, with the districts retaining responsibility for actual development control through local plans. Yet this splitting of functions carries with it inherent risks. In the 1968 Town and Country Planning Act, for example, it was never anticipated that a local plan might not conform to an approved structure plan.

> Clearly, once the responsibility for local plans is allocated to a different authority, this concept breaks down, since the districts may have very different ideas from the counties on the way in which the general policies in a structure plan are to be elaborated in their areas. The scope for conflict is very great, particularly since district authorities are independent political entities which are not subservient to the county. Furthermore, given that the districts are responsible for local plans, there is an inevitable temptation for counties to formulate their "policy and general proposals" in greater detail than would be the case if there were no division of functions. In this way they can keep a tighter rein on the district. (Cullingworth, 1976b, p. 88)

To complicate the picture still further, the planning framework in Scotland now differs in several important respects from that existing in England. Cullingworth suggests that these differences may be ascribed in part to the more drastic nature of Scottish local government reorganization and in part "to a canny move to avoid some of the difficulties which can be expected in the English system" (1976b, p. 91).

In contrast to the Redcliffe-Maud report, the Royal Commission on Local Government in Scotland (the Wheatley Commission, 1969) "rejected the unitary principle as unworkable for Scotland, believing that a single-tier solution would result neither in efficient planning and administration of the major services, nor in satisfactory democratic representation" (House, 1977, p. 40). In broadly following the Wheatley Commission recommendations, the Local Government (Scotland) Act of 1973 created 9 regional and 53 district councils, together with three "all-purpose" island authorities (Table 25, Figure 11). The extent to which local government reform in Scotland was more radical in terms of redefining local government units and creating more "logical economic and social entities" may be seen from the fact that these 65 new units replaced a staggering 430 local authorities (4 cities, 21 large burghs, 176 small burghs, 33 counties, and 196 districts) (House, 1977, p. 40). Only 49 of the new authorities, however, have planning powers (Cullingworth, 1976, p. 67). In practice this means that there exists a two-tier planning system in

Table 25. Administrative Authorities in Scotland (Local Government [Scotland] Act, 1973)

Authority	*Population (1974)*	*No. of Districts*	*Planning Functions*
Regional councils:			
Central	267,029	3	Two-tier structure; planning functions divided between region and district
Fife	337,690	3	
Grampian	447,935	5	
Lothian	758,383	4	
Strathclyde	2,527,129	19	
Tayside	401,183	3	
	4,739,349	37	
Borders	99,105	4	Regions and general planning authorities; districts have no planning responsibility
Dumfries and Galloway	143,711	4	
Highland	178,268	8	
	421,084	16	
Island authorities:			
Orkney	17,462		No districts; single-tier administrative and planning units
Shetland	18,445		
Western Isles	30,060		
	65,967		

the six largest regions (that together contain 90 percent of the population) and a single-tier general planning authority in the less densely populated Borders and Highland regions and island authorities (Table 25).

In view of the size of the Scottish regions, the Local Government (Scotland) Act of 1973 introduced a number of changes in planning procedures. In the first place, a single structure plan covering the entire region that could be continuously updated was clearly inappropriate. Thus regional and general planning authorities may prepare structure plans for different subunits of the region. Indeed, the presumption is that it will not be necessary for the entire region to be covered by a structure plan.

In place of a single structure plan, the first-level units must prepare what is called a regional report. This is intended to be even more flexible and broad-ranging in its content than a structure plan, and there is no formal procedure for its preparation, submission, or approval.

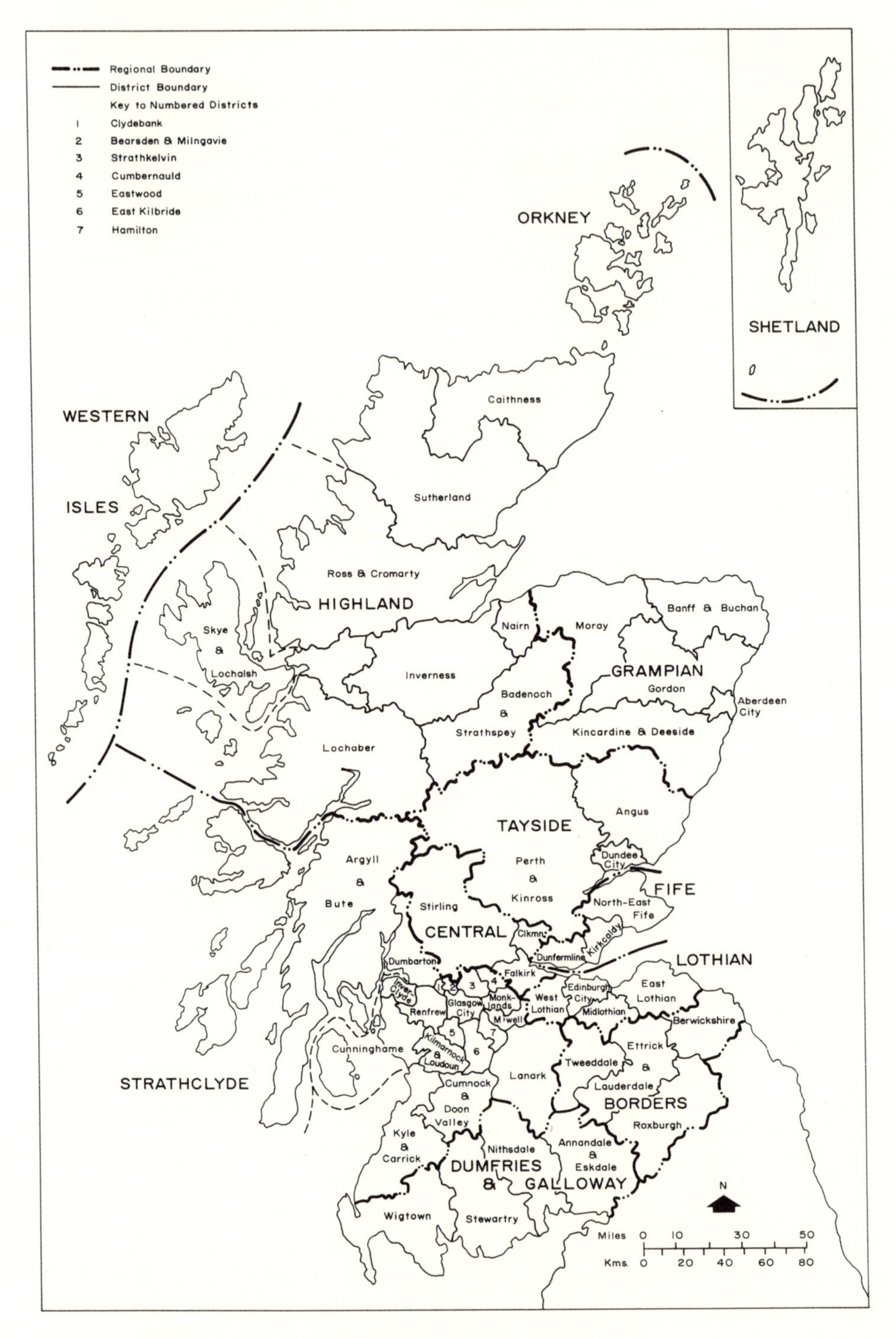

Figure 11. Scotland: Local Government Authorities

> It may be used to provide a basis of discussion between the Secretary of State and a region about general development policy; it may provide a basis of guidance for the preparation or review of structure plans; and, in the absence of structure plan, it may serve as a guide to district planning authorities and developers on planning policies . . . The only requirements are that a regional report shall be based on a survey, that affected local authorities shall be consulted, that it shall be submitted to the Secretary of State (who "shall make observations" on it) and that it shall be published. (Cullingworth, 1976b, p. 91)

The Scottish system thus provides regional planners with a degree of flexibility (i.e., through the regional report and/or structure plan) that does not exist in England. Local plans remain the responsibility of the districts (except in the single-tier planning entities, where they are the responsibility of the general planning authority). Moreover, under the terms of the 1973 act, local plans are mandatory, shall be prepared "as soon as practicable," and may be submitted for approval even in the absence of a structure plan. In these circumstances, local plans have received a much higher priority in Scotland than in England.

This planning structure should provide Scotland with an appropriate framework for dealing with oil-related onshore development. A region's general policies and proposals—covering such key issues as the type of oil development that is to be encouraged and potential housing, recreational, welfare, and transportation needs—would be set out in a regional report or structure plan, while local plans would be primarily concerned with identifying appropriate locations for particular types of development. The opportunity exists to coordinate, at least in theory, planning initiatives and policies and to achieve a balance between development/conservation and national/regional/local interests.

It is perhaps premature to pass judgment on the new structure, but serious weaknesses have already become apparent. These relate particularly to the strains that must inevitably exist within a two-tier planning hierarchy with overlapping functions and responsibilities. Local authorities are particularly sensitive to any attempt to impose policies or plans.

The "call-in" powers of Scottish regional authorities is a particular source of friction. This statutory power first appears in the Local Government (Scotland) Act of 1973. It allows regional authorities to "call-in" local plans for approval.[9] While it is clearly necessary to ensure that the decisions made by the district councils are not prejudicial to the interests of the region as a whole, such powers appear somewhat contrary to the philosophy underlying the reform of the old-style planning system. Similarly, where the issues raised in a particular planning application are considered to be of general significance to the region, the regional authority is empowered to call in the application for determination at the regional level.

Although an attempt was made in the 1973 act to clarify what comprised appropriate regional planning concerns (for example, highways, public transport, education, water) there are still issues where responsibility is unclear. In 1978, for example, the Grampian Regional Council was attempting to call in applications relating to shopping centers and housing estates, arguing that such developments were of regional significance—a view not shared by the district councils. Such issues, which must ultimately be resolved by appeal to the Scottish Secretary, are time-consuming and divisive. Moreover, the attempt to influence may be more subtle. Thus a regional authority provides a strategic brief to "guide" the district authority and must be consulted throughout the preparation of a local plan.

In the aftermath of local government reform, some organizational problems were inevitable. But the protracted and frequently acrimonious "consultations" within the planning hierarchy have undoubtedly contributed to delays in the preparation of structure plans. Shortages of qualified personnel, at a time when planning staffs are being required to consider a broader range of complex and multifaceted policy issues, have compounded this problem. In Scotland, regional reports were prepared by most authorities during 1976, but the first structure plans (for Lothian Region and Shetland Authority) were only submitted to the Secretary of State for approval in mid-1978. In England, where the regional report option is not available to the counties, the situation was so disturbing that the Department of Environment issued a circular "advising" all county authorities to restrict the scope of structure plans by concentrating only on those issues that were of key structural importance to the area concerned (Department of the Environment, 1974b).[10]

In this respect, good intentions notwithstanding, criticism of the delays in preparing and updating the old-style development plans appears to have gone unheeded. It is somewhat ironic to hear developers who were previously critical of the rigidity and inflexibility of development plans now indicting the system for the uncertainty arising from the absence of plans. It should, of course, be emphasized that in this transitional period, decisions on individual planning applications (including virtually all onshore oil-related development) have continued to be based on the old-style development plans with all their admitted imperfections and inadequacies. The influence of the new-style structure plans, however, is becoming increasingly apparent in the new production phase of offshore activity. Such issues as the desirability of downstream facilities in a regional context are particularly appropriate for consideration in a structure plan.

An equally serious long-term concern is the extent to which the "old-style" development plans have influenced the direction and content of the "new-style" local plans. In England, delays in preparing structure plans have meant that local plans have frequently been formulated "lacking context in a wider framework" (House, 1977, p. 43). The situation is similar in Scotland, where higher priority has been given to local plans and where such plans may now be approved in advance (or even in the absence) of structure plans. In such a situation, there is the temptation to draw on the development plan as the basis for the local plan. A key issue is whether local planning is so far ahead as to preempt and constrain structural planning. There seems to be a real possibility that structure plans will represent a rationalization of past actions by local authorities rather than a basis for rational, forward planning in a regional context.

DEVELOPMENT CONTROL

The preceding section outlines the origins and structure of the planning system in the United Kingdom as it has evolved since World War II. It is evident that deficiencies and contradictions continue to exist in the preparation and integration of structure and local plans and in the continued parochialism of some authorities that inhibits effective dialogue. Such criticism has greater force in England than in Scotland. In Scotland, the more radical reorganization of local government units and planning powers and the need to deal with the major onshore impact of North Sea oil development have perhaps combined to contribute to a more flexible response to the challenge of development planning.

In the context of North Sea oil development, the procedures whereby local authorities consider and either approve or deny applications for the siting of particular support facilities (usually referred to as *development control*) are of particular relevance. As noted earlier, any application for planning permission will be considered by the local authority in the context of its development plan or structure/local plan. The actual method of applying for planning permission has remained virtually unchanged since the 1947 Town and Country Planning Act (Figure 12).

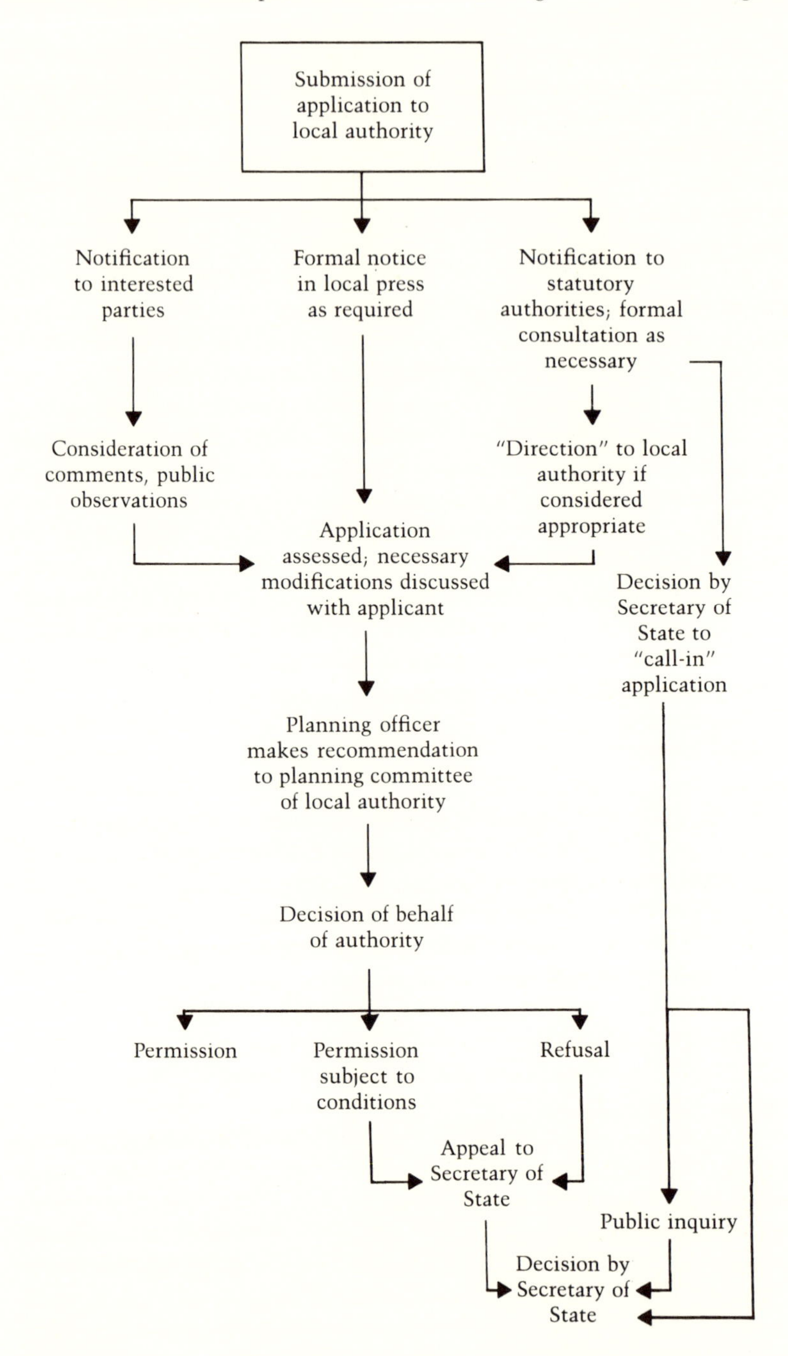

Figure 12. Planning Application Procedures (adapted from Ardill, 1974, pp. 96–97)

Thus all development proposals must initially be submitted to the responsible local authority. If the proposed use of a site is compatible with the development/local plan, no public notice is required, although it appears that such notices are frequently published as a matter of course in the local newspaper. If the proposed use is not in accordance with the development/local plan, the district clerk is required to publish a formal notice summarizing the application and indicating where further information may be obtained. Since a decision on the application must be made within two months of the formal notification, only a comparatively short period of time is available for the public to register objections or submit written comments.[11] Where a developer wishes to obtain an indication of the authority's intention prior to incurring a major financial outlay (for example, prior to the purchasing of land or preparation of detailed plans), "outline" planning approval may be sought. In such instances, plans and particulars relating to siting, appearance, access, and support services need not be submitted in any detail and are left for subsequent approval (known as "reserved matters") in the event that the outline application is approved.

Once outline planning permission has been given, the local authority is committed to allowing the development to take place in one form or another once any imposed conditions or "reserved" matters have been met by the applicant. Thus a situation exists whereby local authorities initially may have to consider a development proposal and assess its likely impact on the basis of the very sketchy information submitted for outline planning consent.[12]

It should perhaps be emphasized that throughout this process, planning permission is not contingent upon "highest and best use of the land" as in the United States. Land is taxed according to current use rather than potential use, and denial of planning permission does not infringe the owner's proprietary rights nor require compensation, providing existing use and value are not impaired (Baldwin and Baldwin, 1975, p. 133).

The basis on which planning permission shall be awarded or denied is clearly stated in the Town and Country Planning Act. Section 29(1) of the 1971 act requires that a planning authority arrive at its decision "having regard to the provisions of the development plan, so far as material to the application, and to any other material considerations." "Other material considerations" is clearly meant to include central government policy. Local planning authorities are well aware of such policies and priorities (for example, the need to expedite development of North Sea oil resources) through notes and circulars from the Scottish Development Department. However, as

Robin Grove-White emphasizes, "in the development control field, departments of central government tend not to *tell* local planning authorities what to do; rather they indicate government policies and leave the planning authorities to interpret them. The fact that a developer can appeal to the Minister if his application is rejected tends to guarantee a broad consistency in planning policies throughout the land" (Grove-White, 1975, p. 8). And, one might add, it ensures that in the final analysis national policies and priorities (as defined and interpreted in London) prevail.

With most applications the planning process is completed when the authority decides whether to grant unconditional permission, to grant permission "subject to such conditions as they see fit," or to reject an application. Where outline or detailed planning approval is withheld, an applicant may appeal to the appropriate minister (in Scotland, the Secretary of State for Scotland). In such circumstances, the minister may order a public inquiry to hear the case for and against the proposal.

It is important to note that these procedures may be circumvented. Thus, under the revised Town and Country Planning Act, the Secretary of State for Scotland retains the power to "call-in" any application for review, i.e., to remove the application from local consideration and to act on it personally. This event is most likely to occur if the application is believed to have important national implications or if there is substantial local opposition.

Should the Secretary of State choose to exercise this option (or indeed whenever a proposed development represents a substantial departure from the approved plan), a public inquiry *may* be held. Holding a public inquiry involves the appointment of an Inspector (or in Scotland, a Reporter), who conducts a formal hearing at which evidence is taken from all interested parties. The Inspector then submits a report (which is based solely on the evidence presented at the hearing) and recommendations to the Secretary. This was the procedure followed when planning permission was sought for a platform fabrication yard at Drumbuie on land owned by the National Trust for Scotland. In this instance (which has become something of a cause célèbre and which is discussed in more detail in a later section) the Secretary of State for Scotland ruled against the proposal, following the recommendations of the Reporter who conducted the inquiry. *But the Secretary is not statutorily bound by the findings of the inquiry and is completely at liberty to overrule the hearing officer's recommendations.* In so doing, the minister may be influenced by government policies that were not raised at the inquiry.

A more recent controversy, for example, surrounded the application by the Cromarty Petroleum Company Limited to construct a refinery and oil storage facilities at Nigg Point on the Cromarty Firth.[13] A public local inquiry was held early in 1975, as a result of which the Reporter, G. W. Maycock, concluded that there were

> major objections to the proposed development . . . because of (a) the impact it will have on the landscape of Easter Ross; (b) the risk it would introduce of a major spill, or a series of minor spills, having a cumulative and perhaps irreversible effect on wildlife in intended National Nature Reserves of international importance; (c) the effects such spills and any related publicity thereto can have on the population in, and visitors to, the area; (d) the short- and long-term effects the development and its associated activity can have on the fishing industry; and (e) the inhibitions and limitations that a privately owned marine oil terminal designed to service but one industrial development and constructed at Nigg Point—which Point is to be considered as a key location in the context of the Cromarty Firth—may impose on the future development of that Firth as a port. These effects would remain at least throughout the working life of the refinery. In addition there would during the construction phase be further disruption of the social fabric and additional strains on the infrastructure of an area already under great stress arising from recent and continuing industrial growth . . . I am not satisfied that there is evidence of a need for the development in this location in either the national or the local interest sufficiently great to warrant overriding the preceding objections. (Maycock, 1975, paragraphs 70–71)

These "final conclusions" of the Reporter have been quoted at some length since it seems unlikely (one hestitates to say inconceivable) that such a strongly worded and forceful official report (perhaps in the context of an environmental impact statement) could have been so easily overturned in the United States. Yet the Reporter's recommendations were rejected by the Scottish Secretary as not being in the national or regional interest (the Highland Regional Council had supported the application), and the plans were approved.

The Minister's decision is final—there can be no appeal except to the courts on a specific point of law. Moreover, there can be little doubt as to the government's priorities with respect to North Sea oil and gas development. In granting outline planning permission for Shell/Esso's petrochemical complex at Mossmorran and Braefoot Bay, for example, the Secretary of State for Scotland noted that

> from the land use point of view, the Moss Morran site is suitable for the development proposed there. On the other hand . . . the site proposed for the related marine terminal at Braefoot Bay would not normally be regarded as an appropriate location for such development in view of the adverse effects on the environment and the area's recreational value. Only an overriding case of need in the national economic interest could justify permitting a marine terminal at Braefoot Bay in face of the amenity objections, even though detailed planning conditions in relation to development there would reduce the adverse environmental effects. The evidence and the Reporter's conclusions have therefore been assessed very carefully in order to determine whether such need for this particular site sufficiently outweighs these objections. No convincing evidence has emerged to support an assumption that a better site for a marine terminal could be found which would adequately serve the needs of the proposed NGL plant. The Moss Morran site is an acceptable location for the building of an NGL plant and no more suitable site has been shown to exist for an associated marine terminal. The Secretary of State therefore feels bound to conclude that the demonstrated need for the marine terminal development at Braefoot Bay decisively outweighs the amenity objections. In reaching this conclusion the Secretary of State has had regard to the need, in the national interest to make provision for NGL separation plant facilities to take advantage of, and avoid wasting, the very large gas resources of the Brent field. (Decision-Letter from Secretary of State to Esso Chemical Ltd. and Shell UK Exploration and Production, August 9, 1979, p. 5)

Even in this brief discussion, a number of questions arise with respect to the effectiveness of a public inquiry in mediating conservation/development and national/local conflicts, in evaluating the technical and policy implications of a proposal, or in considering the suitability of alternative locations. These critical issues are discussed in a subsequent section. For the present, one may note that, given the time and effort involved (the Drumbuie inquiry, for example, lasted from November 1973 to May 1974 and, according to Baldwin and Baldwin [1975, p. 61], involved forty-three days of public testimony at a cost of $360,000), the public inquiry procedure is not frequently invoked. Indeed, it appears to be regarded as something of a "court of last resort" to be used sparingly and only when informal discussions among the developer, the local authority, and contesting groups have failed to produce a compromise solution acceptable to all parties.

The existence of reserve "call-in" powers clearly introduces a degree of uncertainty into development control proceedings at the local level. However, there have been other, more immediate pressures on local authorities in processing oil-related planning applications. In many respects, local planning authorities were woefully ill-prepared for the scale and complexity of oil-related development. Deficiencies in many development plans have already been mentioned. Moreover, in the early exploratory phase of offshore activity, onshore development was deceptively low-key. As a 1975 Scottish Development Department paper acknowledges,

> The early service bases were accommodated within existing harbours (often as permitted development) without significant problems and it was not until the demand for platform construction sites occurred that it became evident that large greenfield sites, population increase and therefore additional residential services would be required. This aspect of oil development was largely unforeseen, probably because the demands of the offshore exploration activity were still modest and tentative, and no comparable pattern had been experienced in relation to the gas fields off East Anglia. (Scottish Development Department, 1975a, p. 7)

The situation changed quite drastically in the early 1970s. Virtually every account of North Sea oil development during this development phase refers to the "overwhelming pressures" on local planning authorities. During 1974, the Zetland County Council was reportedly receiving applications at the rate of a hundred a month, compared with a hundred a year previously (Francis and Swan, 1973, p. 45). Ross and Cromarty County Council temporarily refused to accept further planning applications for the Cromarty Firth area to "take stock" of the situation and ensure that planners "might have time to consider developments to date and so might avoid being stampeded into proposals that they might later regret . . . We found that officials were working under extreme pressure since they were understaffed for the vast increase in work. In 1967, the Council received 550 applications for planning consent; the current rate is 1,700 per year, and complexity increases. This is typical of the planning scene generally up and down the coast" (ibid., pp. 48–49).

Quite apart from the heavy work load is the issue of the expertise available to local planning authorities. Thus Baldwin and Baldwin (1975, p. 60) note, for example, that in 1974 the planning staff of the City Council of Aberdeen (the city that has become the major supply and administrative focus for North Sea oil development) included only three qualified urban planners and two architects. There was no ecologist or environmental scientist to assess the environmental consequences of proposed developments. In the smaller local authorities, human resources were even more limited. In Argyll, the total planning staff in the early 1970s consisted of one full-time person with some part-time support.

Given their limited resources (and the complexity of many applications), planning authorities have frequently been forced into a position of having to rely on the applicant for information about the potential impact of a proposal. In 1971, for example, Brown and Root submitted an application to the Ross and Cromarty County Council to establish a platform fabrication yard at Nigg Bay. In response to a questionnaire prepared by the council, the contractor estimated that the construction force would number between four hundred and six hundred persons—a figure which corresponded very closely to the size of the labor force in the process of being laid off at a recently completed aluminum smelter in Invergordon. The council approved the application unconditionally. By mid-1974, with two platforms under construction, the work force was around three thousand, placing an intolerable strain on the local housing market and service infrastructure. In this instance, the contractor ultimately purchased two passenger vessels and moored them in Nigg Bay to provide temporary accommodation.

Obtaining information about a proposal may prove as difficult as evaluating its quality. As noted earlier, if outline planning permission is sought, prospective developers need only describe a project in the most general terms. As noted by Robin Grove-White,

> Once the developer's planning application is lodged, he is still in a strong position. Only he knows why he needs the site, and what the economics of its exploitation are, for he is under no obligation to make known to the public or even to the local planning authority anything more than the fact that he wants to develop a particular site for a simply defined purpose. And while he may be willing to discuss these matters in more detail with the planning authority when he thinks they may grant planning permission, he may decide not to do so when he calculates that his application will eventually be determined by the Minister. Sometimes it will be to his advantage not to reveal his case until the public inquiry takes place, as this will lessen the chance of its being effectively rebutted by

> the local authority (where the local authority is an objector) or by other objectors to his proposals. (Grove-White, 1975, p. 13)

Assessing the "good faith" of an applicant may be as critical an issue as evaluating the project's potential impact. Many applications relating to North Sea oil development have clearly been speculative. In other cases developers appear to have submitted duplicate applications in the hope that this move would persuade at least one of the authorities involved to act expeditiously. In either case, planners must initially evaluate a proposal without any assurance as to the seriousness of the developer's intentions. The uncertainty thus created in terms of assessing the real demand for land, housing, and community services is reflected in a report by the Chief Planning Officer of the Scottish Development Department: "The promise of new finds has meant that more service and fabrication firms have wished to get into a position from which they can tender for work when it is required at a date often as yet undetermined . . . Not all firms in possession of sites will be awarded work simultaneously or continuously, and there has been some danger of double-counting in the impact these operations may have on housing and community development" (Francis and Swan, 1973, p. 74).

The Dunnet Bay "fiasco" (this description is from a special issue of *Architect's Journal*) illustrates the problem of speculative applications. In 1972 Chicago Bridge Limited, an American-owned construction firm, sought planning permission for a platform construction yard employing four hundred to five hundred persons at Dunnet Bay on the Caithness coast of northern Scotland. The site in question had been zoned as an area of high landscape value in the council's development plan. Moreover, this zoning had been approved as recently as 1969 when the land in question had been purchased from the Forestry Commission on the understanding that it would be retained for recreational purposes. All but a small part of the site had been designated a Site of Special Scientific Interest.[14]

A key concern was the contractor's proposed use of a new building technique—constructing steel platforms in an upright position, thereby greatly increasing the visual intrusion on the landscape. Early in 1973, the Caithness County Council asked the developer to hold a series of consultative meetings in the affected communities. As the application was clearly inconsistent with the local development plan, the council's subsequent decision to approve the request had to be ratified by the Secretary of State for Scotland (technically this is referred to as an Article 8 direction under the Town and Country Planning Act). Although contrary to the wishes of the local authority, the Secretary felt that the circumstances required that a public inquiry be held. At the inquiry, criticism was expressed by several groups including the Countryside Commission for Scotland, the Nature Conservancy, and the North Coast Conservation Group of the Conservation Society. Interestingly the bulk of the opposition came from "outsiders." The county clerk noted that "it would be fair to say . . . that the names and addresses attached to the first petition [from some five thousand persons supporting the proposed development] are almost entirely local while those attached to the counter petition [from some five hundred persons] are virtually all from the central area of Scotland" (Communication of County Clerk to Secretary of State for Scotland, February 27, 1973). In this case, the Secretary endorsed the Reporter's overall conclusion "that economic and employment advantages would be conferred upon Caithness and Sutherland by the establishment of such a site which would outweigh the damage to amenity while the site is in operation, and that satisfactory rehabilitation can be effected so that there will be no permanent damage to amenity" (Communication of Secretary of State for Scotland to County Clerk, Caithness, October 15, 1973).

Thus after detailed consideration, including a public inquiry, the proposal was approved—subject to certain conditions affecting landscaping, dune stabilization, and the restraint that "no site preparation or other work shall be carried out on any part of the site without specific approval of the local planning authority until the applicants have received a firm order for a production platform." The intent, according to the Scottish Development Department, was to ensure that a section of the coastline was not "sterilized" unnecessarily. Moreover, as Chicago Bridge had indicated that after receiving an order it would take up to six months to obtain the steel required for construction, the Scottish Development Department argued that the restraint would not cause unnecessary delay, as this six-month interval could be used for site preparation. Six months later the company announced that it was no longer interested in the site and had sought and received planning permission for another site—in Ireland, where the local authority had imposed no restrictions and had further agreed to bear the cost of site preparation. At the end of the decade, the company had not received an order for a production platform, and the site remained unused.

As recent events at Nigg Point have demonstrated, uncertainty over a developer's intentions may well persist even after construction work has begun. Indeed a developer's own plans will reflect the changing circumstances of North Sea exploration and production. It always seemed probable, for example, that any production plans for the nearshore Beatrice field (discovered in September 1976) would involve additional onshore development around the Cromarty Firth. And as approved by the Department of Energy in 1978, development plans for the field involve the transportation of oil by pipeline to Nigg Point.[15] Even prior to this announcement, however, construction work at the Nigg refinery site had been suspended by Cromarty Petroleum pending a review of project design. This suspension occurred amid considerable local speculation that the tank farm and refinery represented only the first step in a phased development program for the site. As reported in the Aberdeen *Press and Journal*, Cromarty Petroleum was planning for a £750 million petrochemical complex, including an ammonia plant, a gas fractionation plant, and an ethane cracker to supply feedstock for ethylene production, to be constructed over a ten-year period. Subsequently both Dow Chemicals and Highland Hydrocarbons outlined proposals for a petrochemical complex at Nigg Point using feedstock from the new North Sea gas gathering network. As approved by the Department of Energy in 1980, the network will deliver associated gas from a number of fields to St. Fergus for separation into methane and natural gas liquids (NGL). The subsequent disposal and use of the NGL for petrochemical feedstock remains uncertain. Planning priority is to be given to piping the heavy natural gases (particularly ethane) to existing plants at Mossmorran, Grangemouth, and Wilton-on-Teesside, but it would appear from the Secretary of State's announcement that there is a definite commitment to some form of petrochemical development at Nigg based on propane, butane, and condensates. A final decision on additional petrochemical development based on ethane does not appear likely for several months. Whether the local planning authority would have approved Cromarty Petroleum's initial proposal if it had been able to foresee the extent and character of the related downstream facilities now being proposed for the site is open to speculation.

In conclusion, therefore, criticisms of the development control procedure need to be seen in context. There have undoubtedly been undue and unnecessary delays in dealing with applications and appeals. In the case of the oil refinery and storage tanks at Nigg Point, for example, Cromarty Petroleum Company's application was originally submitted in December 1973. Final approval was eventually received in December 1976. Yet such delays were to some extent inevitable given the sheer number of sometimes speculative applications, the increasing number of appeals, and the shortage of qualified staff who were simultaneously grappling with the need to prepare new-style structure and local plans. The national emphasis on rapid exploitation of North Sea oil and hence on the provisions of onshore support facilities has clearly strained to the limit the ability of local planning authorities to evaluate and process applications.

In these circumstances it is not surprising that many planning decisions were essentially reactive. In the early phases in particular, granting applications was often a pro forma matter. The planners and communities lacked the resources and knowledge to evaluate the immediate and long-term implications of a proposal. Whether public inquiries, which are essentially restricted to issues raised by the objectors, would have improved the situation seems questionable, since they would merely have provided the developer with a forum to make an unchallenged expert presentation. There was precious little opportunity for long-range planning designed to ensure balanced, carefully phased development that embraces a "non-oil" future—"the burdens of immediate problems, coping with the boom before the decline, are too great" (Baldwin and Baldwin, 1975, p. 27).

Only now, as the pace of development begins to slacken, can authorities begin to place greater emphasis on forward planning. Yet many of the initial decisions, on such issues as service bases, pipeline landfalls, and tank farms, have tended to narrow the range of choices open to planners and thereby to constrain subsequent development decisions. In more specific terms, the processing of oil-related development applications raises a number of key issues, including (1) the quality and scope of the information available to planners in local authorities, (2) the ability of local planning authorities to adequately assess the impact of proposed developments, (3) the extent to which alternative sites can be identified and considered, and (4) the manner in which local decisions can be coordinated with and reconciled to the national interest.

The Drumbuie Inquiry

The proposal for a platform fabrication yard at Drumbuie on Loch Carron, one of the numerous deepwater lochs on the west coast of Scotland, provides a useful point of departure for a discussion of the reality of planning for offshore oil in the United Kingdom. As noted by MacKay and Mackay, "nothing has happened in fact at Drumbuie but it provides the best example of the range of issues surrounding the impact of large oil developments on the small communities of the Highlands and Islands" (1975, p. 146).

The controversy essentially revolved around the suitability of the Drumbuie location for the construction of offshore production platforms. Such platforms are clearly integral to the development of offshore oil resources. Their construction is closely related to the identification of commercially viable fields, and hence such questions as the number of platforms required and where they should be constructed can be anticipated at an early stage in the development process. While the question of the numbers of platforms required is closely related to the size and distribution of commercial fields, the major factors in locating fabrication yards are likely to be proximity to the offshore field and availability of skilled labor (for example, workers with experience in shipbuilding and knowledge of welding techniques). The precise social impacts of such facilities (in terms of the need for additional housing and services for the labor force) vary from one location to another but are likely to be particularly severe in smaller and more isolated communities. Of particular concern to the local community may be (1) the relatively short "life span" of such facilities in relation to the production cycle for any oil development, and (2) the industry's vulnerability (given the uneven pace of discovery and development) to sudden and unpredictable layoffs, which are particularly serious wherever a large proportion of the local labor force has become dependent upon this single activity.

In these circumstances it seems unlikely that there would have been any serious interest in the remote Drumbuie site had not the deep water and severe weather conditions experienced in the North Sea encouraged some operators to think in terms of a change in platform design and technology. In particular, North Sea conditions encouraged a shift from steel to the use of prestressed concrete in the construction of offshore production platforms.

> It has been suggested that, due to the extremely rapid growth of the oil industry, there is a need to provide structures of a new type, easier to put to work, presenting less hazards in use and with an improved service life. Prestressed concrete is especially suited to offshore production platforms because it offers good corrosion resistance, does not suffer fatigue to the same extent as steel under the stressing conditions of maximum wind and wave forces, and is comparatively easy to build with less demand for skilled workers. (Francis and Swan, 1973, p. 24)

However, the construction of concrete platforms introduces a new locational requirement since, in contrast to steel "jack-ups," which are built on traditional slips, concrete platforms must be built in the upright position and therefore require a protected deepwater site. "Manufacturers of concrete platforms have typically looked for sheltered situations with inshore depths of 400 ft. to 500 ft. which enable the structure to submerge gradually as it is built up and so retain the working area at a practicable height. The fact that construction takes place with the structure in the water means that the onshore site may be less than half the size of that required for steel platforms" (Chapman, 1976, p. 166). Significantly, this new platform technology originated in Norway, where the fjord coastline offered numerous possible construction sites close to the offshore field. In contrast, the east coast of Scotland offered few sites that satisfied the necessary physical requirements. In these circumstances, interest was inevitably focused on the deepwater lochs of the west coast of Scotland.

The controversy was initiated in April 1973, when two British companies, John Mowlem and Taylor Woodrow, submitted separate applications to the Ross and Cromarty County Council for permission to construct concrete production platforms at Port Cam near Drumbuie on Loch Carron (Figure 13). In their applications the two construction companies (which subsequently formed a joint consortium) argued that Drumbuie was the most suitable location on the Scottish coastline, possessing the necessary combination of a sheltered bay, deep water, and adequate level ground onshore, to construct platforms of a new Norwegian design. Three of these "Condeep" platforms were already under construction in Norway. In these circumstances Woodrow/Mowlem argued that the success and popularity of the design had been demonstrated and that further orders from oil companies could be anticipated. The general intention was to carry out the first stages of platform construction in a dry dock at Drumbuie and then float out the platforms for completion in the deeper, protected waters off the Crowlin Islands in the Inner Sound.

Drumbuie, however, was no more than a small hamlet with a total population of twenty-four persons in 1973.

> Traditionally, the people of the region have earned their livelihood through a combination of activities including crofting (small farming with some cultivation but a greater amount of sheep-raising on shared lands), fishing, and providing "bed and breakfast" accommodations for summer tourists. Public-sector employment is also important, with a substantial number of persons working in public utility, education, and transport jobs. Shopping and recreational facilities are almost non-existent. There is a waiting list for all new housing constructed by the local authorities . . . Drumbuie and neighboring communities are steeped in the crofting way of life and in old-fash-

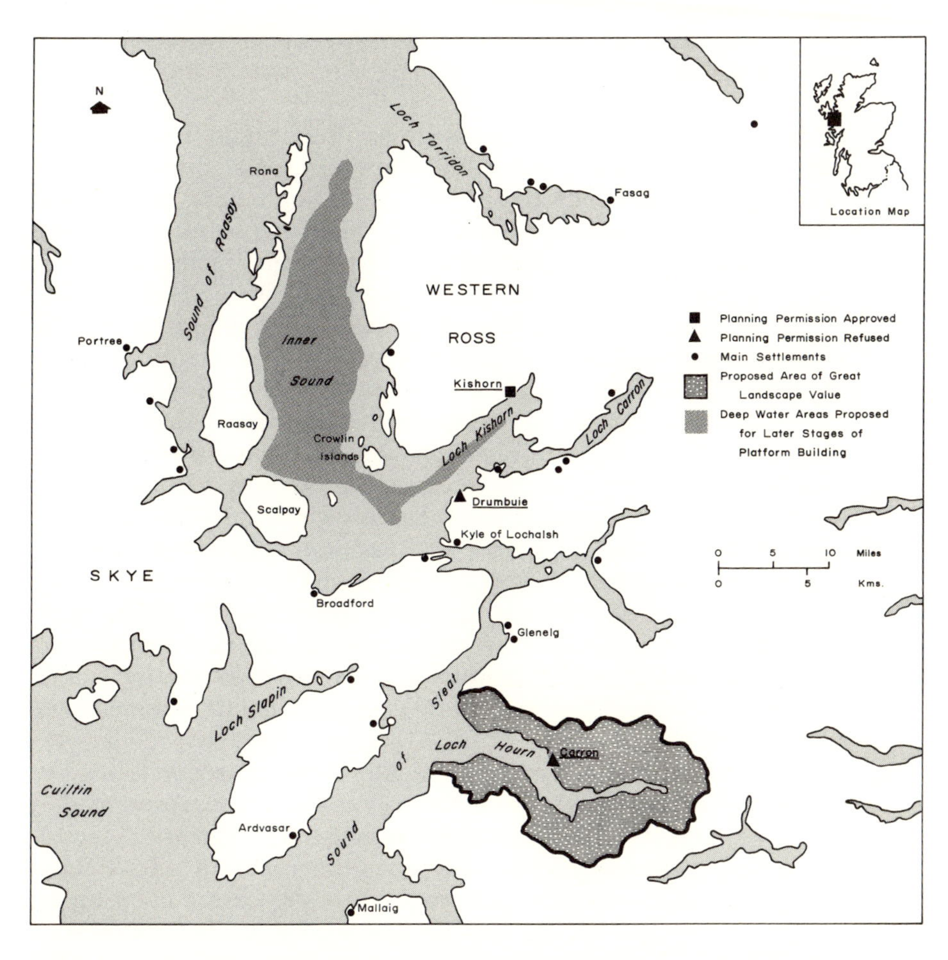

Figure 13. Drumbuie: Proposed Site for Platform Fabrication Yard (data from Hutcheson and Hogg, 1975, p. 107)

ioned traditions of strict religious observance and family and community ties. (Baldwin and Baldwin, 1975, p. 81)

Moreover, Drumbuie's isolation (its communication links take the form of a single-track railway line and a minor road oriented northeastward toward Dingwall and Inverness) would have required the importation of all construction materials by sea and either the installation of generating capacity on site or the erection of a new long-distance transmission line.

Quite apart from the economic and social problems that would have arisen from the influx of a labor force of seven hundred to eight hundred persons (contractor's estimate), the land in question had been left "inalienably" to the National Trust for Scotland, a private organization dedicated to the preservation of land of particular scenic, historic, cultural, or other value. The Trust in turn leased the land (which also happened to be the only arable land in the area) to local crofters. "Since the National Trust strongly opposed platform development [at the Drumbuie site], Parliamentary action would have been required to change the rule of inalienability" (Baldwin and Baldwin, 1975, p. 83). In these circumstances, the Secretary of State for Scotland exercised his rights to "call-in" the planning application from the local authority and announced that a public hearing would be held.

At the hearing, the main advocates of the project were the two construction companies, the British government (as represented by the Department of Trade and Industry) and the Highland and Islands Development Board (a semiofficial governmental body whose objective is to promote development and reverse the pattern of outmigration in the more isolated areas of northwest Scotland). Opponents of the proposal included the National Trust for Scotland, the South West Ross Action Group (a local coalition), the Ross and Cromarty County Council, as well as two national environmental groups, the Conservation Society and the Friends of the Earth. As noted by Baldwin and Baldwin, a notable absentee from the hearings was the oil industry. The Reporter who conducted the inquiry subsequently commented: "I thought it significant that not one representative of any oil company appeared at the Inquiry to state that the [oil extraction] programme would be delayed if the Condeep design could not be built in the UK, or to state how many Condeep-type structures they would require" (Baldwin and Baldwin, 1975, p. 85).

Issues raised at the inquiry included (1) the assumption by the contractors (and the government) that the Condeep design was favored by the oil companies, (2) the level of demand for production platforms as envisioned by the Department of Trade and Industry, and (3) the availability of alternative sites to Drumbuie. That development on such a scale would produce profound changes in the way of life of the community could hardly be disputed.

In calling for approval of the project, the Department of Trade and Industry argued that there would be a need for as many as fifty production platforms in the U.K. sector of the North Sea by the mid-1980s and that about half of these would be of the concrete "gravity" type (gravity in the sense that they are held in position by their own weight rather than by steel pilings). It noted that the Drumbuie yard could reasonably expect to produce two platforms a year between 1975 and 1985, which at $40 million per platform would produce a total revenue of $800 million in orders, with the bulk of the materials, labor, and equipment coming from U.K. sources (MacKay and Mackay, 1975, p. 147). Other industry estimates ranged as high as eighty-one platforms needed between 1975 and 1981. A major concern therefore, from the U.K. government's perspective, was the potential loss of orders for concrete platforms to Norway, a loss which would lead to a further deterioration in the nation's balance-of-payments situation.

However, any estimate of the number of production platforms that will ultimately be required must be treated with caution, contingent as it must be upon such uncertainties as the success of exploration and likely changes in production technologies. Even at the time of the Drumbuie inquiry, it was difficult to reconcile the higher estimates with the conclusions of an independent study undertaken for the Scottish Development Department which anticipated a need for about thirty production platforms but cautioned that with respect to Drumbuie "it is not possible to make any firm predictions as to the numbers of platforms which would be constructed at the site or over what period the construction would continue. The available estimates suggested that the life of the site would be between 10 and 15 years . . . it must be assumed that on occasion two platforms would be under construction at once, and at other times the site would not be in use" (Sphere Environmental Consultants Ltd., 1973, p. 3).

MacKay and Mackay (1975, pp. 147–159) estimated that only thirty to thirty-five platforms would be required by 1980, noting that "the DTI has refuted its own estimates (perhaps unknowingly). In a statement on platform demand and platform sites issued in August 1974, the newly formed Department of Energy said that on average a production platform was expected to produce about 5 million tons of oil per year. Taking into account gas and pumping platforms (reckoned to number between 5 and 10 by 1980), the DTI's evidence at Drumbuie implies a 1980 production rate of 200 million tons of oil, compared with the official estimates of between 70 and 100 million tons at that time" (MacKay and Mackay, 1975, p. 147). MacKay and Mackay were also strongly critical of the Department of Trade and Industry's suggestion that yard capacity should be adequate to meet peak annual demand which "implies that in the non-peak years quite a few sites would be without orders and there would be considerable redundancies . . . such fluctuations in employment are disastrous in the Highlands (though not necessarily so in the Glasgow area or the industrial areas of England) and particularly so in small and isolated communities such as Drumbuie" (ibid., p. 148).

As it became apparent that obtaining planning permission for sites along the northwest coast would be at best a protracted business, there occurred a remarkable growth of interest in alternative sites along the Firth of Clyde, which was suddenly discovered to be deep enough for construction of a variety of concrete platform designs. No fewer than twenty applications were submitted in 1974 to the five counties bordering the Firth of Clyde, the majority (including one by the Mowlem/Woodrow group) being received after the termination of the Drumbuie inquiry. A number of sites were applied for several times over (Hutcheson and Hogg, 1975, pp. 104–105). It must be acknowledged that the concrete platform designed for the Ninian field could probably have been built only on the Loch Kishorn stretch of the Scottish coastline. But this design, which required depths of nearly forty fathoms rather than the thirty fathoms required for most other concrete platform designs, was a subsequent technological advance that followed the Drumbuie inquiry.

It will be noted that at no stage in the planning process is there any requirement for detailed assessment of the social, economic, or environmental impact that may accompany a proposed development. Thus, at the Drumbuie hearings each participant injected separate estimates and concerns regarding the likely effect of the project on both the local community and the scenic amenities of the area. The environmental consultants, Sphere, did review the project for the Scottish Development Department, but their assessment appears to have been far from inclusive. In addition to the issues already discussed, the Sphere report identified as potential causes for concern (1) the likelihood that the sand and gravel for cement production would have to be obtained from local quarries or offshore dredging and (2) noise. In contrast to other opinions expressed at the public hearing, the Sphere report suggested that aesthetically the site was not "outstanding" when compared with other parts of the coastline. Somewhat paradoxically, the consultants also concluded that as dredging and filling would destroy the existing habitat, no survey of the ecology of the area or description of the marine flora and fauna was necessary (Baldwin and Baldwin, 1975, p. 83).

While the public hearing was still in progress, the Conservative government of Edward Heath introduced a bill into the House of Commons which "would enable the Government to acquire, using an accelerated procedure if necessary, land which is urgently needed for certain projects related to the production of oil and gas . . . The accelerated procedures would apply also to planning permissions required for the projects themselves or essential and urgent supporting activities." In effect, the bill would have provided the government with the power to nationalize land, since the government would lease the land acquired through compulsory purchase orders to the developer. This initiative to speed up the processing of planning applications clearly originated within the newly created Ministry of Energy whose Secretary, Lord Carrington, introduced the bill. Equally clearly the bill was intended primarily to provide sites for the construction of production platforms—indeed, it quickly became known as the "Drumbuie legislation." In defending the bill, the Secretary of State for Scotland spoke of the "exceptional circumstances of our national need." He further stated, "It has become clear that, if at least one such site does not become available this summer, we could lose a significant proportion of the oil which we would otherwise be extracting from the North Sea in 1977 and 1978" (House of Commons, 1974b, p. 626).

Reaction to the bill was predictable. Considerable support was expressed by the business community. The *Economist* of London noted that a speeding up of planning applications was long overdue and concluded:

> If rigs and platforms are going to be built [at Drumbuie], this is probably the best way to do it because—the spoiling of the stretch of West Highland countryside apart—the biggest local fears are that a sizeable town would have to be brought into being to support the platform-building enterprise; the town would die when sufficient platforms had been built in 10 years, and possibly sooner, if Mowlem . . . failed to bid competitively for the work available. The Government will presumably take care now of both contingencies. (*Economist*, February 2, 1974)

Even the *Economist*, however, could not forego the observation that "it would have been better in some ways to center the platform-building business on the Clyde, where men are looking for jobs."

Considerable unhappiness with the bill was expressed by Scottish Members of Parliament of the Conservative Party; an even greater sense of outrage was expressed by opposition M.P.'s of the Labour Party, one of whom commented that "the final pretense has been dropped that the Scottish people have any means of expression at the decision-making level through the office of the Secretary of State" (quoted in Baldwin and Baldwin, 1975, p. 86).

The *Economist* observed that the Scottish Nationalist Party had discovered a new asset in Lord Carrington and cautioned that "few politicians in England realize the strong emotions most Scots have when it comes to North Sea oil" (*Economist*, February 9, 1974, p. 25). Such nationalist sentiments were further inflamed when it was reported that the Scottish Secretary of State had deferred to Lord Carrington's desire to see the new executive powers invested in the Ministry of Energy in Thames House, Westminster, rather than in the Scottish Office, St. Andrew's House, Edinburgh. "It seems that the Secretary of State for Scotland stood in the way of [a Scottish-based executive agency to supervise the development of North Sea oil] because he was unwilling to put up any fight against those in the cabinet who wanted North Sea oil directed from London. The creation of the energy ministry has already been interpreted as a blow to the Scottish Office's position. The new bill would be the knock-out punch" (*Economist*, February 9, 1974, p. 25).

With the defeat of the Conservative government early in 1974, the Drumbuie "legislation" became moot. The public inquiry continued, with the Reporter submitting his recommendations to the new Scottish Secretary, William Ross, in May 1974. And on August 12, 1974, it was announced in the House of Commons that "after careful consideration of all the aspects of these applications and of the recommendations of the Reporter, the Secretary of State has concluded that planning permission should not be granted" (*The Scotsman*, August 13, 1974, p. 9).

On the surface at least, it was a victory for common sense and a vindication of the planning process. Yet it was immediately followed by a decision to permit platform construction at Loch Kishorn, an almost identical site only a few miles away from Drumbuie (Figure 13). Thus, the reasons that led the Secretary to reject the Drumbuie application remain obscure. As noted by Baldwin and Baldwin, the Secretary was careful to avoid "precedent-setting statements" or to identify which factors were influential in his decision. "His letter of decision raised all the major objections brought out at the hearings—visual effect, noise problems, strain on the existing economy and social structure, and inalienability of land ownership by the National Trust for Scotland. Yet in almost every case, the Secretary stated that in his view, each particular objection would not by itself warrant a negative decision" (1975, p. 88).

What is clear, however, is that the politically popular Drumbuie decision was inextricably linked with the application submitted by the Howard-Doris group to construct concrete gravity platforms at Loch Kishorn. This application *had* received outline planning approval from the Ross and Cromarty County Council (which had opposed the Drumbuie proposal), a significant factor in the more favorable local attitude being the greater sensitivity to local feelings shown by the Howard-Doris group. "Among other conciliatory moves, the group is paying £25,000 a year into a trust fund to deal with problems that are bound to arise when eventually the platform builders pack up and leave in the second half of the 1980s. The group will also restore the landscape, pay for extra police for the Loch Carron area, and consult the local communities on the various decisions that will affect life in the area" (*Economist*, March 23, 1974).

Moreover, from the Scottish Secretary's point of view, a major advantage of the Loch Kishorn site was that the land in question was not owned by the National Trust, and hence Parliamentary legislation would not be required. Thus, the Scottish Secretary was able to placate public feeling by rejecting the Drumbuie proposal and simultaneously to announce that he was "calling-in" the Loch Kishorn application, noting that another public inquiry would not be necessary, as the essential information had already been debated at the Drumbuie inquiry. And in October 1974, planning permission for the Loch Kishorn project was granted by the Secretary, subject to certain restrictions on working hours and size of the labor force. In November, the Howard-Doris group signed its first order, a concrete platform for British Petroleum's Ninian field.

Yet the attitude of the local council (an admittedly significant factor) and the issue of land ownership apart, it is difficult to see the particular advantages of the Loch Kishorn site over areas such as Clydeside where infrastructure and unemployment abound. The Loch Kishorn site is, if anything, even more remote and isolated than Drumbuie (no rail link, and workers and materials must be ferried across the loch), while its scenic qualities are superior. Housing is in such short supply that the contractor initially housed the workers in a converted liner. "When the Drumbuie applicants discussed Kishorn (along with other alternative sites) during the public inquiry . . . they called it 'unattractive for platform construction.' In addition to certain technical shortcomings, spokesmen mentioned the scenic sacrifice that development of Loch Kishorn would require

Table 26. U.K. Sector: Oil Production Platforms

Field	*Platform Contractor*	*Yard*[a]	*Installation*
Concrete platforms			
Beryl	Norwegian	Stavanger	1975
Brent (B)	Norwegian	Stavanger	1975
Brent (D)	Norwegian	Stavanger	1976
Dunlin	Andoc	Rotterdam	1977
Brent (C)	McAlpine/Aker	*Ardyne Point*/Stord[b]	1978
South Cormorant	McAlpine/Aker	*Ardyne Point*/Stord[b]	1978
Ninian (Central)	Howard-Doris	*Loch Kishorn*	1978
Steel platforms			
Auk	Redpath Dorman Long	*Methil*	1974
Forties (A)	Laing Offshore	Teesside	1974
Forties (C)	Highland Fabricators	*Nigg Bay*	1974
Forties (B)	Laing Offshore	Teesside	1975
Forties (D)	Highland Fabricators	*Nigg Bay*	1975
Montrose	Union Industrielle	Le Havre	1975
Piper	McDermott/Union Industrielle	*Ardersier*/Le Havre	1975
Brent (A)	Redpath Dorman Long	*Methil*	1976
Claymore	Union Industrielle	Cherbourg	1976
Thistle	Laing Offshore	Teesside	1976
Heather	McDermott	*Ardersier*	1977

Ninian (South)	Highland Fabricators	*Nigg Bay*	1977
Ninian (North)	Highland Fabricators	*Nigg Bay*	1978
Murchison	McDermott	*Ardersier*	1979
Tartan	Redpath De Groot Caledonian/Union Industrielle	*Methil*/Cherbourg	1979
Fulmar	Redpath De Groot Caledonian/Highland Fabricators	*Methil/Nigg Bay*[c]	1980
Beatrice (A)	Dragados y Construcciones	Almeria	1980
Beatrice (B)	Dragados y Construcciones	Almeria	1980
North Cormorant	Redpath De Groot Caledonian/Union Industrielle	*Methil*/Cherbourg	1981
Brae	McDermott	*Ardersier*	1982
Magnus	Highland Fabricators	*Nigg Bay*	1982
Maureen	Ayrshire Marine/HDN Offshore	*Hunterston/Loch Kishorn*	1982
Northwest Hutton	McDermott	*Ardersier*	1982
Beryl (North)	Redpath De Groot Caledonian/Union Industrielle	*Methil*/Cherbourg	1983
Tension leg platform			
Hutton	McDermott	*Ardersier*	1983
Converted drilling rigs			
Argyll	Wilson-Watson	Teesside	1975
Buchan	Lewis Offshore	*Stornoway*	1980

[a]Scottish yards are italicized.
[b]Constructed to deck height by McAlpine of Ardyne Point and towed to Stord, Norway, for completion by Aker.
[c]Platform constructed by Highland Fabricators, well head jacket by Redpath De Groot Caledonian.
Source: Department of Energy, 1980, p. 42, updated.

and the heavy traffic that it would create in the village" (Baldwin and Baldwin, 1975, p. 90).

Moreover, approval of the Loch Kishorn site flies in the face of the Scottish Secretary's own guidelines announced in a separate statement on the same day as the Drumbuie decision: ". . . my general approach is to look favorably on applications for technically suitable sites which have access to existing facilities, which can draw on existing sources of labor, and which make use of existing infrastructure and services: correspondingly, difficulties must be expected over applications for sites which lack these" (Statement by the Secretary of State for Scotland as quoted in Baldwin and Baldwin, 1975, p. 89).

To bring the Drumbuie story up to date, it should be noted that the concrete platform for the Ninian field was floated out early in 1978. The contractor at Loch Kishorn, Howard-Doris, had been unable to secure a second order, and the yard was therefore placed on a maintenance basis and the workforce discharged. In many respects therefore, the estimates and claims originally advanced by the contractor and the Department of Trade and Industry at the Drumbuie inquiry have proven to be ill founded. In retrospect, as indicated in Table 26, the official estimates of production platforms required in the first phase of North Sea oil development were considerably inflated. As a result sites have been prepared but not utilized (for example at Portavadie). At Hunterston, where development approval for a platform construction yard was granted in 1975, the contractors finally secured their first order, a steel gravity platform for the Maureen field, in 1979. In other cases yards have been forced to close or lay off workers owing to lack of orders, with all the economic and social disruption that this entails.

More particularly, in the context of the Drumbuie inquiry, concrete platforms have not found universal acceptance in North Sea conditions. Orders have been completed for only seven concrete structures for the U.K. sector. And of these only three have been constructed in British yards. In addition to the Ninian platform, these include Platform C for the Brent field and the Cormorant platform, both constructed by McAlpine Seatank at Ardyne Point on the Firth of Clyde. Planning permission for this McAlpine yard was originally awarded in March 1973, but its development really dates from the post-Drumbuie interest in the Firth of Clyde.[16] As at Loch Kishorn, isolation and poor communications created serious problems. Materials had to be brought in by sea while the bulk of the labor force had to be ferried across the firth from Renfrewshire. The similarity with Loch Kishorn is completed by the enforced closure of the McAlpine yard in 1977 due to lack of orders. Judging by recent orders there is now little enthusiasm for concrete gravity structures in the North Sea. As offshore exploration has moved into deeper waters, interest has focused on other innovative designs, notably the tension leg platform (TLP), which is essentially a floating structure anchored to the seabed by heavy steel cables. The first commercial platform of this design was ordered by Conoco for the Hutton field in 1980. The escalating costs of fixed platform construction, together with the trend toward smaller fields and deeper waters, seem likely to favor newer production concepts (for example subsea completion) in the decade ahead rather than massive concrete platforms. Indeed a recent article on offshore construction trends suggested that concrete gravity platforms might well "go the way of the dinosaur" (*Offshore*, vol. 39, no. 10, 1979, p. 178).

The Drumbuie case has been discussed in considerable detail not simply because it is illustrative of planning procedures in the United Kingdom but because it raises issues that are basic to the entire planning process for offshore oil development. For the present, these issues may be identified as follows:

1. The dispersal or concentration of the impacts, costs, and benefits arising from offshore oil exploitation.

2. The degree of, and the most appropriate forum for, public involvement in the planning process.

3. The rate of oil development, recognizing that the Highland and Islands have experienced repeated "boom-bust" cycles extending from the "Highland Clearances" in the early nineteenth century through the collapse of the wool, herring, and whaling industries.

4. The assessment of impacts arising from particular onshore oil-related facilities and the ability of communities to meet the demands placed upon them.

5. The availability and flow of information between developers, planners, and affected communities.

6. The appropriate relationship between local, Scottish, and national planners and officials, recognizing that pressures for rapid exploitation of offshore oil in the United Kingdom have encouraged growing central government intervention in land use planning and regulation.

7. The difficulty of predicting the nature and scale of oil-related needs in a situation characterized by geological uncertainty and rapid technological change.

These issues are discussed in more detail in subsequent sections.

National Planning Strategies for Onshore Oil Development

As previously discussed, decentralization of planning functions was a basic principle underlying the reforms of the Town and Country Planning Act; the intention was to make central government less intrusive (at least in a planning context) and to make local responsibility for local affairs a reality. Admirable as this principle may be, it should be equally clear from the preceding section that in the early stages of onshore oil activity, local authorities were ill equipped to accept such responsibility. Planning decisions, affecting not only the pace of offshore oil development but also the character of onshore communities, were made on an ad hoc, site-by-site basis, with little attempt to evaluate the impact of a proposed development, to identify alternative sites, or to coordinate land use policies. This situation did not go unrecognized. Thus, the Select Committee on Scottish Affairs issued in 1972 a report entitled *Land Resource Use in Scotland* that emphasized the lack of coordination throughout the planning hierarchy. The committee acknowledged that while the existing machinery might be adequate for "normal circumstances," it was ill equipped to deal with the exceptional demands of North Sea oil development, particularly in the absence of national policy guidelines: "Unless urgent attention is paid to this task, the situation will be characterized by 'decisions-on-the-run' as is clearly the case at present in respect of major industrial developments arising from North Sea oil, which lead either to bad partial decisions or to refusal after lengthy delay and in the absence of clear, considered, and practical alternatives" (Select Committee on Scottish Affairs, 1972, p. 95).

The Select Committee went on to urge that a system of advanced zoning and land requisition for onshore facilities should be implemented, whereby certain areas should be set aside for development. This plan "would enable [oil companies] to be told straight away that sites were immediately available; whereas on [other] sites, they would have to await planning permission which might be delayed or eventually refused. This approach has the advantage that it would do away with the need for developers to purchase an option in advance for land for which planning permission might not be forthcoming" (Francis and Swan, 1973, p. 75).

The emphasis was clearly on speeding up planning procedures and facilitating the development of offshore reserves; there was less emphasis on the need for a careful project-by-project evaluation of the immediate and long-term impacts of oil development on the environment and on local communities. Given the economic environment within which North Sea oil development was occurring, there was inevitably strong pressure on central government planning departments (which in effect meant the Scottish Development Department) to provide "guidance" to local authorities.

In general the guidance has taken the form of background briefing papers, planning guidelines, and technical advice notes. While bolstering the resources of local planning authorities in these ways was clearly necessary, central government "guidance" is by self-definition intrusive; inevitably there will be suspicion in some quarters that such efforts are more concerned with promoting national interests than with strengthening local authorities.

Central government guidelines and initiatives (and their impact on relationships within the local, regional, and national planning hierarchy) are the primary concern of the following section. In addition it should be noted that in a further attempt to improve the quality of oil-related planning at the local level, the Scottish Development Department assisted with the preparation and financing of environmental impact statements. However, the Scottish experience with what is usually referred to as project appraisal is the subject of a separate section.

COASTAL PLANNING GUIDELINES

National policy guidance on oil-related development along the Scottish coastline is summarized in *North Sea Oil and Gas: Coastal Planning Guidelines* circulated by the Scottish Development Department in September 1974 (Scottish Development Department, 1974a). In a circular (No. 61/1974) accompanying the document, local authorities were advised that "the Secretary of State will have regard to these guidelines in taking his decisions on individual planning applications which come before him, and he strongly recommends that planning authorities should take account of them in preparing their development plans and in the exercise of their development control judgements; at the same time the guidelines do not override the provisions of existing development plans, and do not prejudice the decisions of the planning authority or the Secretary of State on individual planning applications" (Scottish Development Department, 1974e).

In general terms the guidelines confirmed the planning framework previously proposed by the Scottish Development Department in a discussion paper entitled *North Sea Oil and Gas: Interim Coastal Planning Framework* circulated to local authorities for their reactions and comments during 1973. Thus the attempt to formulate a coastal planning strategy dates from the early 1970s, with the final guidelines appearing some seven years after the first wildcats were drilled off the Scottish coastline and five years after the Ekofisk and Montrose discoveries.

The main policy guidelines take the form of identifying what are termed *preferred development zones* and *preferred conservation zones* (Figure 14): "Within the overall economic policy of encouraging the development of onshore oil and gas related facilities in Scotland it is considered that there is substantial benefit to be gained from grouping most development into zones, and from restricting major development in areas where conservation is particularly important" (Scottish Development Department, 1974a, p. 7).

The basic principle underlying the coastal planning strategy therefore was concentration rather than dispersal of oil-related developments and their impacts, both beneficial and adverse. The advantages of clustering development are identified in the report as

> (a) avoidance of a scatter of industrial development, affecting many small communities and numerous rural areas;
>
> (b) full use of existing labour pools, housing, and public services;
>
> (c) economic provision of additional services needed for the new developments;
>
> (d) possibility of diversification to cushion any subsequent decline.
>
> In addition there would be possibilities of using new developments to rehabilitate declining areas, and of affording existing businesses the opportunity to adapt to serve new needs. (Scottish Development Department, 1974a, p. 6)

The preferred zones themselves were identified in part by "the need to site new developments in a way that makes the best use of existing labour and infrastructure or minimise the effect of subsequent decline" (ibid.) and in part on the basis of a coastal survey. The published guidelines provide no information on the coastal survey. In 1972, however, the Scottish Development Department's consultant landscape architect had been commissioned to carry out a survey of the coast, working in conjunction with the Countryside Commission for Scotland and the Nature Conservancy Council. "In the course of earlier work for the Department, the consultant had already walked considerable stretches of the coast and the survey involved completing this visual analysis where appropriate, while at the same time drawing together some other available material"

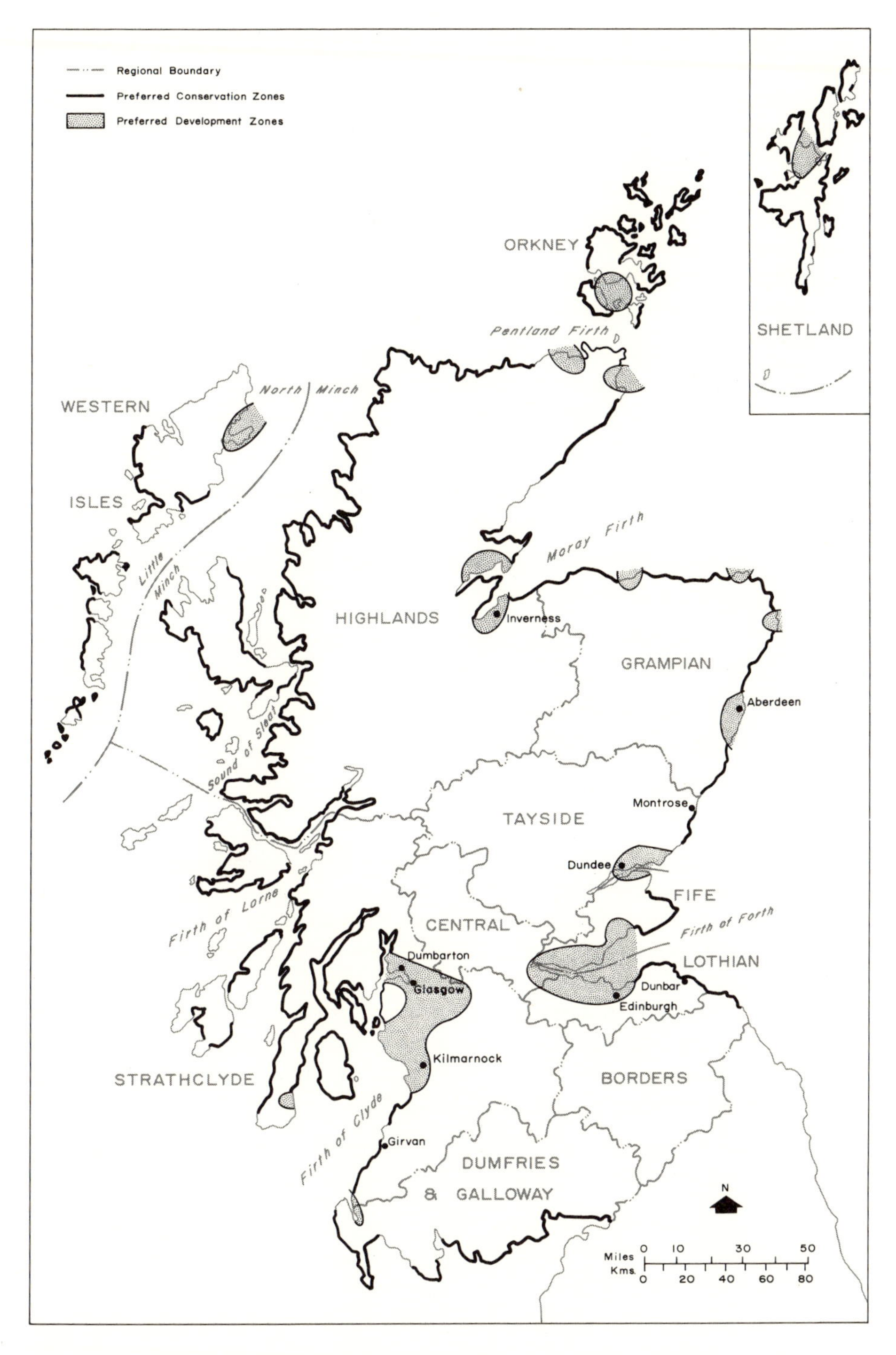

Figure 14. Scotland: Coastal Planning Guidelines (data from Scottish Development Department, 1974a)

(Scottish Development Department, 1975a, p. 7). The basis of the survey would therefore appear to be primarily an individual aesthetic assessment.

On the basis of the survey, the development zones "within which sites for all oil and gas related development seem likely to be appropriate and within which such developments should be encouraged" were identified as areas comprising some or all of the following preferred characteristics:

> (a) a community or series of settlements which can be expanded without incurring the risk of severe economic or social decline as a result of overdependence on one source of employment;
>
> (b) some flat land on the coast and in the hinterland which could absorb major development;
>
> (c) suitable ports and harbours with some potential for developing the dockside land;
>
> (d) existing communications and infrastructure or the potential for improving them economically;
>
> (e) areas in which economic and environmental recovery and rejuvenation are required;[17]
>
> (f) areas in which operations associated with oil and gas could be grouped. (Scottish Development Department, 1974a, p. 6)

Of the sixteen preferred development zones identified, the Central Belt of Scotland (broadly defined as extending from Girvan to Dumbarton on the west coast and from Dunbar to Montrose on the east coast) is singled out as a preferred development zone of the highest priority.

The preferred conservation zones are described as of "such national environmental, scenic and ecological importance that major new developments within them would be inappropriate." More specifically, they are areas containing:

> (a) a coastline with scientific, ecological or scenic features which would be vulnerable to development;
>
> (b) particular sections of the coastline where an existing or proposed use would be incompatible with major oil and gas developments;
>
> (c) areas of the coast containing small scale communities whose expansion might cause serious economic and social problems;

(d) areas of the coast with towns and villages whose character should be protected;

(e) tourist and recreation areas or other places where developments other than major industrial processes should have priority. (Scottish Development Department, 1974a, p. 7)

Developers are specifically warned that in these areas proposals will be carefully scrutinized, that they would have to be justified by "compelling arguments, including a demonstration that no suitable sites existed outwith [*sic*] a preferred conservation zone" (ibid., p. 7) and that a public inquiry would probably be necessary.

The document is careful to emphasize that the guidelines do not constitute a rigid plan. Thus, "it should be noted that [these zones] do not necessarily conform with zonings for industrial development in statutory development plans. Within preferred development zones there are many areas which should be retained for agricultural or recreational use. Similarly there will perhaps be some cases where the acceptance of major industrial development in a preferred conservation zone is justified" (ibid., p. 6). The overall intention is clear, however, and in the summary of policies, the Secretary of State for Scotland specifically directs local planning authorities "to relate their development plans and development control judgement to [these guidelines] and . . . hopes that all concerned will draw them to the attention of prospective developers" (ibid., p. 4).

In general the guidelines would seem to provide a sensible, albeit fairly modest, basis for a planning strategy designed to protect coastal amenities while expediting the exploitation of North Sea oil. In evaluating their effectiveness, however, a number of key issues must be considered. First, when one compares the identified preferred development zones with those areas in which development was *already* occurring as a result of ad hoc, "decision-on-the-run planning," it is difficult to refrain from the conclusion that the guidelines were a post-factum rationalization for what had occurred rather than a framework for forward planning. Second, the authority of the guidelines is clearly undermined when such developments as the platform construction yard at Loch Kishorn are permitted within preferred conservation zones. Third, the basis for defining and delineating the respective zones seems weak. The coastal survey in particular appears to have been highly subjective. Fourth, the guidelines do not provide for an analysis of the impacts of any proposed development or for the evaluation of alternative sites.

Clearly, some of these reservations about the effectiveness of the coastal planning guidelines have greater merit than others. Moreover, these and other criticisms must be seen in context: i.e., the coastal planning guidelines did not stand alone and were intended to complement (or where necessary to encourage) local forward planning initiatives. The issue of *timing*, however, would seem to be crucial. At the time of their publication, it was stated that "the Government believes that the guidelines contained in this document go as far as is practicable at present towards setting out a national strategy for coastal development related to oil and gas exploitation" (Scottish Development Department, 1974a, p. 4).

It seems unfortunate, to say the least, that such broad guidelines could not have been formulated a few years earlier. Had this occurred, it seems inconceivable that the list of preferred development zones would have been as extensive or as broadly delineated. It is instructive, for example, to note that the consultant on the coastal survey was particularly critical of the degree of landscape intrusion associated with the Ardersier platform fabrication yard on the Moray Firth. Yet in this respect the survey was undertaken retroactively, and this area was included in a preferred development zone. It is difficult to accept the view that the information base, in terms both of coastal features per se and of the character of onshore oil-related development, was so deficient as to prevent the identification of suitable development zones (at least at the highly generalized level represented in the coastal planning guidelines) at a much earlier stage in the planning process. A higher degree of specificity could then have been incorporated into the guidelines as more detailed information, both about the ecological and aesthetic attributes of the coastline and about the nature of oil-related needs, became available. In this way, a more carefully elaborated and detailed planning framework might ultimately have emerged that would have served to balance the needs of the oil industry, the priorities of the conservationists, and the interests of individuals and communities most directly affected. Since formulation and application were delayed until the mid-1970s, however, the scale, pace, and complexity of oil development had created a situation whereby many planners dismissed the guidelines as being "too little, too late."

Others have criticized the guidelines' attempt to divert the oil industry

> from its natural predilection for the North of Scotland to locations in the Central Belt . . . It seems that, left to its own devices, the offshore oil industry will increasingly cluster together on the east coast of Scotland, particularly in the Aberdeen area. If, as is the case, the Scottish Office feels it is desirable to decentralise activity, then it is necessary to identify

> very carefully those activities that can be located elsewhere without destroying the generation of external economies which appears crucial to the long-term establishment of the offshore oil industry. There is then a potential conflict between long-run economic efficiency and more short-run social considerations of alleviating high unemployment. (MacKay and Mackay, 1975, pp. 143–145)

The Scottish Development Department is aware of the criticisms and reservations that have been expressed concerning its coastal planning strategy. A more recent working paper, for example, emphasizes that "providing guidance by the use of broad zones in this way does not mean that individual sites of projects can be decided by reference to such guidelines alone. It is only part of the jigsaw . . . Indeed it can be suggested that the national planning part of the jigsaw provides an incomplete picture unless it is followed up by forward (statutory) planning at the regional and local level, and unless it can be regarded as part of a continuing puzzle which admits of revision, addition, and adjustment" (Scottish Development Department, 1975a, p. 8).

It is also evident from this working paper that the Scottish Development Department is extremely sensitive to criticisms of its coastal planning strategy. Thus the report notes that the *Coastal Planning Guidelines* had been accepted by the Oil Development Council and by most local planning authorities and that where criticisms had arisen, "they have in some cases resulted from a misreading of the context. In other cases, in labelling the guidelines superficial, or in seeking more detailed social and economic appraisals of the development zones, there has been an understandable desire to have all the pieces of the jigsaw at once" (Scottish Development Department, 1975a, p. 8). A desire to have all the pieces of the jigsaw at once does not perhaps seem unreasonable, particularly in view of the fact that the criticisms were being voiced nearly ten years after the start of oil exploration off the Scottish coast.

NATIONAL PLANNING GUIDELINES

In addition to developing a broad coastal planning strategy, the Scottish Development Department has attempted to assist local authorities by providing background briefing papers on the likely impacts of particular facilities. Thus, a discussion paper, *North Sea Oil and Gas Pipeline Landfalls*, was distributed by the Scottish Development Department in 1974. This provided some basis on which local authorities could evaluate development applications for pipelines. In the early phases of onshore development, these briefing papers again invite criticism on grounds of superficiality and timing. Baldwin and Baldwin commented that, as late as 1975, "in many critical subject areas no such help has been forthcoming. Two years ago, a document on harbor-based service and supply facilities would have filled a vital local planning need, but no such guide was prepared. Today, one discussion paper on the economic and land-use planning needs for refinery, petrochemical and related oil or gas industries, and another on transportation planning for Scotland, would be equally useful. Unfortunately, the pace of development along the North Sea has seemed to preclude preparation of such timely documents" (Baldwin and Baldwin, 1975, p. 138). The *National Planning Guidelines,* however, provide an effective response to this criticism and suggest that whatever deficiencies have existed in the past, the Scottish Development Department itself is now much better equipped to identify and provide guidelines on the impacts of oil-related development.

The first set of *National Planning Guidelines* (dealing with petrochemical facilities) was issued in May 1977. In a number of key respects the guidelines are less *directive* than the *Coastal Planning Guidelines.* In this sense they are less guidelines than information sheets on the likely characteristics of petrochemical developments. Thus the guideline as published in 1977 "describes briefly and in simplified form petrochemical processes and their planning implications. It is intended as a starting point and suggests where planning officers and planning authorities may go for further information" (Scottish Development Department, 1977a, p. 3).

After describing and noting the wide range of downstream plants that might need to be accommodated, the guideline identifies both the likely demands for labor, land, and services that would be generated by such plants and the safety and pollution issues that would need to be considered in dealing with any such planning application. The guideline is supplemented by a series of planning information notes (Table 27). The "A" series, issued in 1977, deals primarily with institutional interests in petrochemical development, the legislation under which they operate, and the kind of information and advice available to local authorities. One note (No. A11) details the background, aims, and objectives of concerned environmental groups, together with points of contact within each. Each note is accompanied by the names, addresses, and telephone numbers of key individuals as well as references to relevant literature. The "B" series deals with the industrial plants and processes that characterize the petrochemical industry. Again, the intent is to provide background information that will assist local planning authorities in forward planning for petrochemical development.

Table 27. National Planning Guidelines for Petrochemical Developments

Planning Information Notes

"A" Series (issued 1977)

A1	Agriculture
A2	Fisheries
A3	Pollution (General)
	(a) Freshwater Pollution
	(b) Marine Pollution
	(c) Air Pollution
A4	Pollution Inspectorate
A5	Water, Sewerage and Solid Waste Disposal
A6	Health and Safety
A7	Industrial Development Aspects
A8	Land Use Planning
A9	Pipelines
A10	Ports
A11	Forum on the Environment

"B" Series (issued 1978)

B1a	General Characteristics of Basic Hydrocarbon, Primary, and Downstream Petrochemical Plants
B1b	Glossary of Technical Terms
Basic Hydrocarbon Plants:	
B2a	Oil Terminals
B2b	Oil/Gas Separation Plants
B2c	Oil Refineries
B2d	Gas Terminals
B2e	Natural Gas Liquids Plants
Primary Petrochemical Plants:	
B3a	Light Olefins
B3b	Aromatics
B3c	Ammonia/Methanol
Downstream Petrochemical Plants:	
B4a	Low Density Polyethylene
B4b	High Density Polyethylene/Polypropylene
B4c	Vinyl Chloride and PVC
B4d	Styrene and Polystyrene

Note: In addition to the National Planning Guideline and Planning Information notes, a land use summary sheet was prepared on existing developments and issued in 1977.

As in the case of the *Coastal Planning Guidelines*, it is important to note the context within which these *National Planning Guidelines* have been prepared and issued. In a general introduction, the Scottish Development Department notes that while the *Coastal Planning Guidelines* "may serve as an example of the type of national land use guidance it may be appropriate to publish from time to time, there are now two new factors: the production of the regional reports and the disengagement of central government from many local planning matters" (Scottish Development Department, 1977a, p. 1).

The underlying philosophy therefore reflects that of the new-style planning system. A circular (No. 19/1977) to local authorities accompanying the guidelines emphasizes that it is the intention to authorize local planning authorities to grant planning permission for a wide range of developments without prior reference or approval of the Secretary of State for Scotland. "The Secretary of State considers that in the future, apart from development contrary to a new development plan approved by him, he should be notified only of those proposals which raise national issues. Such issues are likely to arise in two ways: from the particular characteristics (such as agricultural quality) of certain land areas which should be safeguarded in the national interest; [and] from particular activities (such as industrial development) having siting requirements which need to be satisfied in the national interest" (Scottish Development Department, 1977c).

As a corollary, therefore, it was thought appropriate to provide guidelines for local authorities that would identify and define the kinds of development that raised national issues relating to land use planning. The guidelines would further serve to indicate the broad national policies that planning authorities should bear in mind in preparing development plans and in exercising their development control functions. In this way the guidelines represent an attempt to balance the need to safeguard national interests (however defined) with the desire to promote local responsibility for development planning and control. As the Scottish Development Department itself noted, "planning guidance at the national level must steer between unhelpful generalizations and unwelcome direction. It should be based on a selection of those issues on which planning authorities, other agencies, and developers might feel in need of guidance on where on balance the national interest lies" (Scottish Development Department, 1977a, p. 1).

In a very real way therefore the national planning guidelines are informational in two senses—in terms of the characteristics of particular developments and in terms of the national interest that must provide the context for planning by regional and district authorities. In this sense the guidelines *are* directive. In the guideline on petrochemical development, for example, it is made clear that "the Secretary of State is anxious that no opportunity for desirable petrochemical development should be lost through lack of preparedness, and equally that individual projects should be considered in more than a narrowly local context. Proper preparation by planning authorities can make an important contribution; if local or structure plans identify sites as being suitable for petrochemical development, companies can hope to satisfy their requirements and establish themselves faster and more easily" (Scottish Development Department, 1977a, p. 2).

In this context, local authorities are required to notify the Secretary of State of any specific proposals "so that he may consider whether to call them in for his own determination" (Scottish Development Department, 1977a, p. 2) and further, even in advance of specific proposals, "to establish the potential for petrochemical developments in their area and to frame their structure plans, local plans and development control policies so that sites can be readily identified if and when required" (ibid.). Recognizing perhaps the reactive, ad hoc nature of early land use decisions relating to North Sea development, many regional authorities have responded quickly and effectively to the mandate contained in the *National Planning Guidelines*. In the Grampian Region, for example, planners have attempted to assess the regional implications of alternative levels of petrochemical development (Grampian Regional Council and Banff and Buchan District Council, 1980). Since the proposed Gas Gathering Pipeline (GGP) will almost certainly terminate at St. Fergus (which already receives and treats associated gas from Brent and natural gas from Frigg), the Grampian Region, and more particularly the Banff and Buchan District, must anticipate further, major proposals relating to the treatment and disposal of gas and feedstock. In these circumstances "it is important that the local authorities examine the nature of these 'petrochemical options' to assess the capacity of the local area to accommodate them and the desirability of attracting and promoting them" (ibid., p. 1). In light of this assessment, a contingency plan was prepared for the region which both summarizes the major planning issues (potential demand for land, employment and social effects, safety risks, pollutant emissions, harbor facilities, and infrastructure needs) and identifies possible locations for each type of development. The intention was not to attempt a detailed impact analysis of the selected sites, since that could be undertaken only after receipt of a firm proposal. Rather the objectives were to (1) draw attention to the major issues that would require detailed and careful consideration at the planning proposal stage and (2) "present an agreed locational strategy which will permit a rapid and coordinated response by the authorities to major industrial proposals" (ibid., p. 2).

OFFSHORE PETROLEUM DEVELOPMENT (SCOTLAND) ACT, 1975

The central government initiatives discussed hitherto represented attempts to provide "guidance" to local planning authorities. They may be viewed as a response to growing pressures (1) to speed up planning procedures, (2) to coordinate development policies at the national and local levels, and (3) to strengthen the information base of local planning authorities. Whether such efforts were too intrusive, indicating directions rather than providing information and hence threatening the desired shift toward a more devolved planning structure, is very much a value judgment that will reflect individual beliefs and priorities about development goals. From the perspective of central government planners, such guidelines were regarded as essential if the national interest in facilitating North Sea oil development was to be realized.

The Offshore Petroleum Development (Scotland) Act of 1975 represents a much more overt central government intrusion into land use planning. In this act, the government took powers to control certain types of oil-related development in "the public interest." Briefly, the legislation gives the Secretary of State for Scotland power to acquire land for "key oil-related activities," if necessary by compulsory purchase. In introducing this legislation, the Secretary of State identified five goals for public ownership:

1. To maximize site use.
2. To make clear to oil companies and platform builders that sites will be available in good time for the best designs.
3. To avoid proliferation of sites.
4. To ensure that only the necessary amount of infrastructure is provided.
5. To enable strict control to be exercised over the development of the production facilities and to ensure the ultimate restoration or adaptation of the sites when they are no longer required for offshore work.

As this list makes clear, although the act refers in general terms to oil-related development, it was primarily intended to provide sites for the construction of production platforms.

The passage of this legislation invites comment in a number of respects. In the first place it is ironic, to say the least, that the act was introduced by a Labour government that had (when in opposition only twelve months earlier) so vociferously denounced similar proposals by the Conservative administration as an "unseemly land grab." To add to the irony, the legislation was introduced into Parliament on August 12, 1974—the same day that the Scottish Secretary chose to announce the Drumbuie decision—by the Minister of Energy.

In the second place, the legislation draws attention to the shortcomings of the government's production platform policies. As noted in an earlier section, estimating the number of platforms that will be required is fraught with difficulty given the uncertainties involved. Yet such calculations have important implications for land use planning, particularly the identification and designation of construction sites. "Under-supply may either delay production schedules or necessitate expensive imports; and over-supply means that one or more yards may be without orders at some point, with all that this implies in terms of economic and social hardships for their labour force" (Chapman, 1976, p. 168).

In practice, pressures to expedite development of North Sea oil combined with an inflated estimate of the numbers of platforms that would be required overrode any chance of a more carefully phased construction program utilizing one or two preferred sites. The Offshore Petroleum Development (Scotland) Act must be seen in the context of the government's fear that production platforms would not be available in the numbers or within the time frame desired if normal planning procedures were followed. The shortcomings of the government's policy are evident in the number of sites for which planning permission was granted but which were never cleared; in those sites which were cleared but never used; and in those sites which have been forced to close down.

The Portavadie site, development of which was expedited under the Offshore Petroleum Development (Scotland) Act, is a classic example of inept planning. In the first place, the site was selected despite its location in a preferred conservation zone and despite a separate consultant's report that Loch Fyne should be declared a conservation area with no fixed sites permitted along its shoreline (Hutcheson and Hogg, 1975, p. 105). In the second place, the land (obtained by the Scottish Office from the Forestry Commission) was cleared at very considerable public expense. The intention was to lease the site, which would thereby remain in the public domain, ensuring that when no longer required the land would be rehabilitated and returned to its former use. Rather than the site's simply being designated, clearance was undertaken in the belief that large numbers of platforms would be required in a comparatively short period of time and that even a six-month delay for site clearance would be unacceptable. Subsequently, the government has been unable to secure an operator for the site, let alone an order—hence the site remains cleared but unused.

While the Offshore Petroleum Development (Scotland) Act demonstrates the difficulty of reconciling competing interests (development/conservation; national/local), it is by no means an isolated example of central government circumvention of normal planning procedures. The case of the outer harbor at Peterhead, known as the Harbour of Refuge, could equally well have been cited. In this instance, Parliament considered and enacted in record speed legislation giving the Secretary of State for Scotland special powers "to develop, maintain and manage, or allow others to do so, harbours in Scotland made or maintained by him."[18] Following government-financed reclamation during 1973, Peterhead has become a major supply base for North Sea oil operations with sites leased to the Peterhead (British Oxygen Company) Base serving the Forties field and to the Aberdeen Service Company's (ASC) South Bay Base serving Occidental, Phillips, and Burmah Oil offshore rigs.[19]

The Harbour of Refuge, opened to commercial development as a result of the Harbours Development (Scotland) Act of 1972, also serves to demonstrate how an early development decision (in this case to relieve congestion in Aberdeen Harbour and ensure adequate facilities for supplying offshore platforms and rigs) may serve to direct the location of subsequent and perhaps less desirable downstream activities. In particular, the expanded harbor facilities were a factor in the North of Scotland Hydro-Electric Board's decision to locate a 1,320-megawatt power station at Boddam, two miles south of Peterhead, and also in Shell Expro's application to construct a natural gas liquids (NGL) separation plant on the outskirts of Peterhead. The latter development involved the separation of NGL from the Brent offshore field into its major constituents and the subsequent transshipment and export of butane and propane by refrigerated tankers from the Harbour of Refuge. As already noted, it became apparent at the public inquiry that the capacity of the harbor—as well as related questions of compatibility with existing and approved uses and of safety in the storing and loading of the hazardous materials—had been inadequately studied. And shortly before the inquiry was due to reconvene, Shell announced that the application was being withdrawn in favor of the Mossmorran site in Fife.

Impact Assessment

The United Kingdom lacks any statutory requirement for project assessment comparable to that existing in the United States under the National Environmental Policy Act of 1969. The general argument, particularly among central government planners, has been that the Town and Country Planning Acts provide a satisfactory framework within which development/conservation interests can be reconciled. More particularly, it is argued that development control procedures are adequate to ensure that particular developments will be sited in such a way as to minimize any adverse impact.

One consequence of this general approach to impact assessment has been an overriding concern with the aesthetic character of development. In the context of oil-related developments, for example, great concern has been expressed over the visual intrusion of such facilities as platform fabrication yards and petroleum refineries on the Scottish coastline. This concern is reflected in the conditions frequently attached to planning permission—landscaping of storage tanks as at Hound Point, relocating the refinery towers to ensure a lower profile against the skyline as at Nigg Bay, site reclamation as at Dunnet Bay, employment of an architect to design and implement a landscape master plan as at Mossmorran. In the early phases of North Sea planning at least, less attention was usually given to the long-term ecological or social consequences of particular developments, let alone to the consideration of "appropriate alternatives to the proposed action" or to "irreversible and irretrievable commitments of resources which would be involved in the proposed action should it be implemented" as is mandatory in the United States under Section 102(2)c of the National Environmental Policy Act.

Given the pace and character of oil development, however, there has been a growing interest in environmental and social impact assessment procedures and techniques in Scotland. Much of the initiative has come from the Scottish Development Department and from local planning authorities. Awareness of the magnitude and frequently irreversible character of oil-induced social and environmental change; realization of the need for independent verification of developers' claims, particularly with respect to the potential hazards of hydrocarbon storage, transportation, and processing facilities; and lack of in-house resources and expertise to deal with the complex technical aspects of oil-related development have combined to create a situation in which more rigorous project assessment (involving where necessary the use of independent consultants) is deemed necessary. Nevertheless, this shift in attitude has come about slowly and unevenly, and project appraisal continues to be undertaken on a somewhat ad hoc basis.

The first official move in the direction of more formal impact assessment came in May 1974, when the Secretary of State for Scotland issued a circular to all local planning authorities requesting notification within two weeks of all major oil-related planning applications, defined as applications for the construction of platform fabrication yards, storage tanks and terminal facilities, refineries, and petrochemical and gas liquefaction plants (Scottish Development Department, 1974d).

While this request may have been intended simply to prod local authorities into processing applications more speedily, it was followed by a "technical advice note" in which local planning authorities were advised as to the appropriate questions to ask in order to ascertain the likely impact of the proposed development on the social, economic, and environmental character of the area. Local authorities were further advised to seek detailed information on the reasons for site selection, the extent to which alternative sites had been considered, and the developer's assumptions about product-demand over time.

While it is easy to deplore the rather belated "official" recognition of the need for a detailed assessment of the impacts of particular oil-related facilities, it must be recognized that institutions and bureaucracies rarely welcome "innovations" that involve a departure from traditional procedures and policies. And what was involved was basically a shift from the traditional emphasis in the British planning system on *development control* to what is now termed *project appraisal*. This shift is acknowledged in the Scottish Development Department's paper, *North Sea Oil and Gas Developments in Scotland: A Physical Planning Resume*: "It is perhaps symptomatic of the unusual type of project involved in oil that the normal handling of the planning application under the heading 'development control' may not be positive enough. Where some unfamiliar operation or process is the subject of a planning application in a rural area, and where it has not been predicted or planned, particularly positive steps are required to find out the full consequences of that application going ahead" (Scottish Development Department, 1975a, p. 10).

Even prior to the Scottish Office's technical note on impact assessment, a number of local authorities had been moving in this direction. As early as January 1973, the Zetland County Council had commissioned a London consulting firm, Livesey and Henderson, to prepare a master plan for Sullom Voe that would "accommodate all the foreseen oil industry and related developments, so as on the one hand to meet the technical requirements of these developments and on the other to cause the least possible damage to agriculture, fishing, and to the social, natural and visual environments of Shetland." More particularly the consultants were directed to assess "the suitability of Sullom Voe as the site for a major industrial complex in Shetland to provide for oil and gas developments. To predict the nature and possible magnitude of industrial requirements, to examine in depth the marine and engineering aspects of the Sullom Voe area for oil industry and related developments, and to assess how such developments can be accommodated with the least disturbance to the Shetland environment, and what complementary infrastructural development will be required" (Scottish Development Department, 1976, p. 13).

On the mainland, the planning department of the Ross and Cromarty County Council undertook an in-house study late in 1973 of the likely physical, economic, and social impacts of a proposed platform fabrication yard at Arnish Point, Stornoway (Ross and Cromarty County Council, 1974). Given the unemployment situation in the Outer Hebrides, the local authority was particularly interested in local employment prospets but concerned about the cultural impact of a major manufacturing employer on an area characterized by small-scale croft farming and fishing and deep-rooted Gaelic influence. The Ross and Cromarty County Council had also commissioned in February 1974 three separate reports on the pollution, noise, and landscape impacts that might be anticipated if planning approval were given for an oil refinery at Nigg Point (from Cremer and Warner, Acoustic Technology Limited, and Architects Design Group). Moreover, the Scottish Development Department had itself employed outside consultants to assess the impact of particular proposals, most notably the proposal for a platform fabrication yard at Drumbuie (Sphere Environmental Consultants Ltd., 1973). In this case, the consultants presented a summary of their analysis at the public inquiry.

Table 28 lists all major environmental impact studies relating to onshore oil development undertaken between 1973 and 1978. While a common element in all of these studies is the attempt to identify and predict the effects of oil-related development on the

Table 28. Environmental Impact Assessment Studies in Scotland, 1973–1978

Location/Project under Appraisal	*Sponsor*	*Consultant*	*Date (Study Period)*	*General Comments*
1. Sullom Voe (master development plan and report on oil industry requirements)[a]	Zetland County Council	Livesey & Henderson (Consulting Engineers)	January 1973 (9 months)	Assessment of suitability of Sullom Voe as the designated location for all oil-related development in Zetland, plus recommendations for guiding development.
2. Loch Carron (Drumbuie) (oil platform construction)[b]	Scottish Development Department	Sphere Environmental Consultants Ltd.	May 1973 (3 months)	Described as the first study of its kind in Scotland. Use made of "the Leopold matrix approach extended to cover social and economic matters."
3. Loch Broom (oil platform construction)	Scottish Development Department	Sphere Environmental Consultants Ltd.	August 1973 (3 months)	Commissioned at the time the Loch Carron study was nearing completion. Similar in approach and methodology. Planning application withdrawn before study completed.
4. Flotta, Orkney (oil handling terminal)[c]	Occidental of Britain Inc.	W. J. Cairns & Associates	August 1973 (10 months)	Commissioned by developer. Objective to identify "all potentially significant environmental effects" and remedial measures. No assessment of social or economic impact.
5. Stornoway (platform construction)[d]	Ross & Cromarty County Council	In-house (technical assistance from Highland & Islands Development Board)	November 1973 (3 months)	Special emphasis on local employment opportunities and housing demand; concern for cultural effect of major manufacturing employer.
6. Firth of Clyde (platform construction sites)[e]	Scottish Development Department/ Department of Energy	Jack Holmes Planning Group; Crouch & Hogg (Consulting Engineers)	January 1974 (6 months)	Examination of several sites on the Clyde Estuary (including those for which planning applications had been received) in terms of social and environmental impact.

7. Nigg Point (oil refinery)[f]	Ross & Cromarty County Council	Cremer & Warner (Consulting Engineers); Architects Design Group; Acoustic Technology Ltd.	February 1974 (7 months)	Consultant's reports added to in-house studies of housing, infrastructure.
8. Loch Carron (comparative analysis of platform construction sites)[g]	Scottish Development Department	Sphere Environmental Consultants Ltd.	February 1974 (1 month)	Undertaken during the Drumbuie inquiry. Ranking of eight alternative sites in Loch Carron area. Comparative engineering study undertaken by Crouch and Hogg for Department of Energy.
9. Loch Eriboll (development feasibility)	Sutherland County Council	Peter Frankel; Economic Consultants Inc.	April 1974 (2 months)	Response to an "obviously speculative development." First stage covered feasibility; second (impact analysis) stage never completed since no possible developments appeared economically viable.
10. Buchan (petrochemical complex)[h]	Aberdeenshire County Council/ Scottish Development Department	Economist Intelligence Unit Ltd.	November 1974 (4 months)	Assessment of demand for sites for petrochemical and other processes of pipeline terminal; appraisal of the combined impact of current and potential industrial developments; preparation of a balanced locational strategy.
11. Peterhead (natural gas liquids separation plant)[i]	Grampian Regional Council	Cremer & Warner (Consulting Engineers & Scientists)	October 1975 (5 months)	Assessment of safety factors, environmental hazards, and "nuisances" associated with proposed natural gas liquids (NGL) separation plant at Peterhead. Influential in alerting authorities to the risk involved in using Peterhead Harbour for the export of propane and butane.

Table 28. Continued

Location/Project under Appraisal	*Sponsor*	*Consultant*	*Date (Study Period)*	*General Comments*
12. St. Fergus (natural gas terminal)[j]	Banff & Buchan District Council	Cremer & Warner (Consulting Engineers & Scientists)	August 1976 (9 months)	Study of proposed Shell gas reception terminal as well as facilities already under construction at St. Fergus for Total Oil Marine and British Gas Corporation. Emphasis on the safety arrangements and pollution control aspects of the combined facilities.
13. Mossmorran (NGL plant & ethylene complex)[k]	Fife Regional Council; Dunfermline District Council; Kirkaldy District Council	Cremer & Warner (Consulting Engineers & Scientists)	1976–1977 (9 months)	Assessment of combined hazards and potential environmental impact of proposed Shell NGL separation plant, proposed Esso ethylene complex at Mossmorran and export facilities at Braefoot Bay. Special emphasis on safety following controversy over Shell's previous application to Grampian Regional Council for site at Peterhead.
14. Moray Firth (offshore field)[l]	Mesa (U.K.) Ltd.	Sphere Environmental Consultants Ltd.	May 1977 (2 months)	Commissioned by operator of prospective Beatrice oil field (located some 12 miles off the Moray Firth coastline) at request of Department of Energy. Proximity to coast, heavy nature of the crude oil, and importance of Moray Firth for commercial fishing raised significantly different environmental issues from other offshore developments.

15. Highland (petrochemical complex)[m]	Highland Regional Council	Cremer & Warner (Consulting Engineers & Scientists)	1977–1978	Development of guidelines for layout and safety zones in petrochemical developments. Criteria then used to assess suitability of a number of sites in the Highland Region for a petrochemical complex.
16. Strathclyde (petrochemical complex)[n]	Strathclyde Regional Council	Department of Physical Planning, Strathclyde Regional Council	September 1977 (3 months)	Examination of alternative sites for petrochemical complex.

[a]Livesey and Henderson, "Sullom Voe and Swarbacks Minn Area: Master Development Plan and Report," 1973.

[b]Sphere Environmental Consultants Ltd., "Impact Analysis: Oil Platform Construction at Loch Carron," August 1973.

[c]W. J. Cairns and Associates, "Flotta, Orkney: Oil Handling Terminal," December 1973.

[d]Ross and Cromarty County Council, "Impact Study: Planning Application by Messrs. Fred Olsen Ltd. at Arnish Point, Stornoway," 1974.

[e]Jack Holmes Planning Group (in collaboration with Crouch and Hogg, Consulting Engineers), "An Examination of Sites for Gravity Construction and the Clyde Estuary," July 1974.

[f]Cremer and Warner, "Environmental Feasibility Report," June 1974; Acoustic Technology Limited, "Environmental Noise and Vibration Study," 1974; Architects Design Group, "Oil Refinery at Nigg Point," 1974; Ross and Cromarty County Council, "Impact Study of Proposed Refinery," June 1974.

[g]Sphere Environmental Consultants Ltd., "Loch Carron Area: Comparative Analysis of Platform Construction Sites," March 1974; Crouch and Hogg, "Engineering Analysis of Alternative Sites at Loch Carron," March 1974.

[h]Economist Intelligence Unit Ltd., "Buchan Impact Study" (Parts 1 and 2), Aberdeen County Council, 1975.

[i]Cremer and Warner, "Report on the Environmental Impact of the Proposed Shell UK (Expro) NGL Plant at Peterhead," February 1976.

[j]Cremer and Warner, "The Environmental Impact of the Natural Gas Terminal at St. Fergus," May 1977.

[k]Cremer and Warner, "The Hazard and Environmental Impact of the Proposed Shell NGL Plant and Esso Ethylene Plant at Mossmorran, and Export Facilities at Braefoot Bay," May 1977.

[l]Sphere Environmental Consultants, "Environmental Impact Analysis: Development of Beatrice Field Block 11/30," July 1977.

[m]Cremer and Warner, "Guidelines for Layout and Safety Zones in Petrochemical Developments," February 1978.

[n]Strathclyde Regional Council, Department of Physical Planning, "Environmental Impact Appraisal: Preliminary Assessment of Sites for a Petrochemical Complex," December 1977.

physical, social, and economic make-up of the impacted area, it will be noted that these early attempts at environmental impact assessment vary greatly in the manner of their initiation, in intent, in scope, and in methodology. In a number of instances, the Scottish Development Department commissioned studies; in other cases, the local planning authorities called in consultants (with occasional technical and financial assistance from the Scottish Development Department and/or the Highland and Islands Development Board); in yet other cases, studies were commissioned by the prospective developer. More significant differences emerge in terms of the scope and intent of the impact studies. Several (Table 28: items 2, 3, 5, 7, 11, 12, and 13) relate to development proposals affecting specific sites.

> These studies were commissioned by the public authorities, *primarily to assist in reaching a decision on the planning application for the project* . . . The work . . . consisted of a detailed description of each proposal, a detailed description of the areas involved, and a systematic analysis of the effect of each project on the area in which it would be located. In most cases *the reports included suggestions as to how the unsatisfactory aspects of the projects might be avoided or minimised if the project came into operation. These studies were confined to the analysis of impacts rather than detailed planning.* (Scottish Development Department, 1976, p. 2; emphasis added)

The application by Cromarty Firth Petroleum Company for permission to construct an oil refinery at Nigg Point may be cited as an example of the way in which many local planning authorities view environmental impact assessment. On the basis of consultant studies, the Ross and Cromarty County Council granted overall planning permission but identified a number of changes that would have to be made in refinery design (most notably a reduction in the height of the refinery towers in the interest of preserving the scenic character of the area) before final planning approval would be granted.

Safety considerations have become a matter of increasing concern to local authorities as oil activities have shifted from the development to the production phase. In addition to a broadly defined environmental assessment, consultants have specifically been asked in several instances to appraise the degree of risk posed by hydrocarbon (particularly gas) storage, transportation, and processing facilities. The Cremer and Warner report on the proposed Shell liquefied natural gas (NGL) separation plant at Peterhead was instrumental in alerting the local authority to the hazards of the project as proposed and to the inadequacies of Peterhead Harbour in safely accommodating existing shipping movements and the export of propane and butane from the NGL plant.

In contrast to the single-development, site-specific impact assessment studies cited above are the reports relating to Sullom Voe (Zetland) and Flotta (Orkney) that are more concerned with working out the details of an approved project than with contributing to a decision in principle on a planning application (Table 28, items 1 and 4). In both instances, the local planning authorities had already determined that oil storage and transshipment facilities would be required for the islands. The studies therefore presumed that the projects would be implemented and were primarily concerned with detailed planning for the location of on-site activities, access arrangements, landscaping, and off-site infrastructure needs. A difference between the two studies was that the Flotta site was identified by the prospective developer, who also commissioned and paid for the report, whereas the Sullom Voe site in Shetland had been selected by the local council. Moreover, the appraisal of the site was commissioned by the Zetland County Council, which from the beginning intended that Sullom Voe should serve a large number of users.

A third group of studies has examined the relative merits of a number of different sites for a particular type of development (Table 28, items 6, 8, and 15). This approach combines elements of project appraisal (albeit at a fairly general level in the absence of *specific* proposals) with forward planning. The Firth of Clyde and Loch Carron studies examined alternative sites for platform construction. The emphasis in these studies was on a comparison of the impacts at each potential site and an analysis of the best combination of sites under varying assumptions about the total number of sites that would be required. The Cremer and Warner study for the Highland Regional Council sought to establish guidelines for the protection of a community from the hazards posed by petrochemical developments. The guidelines essentially took the form of strict controls on plant layout, separation distances, and safety zones. In the light of these requirements, a number of sites were evaluated to see if they could safely accommodate various hypothetical combinations of petrochemical plants.

A fourth group of impact assessment studies has addressed other issues and has been regional in its focus. The Loch Eriboll and Buchan studies involved an assessment of the impact of several major development proposals within a relatively small area (Table 28, items 9 and 10). In the case of Loch Eriboll, the Sutherland County

Council had received several proposals for an oil terminal, refinery, and platform construction yard on neighboring sites in an isolated and extremely scenic area. (In this context, it is worth noting that not only does Loch Eriboll lie within a preferred conservation zone as defined by the Scottish Development Department, but this particular region is singled out as being of outstanding ecological significance and scenic importance. See Scottish Development Department, 1974a, p. 12) The local authorities considered the applications to be purely speculative but engaged consultants to examine the physical and economic feasibility of the proposed development. This analysis was to have been followed by an impact study of those developments that seemed feasible, but none of the proposals seemed to be sufficiently likely to justify more detailed examination. In the Buchan case, the Aberdeenshire County Council had already granted permission for a natural gas pipeline landfall. The purpose of the study was to examine the type of secondary onshore development that might be attracted to the area (for example, ammonia and ethylene production) and to assess the likelihood and implications of such development. A second phase of the study indicated possible locations for the various plants and identified various constraints on the level and type of development that could be accommodated.

Finally, in view of the location of the Beatrice field, the Department of Energy asked the field operator to undertake an assessment of alternative development plans with particular reference to their implications for the fisheries and amenities of the Moray Firth (Table 28, item 14). This appears to have established a precedent in that in 1980 the Secretary of State indicated that "as part of the process for preparing a development plan for a discovery near to shore [I will expect that] the licensee will normally carry out a study into the implications of the proposed development on the marine environment, on other users of the sea, and on local coastal areas, consulting the relevant local authority as appropriate" (Department of Energy, 1980, p. 49).

Many of these studies have been severely criticized in terms familiar to those who have worked with environmental impact statements in the United States—descriptive, superficial, inadequate, incomplete. A critique of the reports prepared for the Scottish Development Department by Sphere Environmental Consultants Limited, for example, concludes as follows:

> Both reports are in many instances superficial and inadequate. The defense against these criticisms would be the lack of time available for the studies, and for the preparation of the reports, and the assertion that because data is limited it does not mean

> that it is wrong . . . It could be argued, however, that a responsible firm of consultants should not elect to undertake a study for which there is clearly inadequate time, or (and this is obviously a more practical suggestion) they should . . . specify what should also have been done had time been available . . . so that the reader can appreciate that on some aspects at least, there is much more to be considered than appears in print. (Project Appraisal for Development Control, 1974, pp. 7–8)

This critique was particularly concerned about (1) the failure to provide any indication of the relative significance of identified impacts: "there would seem to be a need therefore to try and develop some form of evaluation matrix, in which the relative importance of the different impact effects can be compared with one another" (p. 3); (2) inaccuracies in the evaluation of potential economic impacts, most notably a serious "underestimate" of "the numbers of local people who would leave their jobs to join the platform builders and related construction work" and an "overestimate" of the amount of subcontracting work that would be done locally (pp. 3–4); and (3) serious inadequacies in several parts of the environmental section:

> It is extraordinary that Sphere should feel that it is enough to have discovered that there are no Sites of Special Scientific Interest (SSSI) or National Nature Reserves (NNR) in the immediate neighborhood, and that the Nature Conservancy was unable to raise any objection on ecological ground to the development. They have completely missed the point. They assume that the lack of any designated SSSI's or NNR's in the area indicates the absence of any features of ecological interest. This is not necessarily the case and is a dangerous precedent. Many areas in the British Isles have not yet been assessed for the ecological potential, as the Nature Conservancy has been obliged to give priority to sites where changes in land use were most imminent . . . These sites may well be of no special ecological interest but a survey of the main vegetation types, the wildlife and the land use is necessary if only to demonstrate this . . . [Moreover] if there is to be a clause for rehabilitation of the area after the industry has come to an end, then it will be necessary to have some idea of the site as it is now. (Project Appraisal for Development Control, 1974, pp. 5–6)

Similar criticisms have been made of the quality of other reports. Despite the limited time often available to consultants, it is something of an indictment to learn that a senior planner was once asked by a consultant on the day before a report was due to be submitted whether there were sites of particular scientific and/or ecological importance within the affected area.

In evaluating the Scottish experience with environmental impact reports, the Scottish Development Department (1976, pp. 3–5) drew a number of general conclusions. In the first instance, it was suggested that both the scope of the impact analysis and the method of carrying it out had to be tailored to suit individual circumstances and different stages in the planning process:

> There are significant variations in the type of information that is relevant according to the stage in the development process that has been reached. Prior to the decision in principle, the main concern of the planning authority is *to identify the main implications of the project* (both advantages and disadvantages) and to understand them sufficiently to compare them for decision purposes. At this point, planning permission may be refused and no further work is necessary unless an appeal is lodged. If it is intended to approve a major project, there will usually be reserved matters and other conditions affecting the design, size, and operation of the development, and *a rather more detailed analysis may be needed* to specify these in adequate detail. If the project seems likely to proceed, *then a third level of work* is needed to evolve the details of the project . . . in a way that maximises the benefit and minimises the disadvantages . . . A recognition of these differences and their varying information requirements make it easier to select the type and depth of information that is needed at each stage. (Scottish Development Department, 1976, pp. 3–4; emphasis added)

This general view appears to have contributed to some suspicion within the Scottish Development Department as to the value of any form of standardized impact assessment procedure or format. This view is explicitly expressed in a 1975 working paper: "standardised methods may inhibit the achievement of the appropriate depth of analysis and *the rapid response required*" (Scottish Development Department, 1975a, p. 10; emphasis added). A flexible approach that relates the depth of analysis to the stage of the development process and that allows for a rapid response in the early stages may have some advantages, yet it must be recognized that once overall planning permission has been given, the local planning authority is statutorily bound to grant full planning permission once any reserved matters or constraints have been resolved.

In these circumstances, might it not be preferable to undertake a thorough assessment of impacts (at least as understood in the United States) prior to the granting of overall planning permission? It seems a little strange, for example, to defer a consideration of infrastructure needs (the example cited by the Scottish Development Department, 1976, p. 4) to the "third level of work" that would *follow* the granting of final planning permission. A more significant omission would be the lack of opportunity within the three assessment levels described in the Scottish Development Department paper to identify and assess the impact of alternatives to the proposed action.

An excellent example of forward planning in this context is the Strathclyde Regional Council's study of possible sites for a petrochemical complex (Strathclyde Regional Council, 1977). This study focused (1) on the suitability of seven sites for petrochemical development from an engineering and physical point of view, and (2) on the major impacts (environmental and social) that could be anticipated at each site if development were allowed. The final section identified those sites which could accommodate different levels of development with the least environmental and social disruption. Unfortunately, the study's conclusions were severely weakened by its restriction to "sites with a history of industrial promotion," but the general approach seems a valuable way of combining forward planning responsibilities with environmental assessment. It is a means to identify the most suitable sites and eliminate unsuitable sites. As already noted, a very similar approach has also been used in the Grampian Region to develop a contingency plan for future petrochemical development. As in Strathclyde, various sites were evaluated to assess their potential for accommodating different types of petrochemical plants in terms of critical criteria and constraints (for example, distance from residential population, availability and quality of land, accessibility). The rationale was to undertake as much forward planning as feasible in order that delays in dealing with specific proposals might be avoided. Thus, in the foreword to the plan, the hope is expressed that "having read this report [potential operators and developers] may feel encouraged to investigate the advantages of locating their plant in Grampian, and feel confident that their proposals will be dealt with expeditiously and efficiently" (Grampian Regional Council and Banff and Buchan District Council, 1980, p. i). In effect, without prejudging individual proposals that would still need to be subjected to detailed appraisal, these types of forward planning exercise should help both developers and planners to expedite the processing of development applications.

There appear to be several practical reasons why the Scottish Development Department is opposed to detailed project appraisal when planning permission is first sought. The first reason is the length of time it might take to undertake such a study. In the narrow legal context, a local planning authority is statutorily required to decide on a planning application within two months of its receipt; in the broader national context, such a policy would undoubtedly slow down the processing of applications to develop onshore support facilities for the oil industry.

A second practical problem relates to the very limited information presently required of an applicant when planning permission is first sought. This procedural point is raised in the critique of the Sphere report referred to earlier which seriously questioned whether "any meaningful and detailed appraisal of impact can be based on information submitted at the outline planning application stage" (Project Appraisal for Development Control, 1974, p. 3). Thus a situation appears to exist whereby the advantage of the fullest possible appraisal of project impacts as early as possible (and it is worth emphasizing that Environmental Impact Statements in the United States have been criticized for appearing at the end of the planning process rather than being an integral part of planning from start to finish) is precluded by the need for speed and by the pre-existing structure of the U.K. planning system. The situation is then rationalized as being preferable in the interests of "flexibility."

A further "reservation" expressed by the Scottish Development Department with regard to impact assessment related to the difficulty of evaluating the relative significance of impacts, recognizing that widely differing priorities and attitudes exist with respect to oil development: "It is probably fair to say that little has emerged from the Scottish studies to clarify and formalize this stage of the process. Various techniques of ranking, weighting, and scoring have been used and have been of assistance but none has emerged as an accepted tool (it should be noted that none of the studies involved a formal cost/benefit analysis)" (Scottish Development Department, 1976, p. 4).

Again, however, the concern expressed is directly related to the "national interest" in facilitating oil development: "The question of priorities has been particularly difficult in the Scottish oil cases because several of them have had a direct bearing on the timing and quantity of offshore oil production, a matter of considerable national interest. This national economic benefit is normally beyond the terms of reference for a local impact analysis [yet] is one of several items that enter the decision making process alongside the description of the local implications of the project" (Scottish Development Department, 1976, p. 4).

A final reservation is related to "the wide variety of analysis techniques that can be used to examine impacts." As a result "it is difficult to define exactly what is adequate in any given case." More caustically, the Scottish Development Department noted that "the number of experts, methods, and differing interpretations of the same facts seems to be limitless." In the absence of a more "exact science" of impact assessment that would perhaps have satisfied the department's civil servants, there was the "need for a commonsense approach, taking account of conditions in the area, and based on an attempt to limit the analysis to a relatively few issues which are accepted as important for the decision in principle" (Scottish Development Department, 1976, p. 4).

Despite what appears to be a rather cool attitude toward the concept of project appraisal (and one is struck by the reiterated theme, sometimes implicit, sometimes explicit, that detailed appraisal early in the planning process will impede rapid development and decision-making), the Scottish Development Department commissioned a group of geographers at Aberdeen University to develop an impact manual that would "help planning authorities to develop expertise in analysing complex planning proposals in a methodical way." As published in 1976 (Department of the Environment, 1976a), the report emphasizes the type of information required for effective impact assessment. An innovative feature is the project specification report that must be completed by the developer and submitted with the application. In this way, the manual places a major responsibility on the developer to provide information as in input into the planning process.

The project specification report covers details of the proposed plant and its processes, the physical characteristics of the proposed site, reasons for site selection in relation to alternative locations, employment characteristics during construction and operation, infrastructural requirements, factors of environmental significance, and the nature of emergency services. The manual further recommends that developers should be advised that planning approval and conditions would reflect the information provided in the project specification report. In this way it was felt that developers would be more likely to analyze the needs and outputs of a proposed project. "Developers acknowledge their responsibility to society as a factor in policy decisions and in any case will appreciate that it is not in their interest to mislead the planning authority, particularly if subsequent applications are envisaged. Speculative applicants are likely to be less conscientious and a useful side effect of the requirement to submit a completed Project Specification Report may be to discourage such applications" (Department of Environment, 1976a, p. 14).

However, the manual does recommend that local authorities obtain independent verification of the data provided by developers. Moreover, the manual suggests that the information received by the local authority should be made available for public inspection.

> Should developers refuse to provide information on the grounds that release could damage their commercial interests, then the local authority should indicate that analysis may have to proceed based on "worst case" assumptions. In addition, should the application go to public inquiry, subjects for which the developer has failed to provide information may be raised as issues meriting special attention, particularly if the planning authority has considered these subjects to be key issues and had brought them to the attention of the Secretary of State. In such circumstances the non-availability of information considered critical might seriously damage the appellant's case. (Department of Environment, 1976a, p. 15)

If these various procedures were implemented, and if local authorities had sufficient lead time before a decision in principle had to be made, the impact assessment procedure would undoubtedly be more credible than at present. Moreover, by providing the information necessary for detailed project appraisal at the outset of the development control process, the developer should benefit as a result of the more rapid and efficient processing of the application. In this context, appendices to the manual provide step-by-step guidance to local authorities on impact assessment methodology, for example, estimating local employment impacts and immigrant flows. In general, these fall short of giving full weight to various modeling and impact analysis procedures that have been developed in the United States. Moreover in testing the manual the lengthy and detailed questionnaire used to solicit information on a proposed development has drawn criticism.

Although the report expressed the hope that "planning authorities will make use of the impact procedure wherever appropriate in their work," there appears to have been little real progress as yet toward incorporating any form of impact appraisal into the statutory procedure for development control. Nor, by and large, have the cumulative impacts of phased development activities (as at Nigg Bay) received the attention they deserve, with projects continuing to be considered, evaluated and either approved or rejected on an individual basis.

A more immediate constraint, however, is the two-month interval between formal receipt of a development application and a planning decision presently permitted local authorities under the Town and Country Planning Act. Detailed project appraisal within such a time-frame is clearly impossible. "Unless the two month period is substantially lengthened, waived or otherwise modified, the planners in Scotland must attempt to anticipate the likely effects of new oil-related development *before* receiving applications for planning permission" (Baldwin and Baldwin, 1975, p. 157). In such circumstances, the combined forward-planning/environmental-assessment approach being attempted by the Strathclyde and Grampian Regional Councils is of particular significance.

Public Participation

The matter of public involvement in environmental planning and resource allocation has received increased emphasis in recent years. O'Riordan identifies four major reasons that have contributed (at least in the United States) to the growing demand for a "participatory strategy."

> First, resource allocation in the face of widespread externalities and conflicting political interests requires the intervention of adversely affected groups. Second, because people are better educated, more informed, and more willing to participate in community affairs, they are more responsive to participatory strategies. Third, many people enjoy a sufficiently high standard of life to be concerned about deteriorations in environmental quality which they find it increasingly hard to avoid. Fourth, the days of the paternalistic administrator, infallible expert, and trustee politician are over; people want to play a part in shaping their own surroundings. (O'Riordan, 1976, pp. 257–258)

Collectively the thrust is toward a sharing of decision-making powers and a reform of institutional structures to allow greater public involvement in environmental policy-making, particularly at the local level where issues of environmental quality and social well-being are of immediate concern.

It must be acknowledged, however, that the environmentalist

tradition in the United Kingdom is somewhat different. In the first place, the level of activism of environmental groups, itself an outgrowth of the grass-roots/town-meeting activism endemic in the American political scene, is lower in the United Kingdom (O'Riordan, 1976, pp. 231–232). There are, of course, numerous conservation organizations, but they tend to be more deeply entrenched within the established political system. "It might even be argued that it is precisely because such organisations are so embedded in the political fabric (and because many of them are patronised by leading public figures) that activism in the sense of grass roots participation is less common"(ibid., p. 231).

Some conservation organizations have a quasi-official and/or advisory status with a statutorily defined role in environmental policy-making. Thus the Nature Conservancy Council and the Countryside Commission are responsible for identifying areas suitable for designation as Areas of Outstanding Natural Beauty, Nature Reserves, and Sites of Special Scientific Interest. Local authorities have an obligation to inform these organizations of any development that might affect a designated site. Even societies without statutory responsibilities are highly influential—but their influence is derived from the prestige and status of executive members and the way in which they operate within the political system. In analyzing the membership and influence of the Council for the Protection of Rural England, L. Allison (1975, pp. 115–123) draws attention to the social status of the governing body who rely on "contacts and expertise . . . to communicate evaluations and ideas through effective channels and have them taken seriously" (ibid., p. 123).

The corollary of this lack of grass-roots environmental activism in the United Kingdom is the greater degree of discretion enjoyed by policy-makers and administrators. "All kinds of regulatory policy making—the setting of ambient standards, determining consents for waste discharges, arriving at performance standards for road and building construction, deciding on what buildings or landscapes should be included in urban or rural conservation areas etc.,—are executed by selective consultation with particular interests, but with no requirement to inform the general public" (O'Riordan, 1976, p. 233).

The scope of administrative discretion in the United Kingdom (which O'Riordan describes as an indication of how far society puts its faith in professional competence and political integrity) should have become evident in the preceding discussion of planning procedures. Thus in Scotland, the Secretary of State is free to decide whether or not to "call-in" a planning application, to determine whether a public inquiry should be ordered, to define the scope of the inquiry, and to accept or reject the conclusions of the appointed Reporter.

An important element in the discretion enjoyed by decision-makers and administrators in determining and implementing environmental policy are the restrictions on public access to information. It is, for example, impossible in the United Kingdom to obtain any data on actual effluent discharges into the nation's atmosphere and waterways, even though, as a neighboring property owner, one may be immediately and directly affected by the discharge. "Public access to all kinds of relevant facts in Britain is not an automatic right, since the publication of information is either controlled by statute or limited by ministerial discretion" (O'Riordan, 1976, p. 233). This extends even to issues of public health and safety. "The Alkali Inspectorate withhold findings regarding fluoride poisoning. The Health and Safety Executive report on the safety of Canvey Island suppressed information on liquefied natural gas which local residents needed to assess whether they risked a holocaust like the Bantry Bay tanker explosion" (Meacher, 1979, p. 14). Yet, as Michael Meacher observes on many issues, "not only do there seem no justifiable grounds for concealment, there are surely very positive reasons why the public *should* know these things." There is no legislation comparable to the Freedom of Information Act in the United States. Again, in the United States, public awareness and involvement are facilitated by the National Environmental Policy Act, which requires a full public disclosure of the environmental impact of any federal project. No such legislation exists in the United Kingdom. The discretion enjoyed by policy-makers and administrators and restrictions on the flow of information clearly weaken the opportunity for effective public involvement in resource management.

In these circumstances, the emphasis in British environmental policy-making is on consultation and compromise rather than on confrontation and bargaining. This general approach is exemplified by the controversy that arose over the proposed site for the landfall of the Frigg gas pipeline. The original site for which planning permission was sought was the Loch of Strathbeg. The ecological significance of this site (particularly as a habitat for wildfowl) resulted in considerable opposition from environmental groups. As described by George Dunnet, a founding member of the Environmental Liaison Group,

> Discussions took place between the developers and several organizations, but perhaps the most detailed and critical dis-

> cussions took place in private under the chairmanship of Aberdeenshire's county clerk between the Environmental Liaison Group and the two developers, Total Oil Marine and the British Gas Corporation. These discussions enabled ecologists from different disciplines to discuss critically with the developers the most likely impact of their plans. We were unyielding on the principle that this highly important ecological site should not be put at risk until everybody was satisfied that suitable alternative sites could not be found. Such a site was eventually found at St. Fergus, a few miles further south. (Quoted in Baldwin and Baldwin, 1975, pp. 102–103)

Whether a fundamental change is occurring in the character of the British environmental movement remains open to debate. From discussion with planners and developers, however, it is clearly their impression that there has been a shift to more activist and adversary form of public involvement in environmental policy-making. Whether the pace and scale of North Sea oil development and the complex range of trade-offs involved have served to encourage such a shift is again unclear. Certainly there appears to be a greater willingness to challenge decisions of administrators, to question expert testimony, and to demand a more accurate accounting of the costs and benefits of proposed developments. Inability to influence decisions directly affecting individual and community welfare and exclusion from the decision-making process are frequently repeated themes. "Every community, particularly the smaller towns and villages, seemed to be afraid of being engulfed by the speed, size and complexity of events, and scarcely any felt that they were being given sufficient information in time for them in any way to influence events" (Francis and Swan, 1973, p. 57).

As described earlier, there are in fact three opportunities for public participation in the planning process. The first occurs when a structure/local plan is being formulated. The Town and Country Planning Acts make public participation in the preparation of structure plans a statutory requirement.

> The main stimulus for this came, not from the grass-roots (still less from local government), but from central government themselves. Under the old development plan system, the department(s) were becoming crippled by what a former Permanent Secretary called "a crushing burden of casework." . . . A new system was therefore required which would remove much of the detailed planning work—including approval of local plans—from central to local government. But this necessitates

> public confidence in local government: hence the importance of public participation. This now becomes, not a desirable adjunct to the planning process, but a fundamental basis. (Cullingworth, 1976a, pp. 257–258)

In practice the response to this opportunity, at least in Scotland, has not been encouraging. Thus the Grampian Regional Council went to considerable length to publicize a preliminary version of its structure plan, holding workshops and mounting a mobile exhibition. Attendance, however, is said to have been extremely disappointing. An important distinction must, of course, be made between the provision of information and involvement. As the Skeffington Report, *People and Planning*, commented,

> most authorities have been far more successful in informing the public than in involving them. Publicity—the first step—is comparatively easy. To secure effective participation is much more difficult. Secondly, some of the authorities who have made intensive efforts to publicise their proposals have done so when their proposals were cut and dried. At that stage, those who have prepared the plan are deeply committed to it. There is a strong disinclination to alter proposals which have been taken so far; but from the public's point of view, the opportunity to comment has come so late that it can only be an opportunity to object. The authority is then regarded more as an antagonist than as the representative of the community, and what was started in good will has ended in acrimony. (Committee on Public Participation in Planning, 1969)

The second opportunity for public involvement occurs during the development control process. As described earlier, any planning application that does not conform to a development plan must be advertised in the local newspaper and on the site. The concerned individual (or conservation group) then has the opportunity to register appropriate comments or objections with the local authority before the application is heard. A controversial proposal is likely to become the subject of a public inquiry. It is this forum therefore that tends to become the primary focus for public involvement in planning issues. As described earlier, such inquiries are ordered at the discretion of the Secretary of State. This event is most likely to occur when (1) a developer appeals against a local authority's refusal to grant planning permission, (2) there is considerable local opposition, or (3) an application is considered to raise issues of national or regional significance and is "called-in."

The general format of a public inquiry should be evident from previous discussion of the Drumbuie episode. An inquiry is held in the vicinity of the site proposed for development—indeed, a major purpose is to ascertain the strength of local feeling toward the proposal. The proceedings are intended to be as flexible and informal as possible "so that the ordinary interested person does not feel inhibited from making a contribution without professional representation" (Scottish Development Department, 1975c). Nevertheless, it is usual for the developer and principal objectors (if they can afford the cost) to be represented by counsel familiar with planning law. The scope of the inquiry is defined in advance by the Secretary of State.

In the past, inquiries have usually been restricted to a consideration of the issues raised by objectors, "criticism of government policy is in general held to be beyond the terms of reference of an inquiry" (Grove-White, 1975, p. 20). It is often the case, of course, that full consideration of the public interest cannot be neatly constrained in this manner. And in recent years there does appear to have been a real effort to broaden the basis of inquiries and ensure that all relevant issues are explored. Thus at the crucial inquiry into Shell Expro's proposal for an NGL separation plant at Peterhead, the Reporter refused to divorce considerations of the site itself from discussion of Peterhead's harbor facilities, even though the latter were not at that time the subject of a planning application.

The general rationale underlying the inquiry is to gather information that will assist the Secretary of State in arriving at a final decision. To this end, testimony is presented by various "expert" witnesses, who are subject to questioning by any of the interested parties and by the Inspector. A technical assessor is usually appointed to help the Inspector evaluate the information presented by the different witnesses. As Robin Grove-White points out, however, despite the "information-gathering" objective, it is almost impossible to avoid polarization—both sides want to win.

> Once the inquiry has begun, a familiar exchange of arguments unfolds. The developer is concerned to show that the benefits his project will bring to an area outweigh the costs to amenity, while those opposing will argue the reverse—that the social and environmental costs they and the public at large are being asked to bear are too high to be compensated by the exclusively economic benefits the project will bring. Both sides are by now firmly convinced that right is on their side. The developer is determined to add as many non-economic benefits to his side of the equation as he can reasonably manage—perhaps arguing that the area is not as beautiful as the objectors claim,

> or that his project will create, paradoxically, a new focus for tourist interest. Similarly, the objectors, to have any real chance of success, must seek to undermine the economic certitudes which are the developer's real strength. (Grove-White, 1975, p. 15)

The Inspector's findings and recommendations are submitted to the Secretary of State, who, as has been noted, makes the final decision.

It has been argued that a public inquiry is well suited to airing the issues surrounding a controversial proposal, to eliciting facts, and to ensuring that the public interest is safeguarded. While it is true that an inquiry does represent one of the few occasions on which opposing views can be publicly expressed before an independent arbitrator, it seems that the odds in such a procedure tend to favor the developer rather than the objector.

The developer starts with a distinct informational advantage. Moreover, a developer can usually afford to buttress a case with the expert testimony of engineering and economic consultants. And it may prove difficult to convincingly refute a developer's case.

> The difficulty for objectors is that they start from an a priori dislike of the project. There may be therefore in their attacks on the scheme's economics an element of justifying a prejudice. They must cast around to find alternatives and then explain convincingly both to the developer and the Inspector why it is that the former has not preferred one of these, and why he should do so now . . . The success of such a ploy, in an Inspector's eyes, would rest on its being demonstrated beyond all doubt that such an alternative was realistic. However, this evidence would have to be produced by the objectors themselves; to be effective it would have to be supported by the testimony of expert witnesses. In this respect as in others, a truly convincing case needs money behind it. (Grove-White, 1975, p. 15)

In part therefore the objector's problem is to be able to find (and afford) experts to challenge a developer's case. In these circumstances, the inquiry—as an adversary proceeding—breaks down, since the contesting groups have unequal access to resources and information. This weakness is likely to become more serious as the issues raised at an inquiry increase in technical complexity.

In the second place, whereas the developer is motivated by a simple, single goal, objectors are likely to embrace a rather diverse range of philosophies and goals and are therefore less likely to present a "united front." Many objections are likely to come from members of the public who are concerned not about the merits of the proposal per se but about mitigating any possible damage to their interests. Perhaps only a minority are opposed to the entire proposal. This variation in attitude is perhaps worth emphasizing in the context of onshore oil development in Scotland.

It is quite incorrect to suggest that disagreements over North Sea development can be interpreted as a straightforward national/local conflict. In many instances, proposals *have* had strong support from both local authorities and local communities. The promise of employment opportunities in remote rural areas characterized by high rates of unemployment and out-migration is extremely attractive (and remains so despite the growing body of evidence that the local employment benefits of such projects tend to be overstated). The application by Chicago Bridge for permission to construct production platforms at Dunnet Bay, it will be recalled, was strongly endorsed by a petition from local residents. Opposition came from outside the region, from the group frequently referred to rather derogatorily as "white settlers." (This name has been given by local residents to "incomers," those wealthy, usually English, individuals who have purchased property—farms, vacation or retirement homes, shooting estates—in rural Scotland.)

A similar situation existed at Drumbuie. In this instance, partly because of opposition from the local authority, the proceedings were represented in the national media as an attempt to impose a large-scale development on an unwilling community. In reality, apart from Drumbuie itself, local opinion was considerably divided on the issue. "Those in favour tended to view industrialisation as the answer to the more deep-seated problems of the area, particularly as a means of stemming the drift away of the younger elements of the population. Those against argued that the size of the project constituted a serious threat to the existing social pattern and would not in the long term resolve the employment situation" (Hutcheson and Hogg, 1975, p. 116). In these circumstances there was a somewhat strange alliance between local opponents of the proposal, the "white settler"/outsider group, and the Scottish Nationalist Party!

It is important to recognize, however, that public opinion may be extremely fluid. Indeed, one sympathizes with the predicament of an administrator weighing alternative courses of action and confronted with vague, conflicting, and shifting expressions of public preferences. At the public inquiry over Shell Expro's application to site an NGL separation plant at Peterhead, there appeared to be a certain consensus locally that after the expansion of the harbor for supply bases, the construction of a new power plant, and the location of the Scanitro liquid ammonia plant, "enough was enough." Yet when Shell subsequently withdrew its application, local planning authorities were severely criticized for "losing the plant," and the sentiment as expressed in editorials and letters was that "the boom is over and we missed out." Should Occidental Petroleum go ahead with its proposal for an ethane cracker at Peterhead, the local community will have a further opportunity to express its views as to the desirability of petrochemical development.

Quite apart from the question of whether or not an inquiry can provide an adequate forum for public involvement, the proceedings are weakened by a tendency to focus on the merits (or otherwise) of a particular site. In an attempt to remedy this weakness, the revised Town and Country Planning Act did provide an alternative to the public local inquiry—namely, a Planning Inquiry Commission. Such a commission would be appointed to hear planning issues of an unusual or complex nature and would enjoy much broader terms of reference than a conventional local inquiry. In particular, it would be authorized to examine the implications of alternative locations as well as matters of unusual scientific or technical character. Clearly, Planning Inquiry Commissions are likely to be extremely time-consuming, and they have in fact not been employed in the United Kingdom other than for the third London Airport at Maplin. Yet perhaps a Planning Inquiry Commission is precisely what is required to assess the implications—local, regional, national—of the various proposals now being advanced for additional petrochemical investment based on hydrocarbon feedstocks.

All in all, the local public inquiry is an awkward procedure for handling controversial applications. But, what is perhaps most critical, the feeling appears to be growing that it is irrelevant. This attitude is well expressed with respect to oil development in Scotland by the conservationist Bill Bourne: "With great concern for 'democracy,' current legislation provides for local public inquiries to be held into any disputed development, after which the ultimate decision is reserved for the appropriate minister. This means that local people who feel strongly about the issue are able to expend all their energy and cash arguing about it in some backwater, while the real decision over the issue is made quietly elsewhere" (quoted in Baldwin and Baldwin, 1975, p. 142). Was the Drumbuie inquiry a useful exercise? Certainly it provided a forum in which the issues were debated, but one cannot help feeling that the final decision was determined not by the merits of the case but by the granting of local planning permission at neighboring Loch Kishorn—despite the fact that such approval was directly contrary to the evidence presented and apparently accepted at the Drumbuie inquiry. Was the Mossmorran inquiry a useful exercise? Certainly the aborted Peterhead inquiry had alerted local authorities to the need for a careful, independent assessment of the risks involved in shipping and processing natural gas liquids. Yet it would appear that the suitability of the Mossmorran site pre-empted the decision with respect to locating terminal facilities at Braefoot Bay, despite admitted adverse effects on the environment and on the area's recreational value. Moreover the Mossmorran decision illustrates yet again the extraordinary discretionary powers enjoyed by the Scottish Secretary. As at Peterhead, opponents of the proposal had expressed particular concern over the safety of both the gas separation plant at Mossmorran and the storage and shipment facilities at Braefoot Bay, which lies within one mile of Aberdour and Dalgetty Bay (population approximately 6,000). In rejecting these concerns the Reporter concluded that "the weight of evidence suggests . . . that the maximum credible spills at the loading jetty (the most likely point at which spills could occur) will be considerably less than the quantity required to result in an unconfined explosion" (Decision Letter from Secretary of State for Scotland to Esso Chemical Ltd and Shell UK Exploration and Production, August 9, 1979, p. 2). The risk of spills from the Mossmorran–Braefoot Bay Pipeline was considered to be "very low" and accordingly "the doubts raised on safety factors are not sufficient to cause the Reporter to believe that these plants cannot be designed to operate within an acceptable level of hazard."[20] Yet the Reporter's conclusions were reached without reference to a special study of the safety issues involved in transporting and storing liquefied energy gases (NGL, propane, butane) undertaken by the U.S. General Accounting Office (GAO). This report (published in July 1978 four months after the Reporter's recommendations had been submitted to the Secretary of State) concluded that the level of risk to the public warranted "the immediate attention of the Congress and the cognizant Federal agencies," and recommended "that future, large-scale liquefied energy gases facilities should be located away from densely populated areas; that any such existing facilities should not be permitted to expand, in size or in use; and that present urban facilities should be carefully evaluated to ensure that they do not pose undue risk to the public" (U.S. General Accounting Office, 1978, p. 1). By implication the report suggested that industrial self-regulation would not be adequate to limit the risk of harm to affected populations to a politically acceptable level. Whether these conclusions would have influenced the Reporter's recommendation on Mossmorran is, of course, open to speculation.

What is clear, however, is that the Secretary of State is not required to reconvene an inquiry or even necessarily to consider additional evidence that was not presented at the original inquiry. Given the GAO report it seems unlikely that a Mossmorran-type proposal would be approved for a similar location in the United States.

It should be evident from this discussion that the constitutionally defined role of the courts in the United Kingdom is markedly different from that in the United States. In the United States the separation of powers between the legislature, the executive, and the judiciary is constitutionally defined, enabling (indeed requiring) the courts to assess the appropriateness of politically determined policies and the correctness of administrative actions and procedures (O'Riordan, 1976, p. 240). Moreover, "judicial legislation [in the United States] is based on adversary relationships and so is grounded in conflict. Perhaps this is one reason why American environmental politics is so bound up with the law (or vice versa)" (ibid.).

In the United Kingdom, by way of contrast, a single institution—Parliament—stands alone at the apex of power, combining legislative, executive, and judicial roles and with ultimate responsibility for defining and delegating "political power," including the duties and functions of the courts (O'Riordan, 1976, p. 241). J. N. Lyon notes that "the courts take no initiatives in important public issues such as those relating to the environment [and] are wary of intruding in areas of responsibility that they regard as belonging properly to the legislative or executive branch" (quoted by O'Riordan, 1976, p. 241). The manner in which Parliament may choose to exercise its power is well illustrated in relation to North Sea oil development by the Offshore Petroleum Development (Scotland) Act. Moreover, again in contrast to the United States, the implicit intention throughout the planning process in the United Kingdom is to *avoid* adversary proceedings. Thus, the public cannot look to the courts for the kind of environmental "relief" and policy innovation that has occurred in the United States.

A standard response (particularly from central government administrators) is that this is as it should be. From this perspective, Parliament, as the elected representative body, is the only appropriate forum for a challenge to government policy. If offshore oil development and onshore land use planning policies are to be changed, it is Parliament that must change them. Yet this view tends to oversimplify the real dilemma—that is, the extent to which the "need" for rapid energy development intensifies central government intervention in land use planning, thereby constraining the opportunity for a more localized and participatory approach to policy formulation. In these circumstances, it is hardly surprising that "many local planning officials and citizens in Scotland evince a reluctant fatalism about central-government control and policy directions" (Baldwin and Baldwin, 1975, p. 143).

O'Riordan (1976, p. 258) suggests several criteria for use in assessing the opportunity for effective public involvement. These include (1) the scope of citizens' rights, more particularly their standing in the courts; (2) the nature of the participatory strategy (ballots, workshops, inquiries) and particularly the degree of interaction and the representativeness of the participants within the strategy; (3) the amount of media publicity given to environmental issues; (4) the degree of public access to information; and (5) the nature, scope, and political effectiveness of environmental impact statements. On the basis of these criteria, it is clear that British planning procedures are not conducive to effective public participation.

These differences in what O'Riordan (1976, p. 230) calls the "political culture" of decision-making in the United States and the United Kingdom have been emphasized because it may well be that, when harnessed to the economic imperatives of correcting chronic balance-of-payments deficits and achieving energy self-sufficiency, they help explain the speed with which the United Kingdom has expedited the development of North Sea oil.

The Shetland Experience

The preceding sections have dealt with the framework for locating and accommodating the impacts associated with the onshore facilities needed to support offshore exploration and production. The focus has been on mainland Scotland, with particular reference to (1) the tensions between conservation/development and local/national interests, and (2) the nature and effectiveness of those measures (such as public involvement, impact assessment, and forward planning initiatives) that might serve to reduce the level of conflict. In Scotland, the efforts of planners have attracted considerable criticism and are sometimes contrasted with the more imaginative and effective planning initiatives adopted in the Shetland Islands. MacKay and Mackay exemplify this view and suggest that the "county council has demonstrated a farsighted and vigorous attitude to the marriage of oil interests with those of Shetland. It appears to have provided a model framework for local authorities to use in such situations" (1975, p. 130). Although the "model framework" has yet to stand the test of time, several facets of the Shetland experience—notably the identification, assessment, and accommodation of social and environmental impacts, the provision and financing of the necessary infrastructure for onshore facilities, and the active role of the local authority—contrast sharply with the mainland experience.

The Shetland Islands comprise a group of approximately one hundred islands with a total permanent population of around twenty thousand.[21] The only settlement of any significant size is Lerwick (about one-third of the total population), which acts as the main employment, commercial, and administrative center. The remaining population is widely dispersed in small crofting and fishing communities. It seems almost mandatory for authors to describe the Shetlanders as "fiercely independent." (After centuries of Norwegian and Danish rule the Shetland and Orkney Islands became part of Scotland in 1469, when the then King of Denmark, Norway, and Sweden defaulted on the payment of his daughter's dowry.) Most accounts contain further references to the islands' remoteness and to the islanders' "spirit of satisfaction with the life they have led for many years" (Baldwin and Baldwin, 1975, p. 106).

Documentation of the islands' geographic location and historical experience is somewhat easier to provide than of the islanders' sense of well-being. The last in particular is difficult to reconcile with the pace of out-migration and the economic history of the islands during the twentieth century. Nevertheless, the islanders clearly have a distinctive and strong sense of cultural identity. The structure plan prepared by the Shetland Islands Council has attempted to define the essence of this identity:

> A predominant feature in the Islands is the awareness of a distinctive identity and lifestyle, the main characteristics of which [may be defined] as follows:
>
> - Communities where the individual feels he matters and has a sense of belonging;
> - Strong family ties;
> - Tolerance through not having been exposed to or confronted by strong social pressures;
> - An absence of serious crime: little parental fear for the safety of young ones;
> - Religious tolerance;
> - The realisation that the continuation of a native dialect affords an enriched means of communication.
>
> Underlying and moulding these characteristics are Shetland's remote location; its small size and small communities; its eco-

> nomic tradition—which allows for personal independence (e.g. crofting and fishing) but has also necessitated community commitment to overcoming hardship—and its Scandinavian connection, still apparent in place names, dialect and traditions. (Shetland Islands Council, 1976, vol. 1, pp. 8–9)

The prospect of substantial oil-related development undoubtedly raised a number of fears within the Shetland community. In the early phases of offshore activity the most significant concerns appear to have related to the potential economic and social consequences of oil-related development. In economic terms, the major fear was that traditional industries (which represent a central element in the community's lifestyle) would be further undermined as a result of labor losses to industrial development. "A balance needs to be struck between too great and rapid a population build-up and facilitating the labour supply situation through immigration" (Shetland Islands Council, 1976, vol. 1, p. 9). In general it appears that the loss of labor to oil-related activities has so far been less serious than anticipated. Indeed, the oil companies often provide a convenient "whipping boy" for problems really attributable to other factors. Thus the structure plan notes that "loss of labour [from the fishing industry] does not seem likely in view of tradition, job loyalty and the now almost universal system of shared ownership of boats . . . employment has contracted but this has been due not so much to oil-related activities as to the fall in catches and the generally depressed state of the market" (ibid., pp. 21–22). More crucial to the future of the fishing industry appear to be such issues as the extension of EEC fishing limits, the conservation of fish stocks, rising capital costs, and the maintenance of adequate processing capacity in the Shetlands. Nevertheless, the high earnings available in the oil sector will undoubtedly continue to tempt those employed in indigenous industries, and "if these industries should go into recession for non oil-related reasons, the oil industry will attract resources from them which would frustrate any subsequent recovery" (McNicoll, 1977c, p. 9).

From the social point of view, the greatest apprehension related to the impact of any large-scale in-migration (particularly of construction workers) that would threaten social values and customs.[22] "Many people feel that the incoming population will have little sympathy with the indigenous community's lifestyle and little concern with the Islands' long-term interests: many may be only transitory residents of one or two years duration" (Shetland Islands Council, 1976, vol. 1, p. 9).

Inherent in these concerns is a clear recognition of the transitory character of much oil-related activity and the need to look to the long-term interests of the islanders. In the opening pages of the structure plan this aspect of oil activity is referred to repeatedly:

> Oil is a limited resource but, while it lasts it will dominate . . . The traditional industries must be safeguarded against the distortion which this single, massive, short-term element could create . . . every new job created in Shetland arising out of North Sea oil will need to be replaced at some future date, or the prospect of emigration and decline must once again be faced . . . the principal objective [of the structure plan] must therefore be to balance the long-term needs of the Shetland community with the relatively short-term requirements of the oil industry. (Shetland Islands Council, 1976, vol. 3, pp. 6–8)

As noted by Keith Chapman, these fears appear to have produced a somewhat equivocal attitude to the immediate benefits of oil development: "These attitudes are coloured by the history of the Shetlands, which have experienced a series of economic disasters extending from the effects of the "Highland Clearances" upon crofting in the mid-nineteenth century to the collapse of the herring and then whaling industries. Fears of a repetition of this cycle have coloured the oil industry as an unwanted intruder rather than a catalyst of growth" (Chapman, 1976, p. 174).

These various apprehensions undoubtedly influenced both the general approach and the particular policies adopted by the local authorities with respect to oil development. Moreover in the late 1960s the Shetland Islanders were enjoying a brief period of relative prosperity. Employment in the traditional industries of fishing, fish processing, and textile manufacturing had increased, in part through the efforts of the Highland and Islands Development Board to modernize and rationalize each of these activities (McNicoll, 1977c). The official rate of unemployment was less than 2 percent in 1971, real incomes were high relative to those on the mainland, and there had been a reversal of the general pattern of out-migration. In such circumstances it has been suggested that the local authority could negotiate from a position of strength. "Shetland's success thus far in meeting oil head-on and remaining firmly in the driver's seat is due in large part to an unintimidated attitude of local officials towards the oil industry. It was apparent to oil men early in their discussions that Shetland was not to be cowed by the money and power of multinational oil companies" (Baldwin and Baldwin, 1975, pp. 116–117). The implication is that the Shetlanders would have been quite happy to do without oil.

However, Hance D. Smith has observed that "sensitivity to the vulnerability of the traditional basic industries, especially fishing and knitwear—the latter particularly subject to fluctuations in demand—has already *encouraged* substantial and perhaps increasing support for industrial development among the public at large" (1975, p. 70; emphasis added). In support of this contention, Smith cites the results of a number of local referenda in which a majority of the local population favored the establishment of oil-related facilities. Such pressures for oil development were presumably reinforced during the early 1970s, when traditional industries began "suffering, in the same way as the rest of Britain, because of recession, EEC policies, quotas, and the like" (Shetland Islands Council, 1976, vol. 1, p. 5).

Against this background, it might perhaps be more correct to interpret the Shetland Council's policy as being based on the premise that some oil-related development was inevitable (even desirable) but that the pace, scale, and character of that development would need to be strictly controlled if the long-term interests of the Shetlanders were to be ensured. As summarized in the structure plan, "if oil was to come, Shetland was to be armed and ready" (Shetland Islands Council, 1976, vol. 1, p. 5).

Such a policy was undoubtedly pragmatic in the sense that it seems unlikely, given the Shetlands' location and the national interest in expediting oil production, that a local decision to actively avoid oil-related development would have been tolerated in Edinburgh and London. Moreover, it undoubtedly did not escape the islanders' attention that in formulating and determinedly implementing such a policy it might be possible to secure a greater degree of political autonomy for a community "constantly threatened by the remoteness of much of the decision-making that affects it" (Shetland Islands Council, 1976, vol. 2, p. 13).

OIL DEVELOPMENT

The physical factors of location relative to oil discoveries in the East Shetland Basin and the existence of natural harbors suitable for the siting of supply bases made it inevitable that the Shetlands would be affected by the oil boom. At the time of the third licensing round, when the United Kingdom first experimented with a system of bids for offshore blocks, Shell paid the record sum of nearly £21 million for a tract approximately one hundred miles due east of the Shetlands. This bid provided the first real indication of the potential of the East Shetland Basin. "Until this time, the only hard evidence of oil off Shetland was the increase in helicopter traffic at Sumburgh airport and the rig crews passing through" (Shetland Islands Council, 1976, vol. 1, p. 5).

In the early exploratory phase of offshore activity, many small "forward" bases were located in the Shetlands. Before 1974, however, the servicing activity appears to have been fairly low-keyed. The bulk of the supplies for the drilling rigs continued to come from Aberdeen and other mainland ports, the role of the forward bases in Shetland being limited to transshipment. This activity was largely confined to Lerwick, which as the major fishing port could provide the necessary shore facilities and services.[23] Since 1974, however, there has been a significant build-up in oil-related activity, including the construction of several purpose-built bases.

By and large it would appear that the impact of supply and service activities has not been disruptive. T. M. Lewis and I. H. McNicoll (1978, pp. 116–117) estimate that when fully operational (it was assumed that the level of service activity would have peaked by 1981) oil supply bases might employ between 210 and 330 persons. Although this is not insignificant, representing between 3 and 5 percent of all Shetland employment in the pre-oil period, it is clear that oil supply bases will have far less impact (both positive in terms of generating income and employment and negative in terms of diverting labor from traditional industries and social infrastructure needs) than construction of the Sullom Voe complex (McNicoll, 1977a). Given the decline of the fishing industry, even the clash between fishing and oil interests over harbor space has been minimal. As noted by Smith, the large natural harbor at Lerwick is particularly "well-suited to industrial expansion without further physical encroachment upon the existing community. The harbour already bears witness to former extensive industrial development in the shape of many areas of derelict herring stations dating from the heyday of herring fishing prior to 1939. These are currently being obliterated by reclamation schemes associated with the building of service bases and a roll-on/roll-off ferry terminal" (H. D. Smith, 1975, p. 70).

Of much greater significance for the Shetlands have been the *production* plans for the fields that have been discovered in the East Shetland Basin. Two major pipeline systems now link these fields with the Sullom Voe oil terminal. The first system, a thirty-six-inch trunk pipeline from South Cormorant to Sullom Voe with feeder lines to the Brent, Dunlin, and Thistle fields, began operating in November 1978 when the first oil was pumped ashore from Dunlin. The second system, involving a separate pipeline from the Ninian and Heather fields was also completed during 1978. By the end of that year the terminal at Sullom Voe was receiving over 200,000 barrels per day from Thistle, Dunlin, and Ninian (Department of Energy, 1979, p. 10). Construction of the terminal facilities, including tanker jetties, storage tanks, and a gas separation plant, fell far behind schedule; it now appears that the complex will become fully operational in 1982. When the current design capacity of 1.4 million barrels per day is finally achieved, the terminal will be handling over half of the oil produced from the North Sea. Gas recovery facilities will also allow the recovery of associated petroleum gases currently being flared off or reinjected. Covering an 1,100-acre site, the cost of the terminal has escalated from an initial figure of £250 million to the current estimate of between £1.3 and £1.5 billion. Similar difficulties have been encountered in forecasting the size of the construction work force, which currently exceeds 5,000 persons. In a small, isolated community where, as MacKay and Mackay note, "the historical process of decline and emigration has produced a situation in which much of the public infrastructure requires renewal" (1975, p. 150), absorbing the direct and indirect impacts of a development on the scale of the Sullom Voe oil terminal complex has proved to be a formidable challenge (Figure 15).

LOCAL AUTHORITY INITIATIVES

In contrast to the situation on the mainland, the local authority in the Shetlands had had little previous experience with land use planning through development control. Indeed the Zetland County Council (predecessor to the Shetland Islands Council) was one of those authorities that had still to develop a county development plan, lacking both a planning officer and a planning department. In retrospect this may well have proved an advantage, freeing the islanders from the constraints of a rigid development plan and the more orthodox approach of mainland planners.

Confronted by evidence of the extent of hydrocarbon reserves in the East Shetland Basin and by the inevitable deluge of oil-related planning applications, the Zetland County Council in 1972 commissioned a port and terminal siting study. Consultants were asked to evaluate five potential deep-water sites in the Shetlands and to identify the most appropriate site from both a scenic/amenity and a maritime/operational point of view. This early expression of a policy of "concentration" reflected the underlying fear of the council that oil companies, "if uncontrolled, would be attracted to any number of the natural harbours provided around the Shetland coastline" (Zetland County Council, 1974, p. 3). Thus the intention from the outset was to avoid a dispersal and proliferation of sites by concentrating oil-related activities in a single, specified location.

The consultant's report, submitted in July 1972, identified two

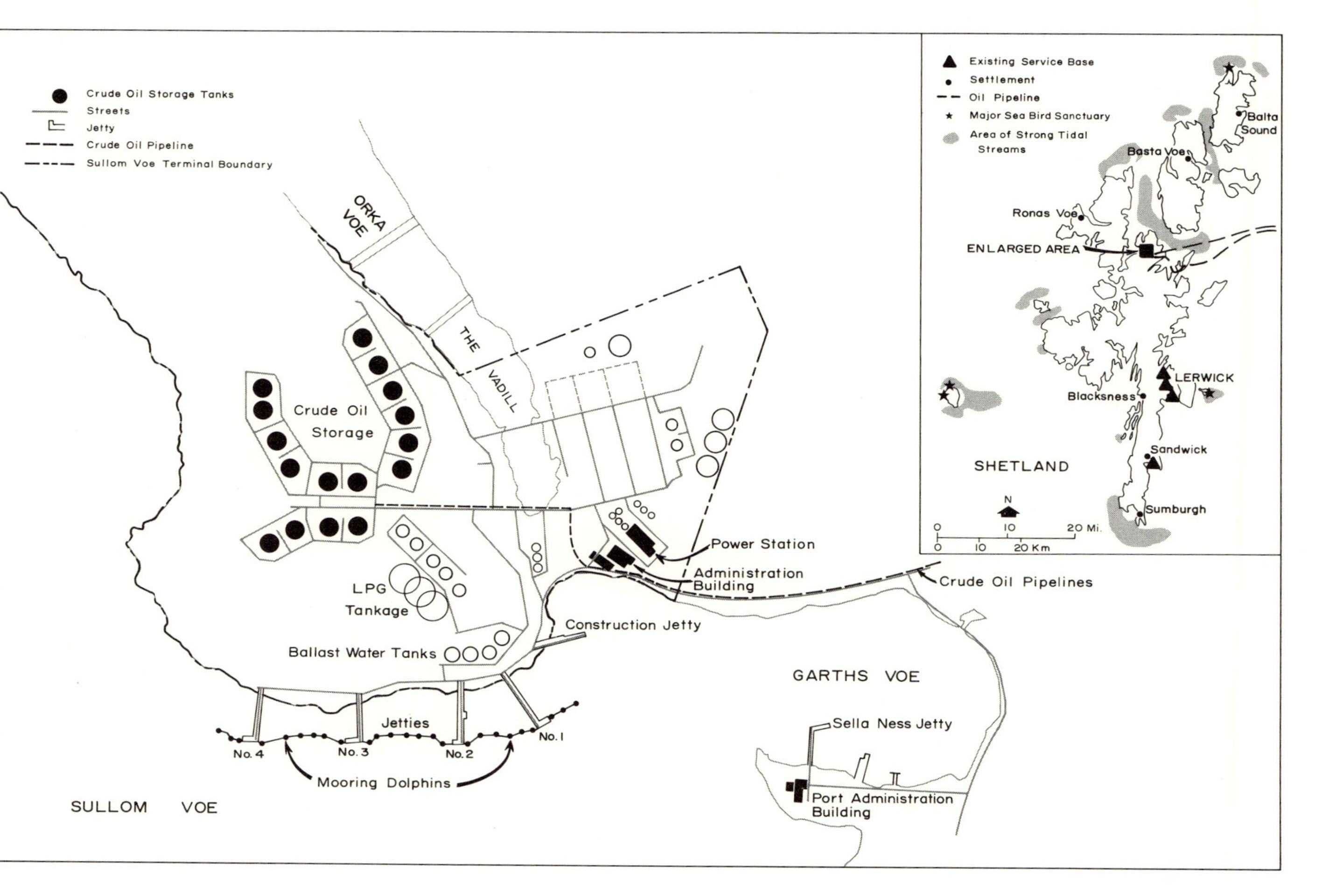

Figure 15. Shetland and Sullom Voe

potential sites, Sullom Voe and Basta Voe (on the island of Yell). Although Sullom Voe ranked second because of occasionally adverse wind and water conditions, its superior onshore features—notably the availability of flat land suitable for industrial development and the absence of any outstanding amenity features—resulted in its designation by the Zetland County Council as the site for a multi-user terminal complex. Moving with some speed, the council in November 1972 sought provisional authorization to acquire land in the Sullom Voe area (through compulsory purchase where necessary), to establish the principle of joint use of facilities, and to acquire appropriate port and harbor regulatory powers.

Believing that its authority (under the Town and Country Planning [Scotland] Act) was inadequate to implement its oil policies, the Zetland County Council took the unusual step of promoting a special order (later to become an Act of Parliament) to acquire additional planning powers. In the process, the council also sought authority (1) to actively participate in onshore development through acquisition of a financial interest in any venture established within the area covered by the order, and (2) to establish a reserve fund that would "provide both during and after the oil era, the means for the Council to take steps which they consider to be in the interests of the Shetland community, particularly in safeguarding the traditional industries or in promoting ventures to diversify the economy" (Shetland Islands Council, 1976, vol. 1, p. 82).

As enacted in 1974, the Zetland County Council Act "empowers the Council and its successor, the Shetland Islands Council, to have harbour authority status over the [Sullom Voe] area; and further, outwith these specific areas to license works and dredging, and to be created harbour authority in areas surrounding such work; to acquire specific lands; to invest in securities of bodies corporate, and to create a reserve fund" (H. D. Smith, 1975, p. 113). The act specifies that the terminal will be run by a Sullom Voe Association, a non-profit-making body on which the council and the oil industry are equally represented. The terminal is actually being built by British Petroleum for a consortium of thirty-four companies, but at least in theory the association has final responsibility for the design, construction, and operation of the terminal. Following passage of the act, the council created the Zetland Finance Company to raise funds for the construction of the terminal and has also entered into partnerships with several companies to provide marine services (tuggage, lighterage, and bunkering facilities) and housing for terminal workers. Income accruing to the council from its commercial undertakings is to be applied to the reserve fund.

Apart from its particular application within Shetland, the Zetland County Council Act raises some fundamental issues regarding control of North Sea developments (MacKay and Mackay, 1975, p. 129). It seems unlikely that the act will set a precedent for similar action by other local authorities. Indeed, such sweeping powers could probably only have been obtained from a Parliament anxious (in the aftermath of the oil price rises and embargo of late 1973) to expedite production of North Sea oil. But in contrast to local authorities on the mainland, the Zetland County Council acted much more decisively and now possesses powers over oil and gas development available to no other local authority. "From having been behind the average local authority in planning matters the Council has pioneered ways in which small local authorities in remote rural areas can cope with massive industrial development" (H. D. Smith, 1975, p. 72).

The considerable control exercised by the council has been further reinforced by an agreement with the oil companies for a tax on each barrel of oil flowing through the Sullom Voe terminal. This barrelage tax, authorized in the Zetland County Council Act, is envisaged as "compensation" to the islanders for disturbance and inconvenience caused by the oil industry and will undoubtedly reduce the islanders' dependence upon central government support. According to one estimate, the local authority's direct revenues from oil (in the form of taxes, rent, "disturbance payments," and joint-venture profits) may well exceed £30 million a year when the terminal is fully operational, a substantial sum when compared with total pre-oil revenues of around £2.5 million a year (McNicoll, 1977c). In addition to meeting infrastructure needs, these revenues are to be used to establish a regional development fund that will hopefully provide the means (1) to diversify and stabilize the island's economic base, and (2) to protect traditional industries, thereby ensuring a smooth transition to a non-oil future.[24] "In the short term certainly, the Act enabled the local authority to reduce the number of speculative land deals and obtain information on the type of oil development being pursued, but in the longer term, the major benefit to the region from the legislation is likely to be the fund raising powers of the Shetland Islands Council . . . from the oil operators" (McNicoll, 1977c, p. 5). Moreover, the council, aware that its capital needs in constructing the terminal and providing the required infrastructure would precede income from its various undertakings, sought and received advance payment of the barrelage tax. "This again raises a precedent, the implications of which will not become clear in the immediate future, but which may be followed by other local authorities" (MacKay and Mackay, 1975, p. 129). Indeed, in the absence of any national program comparable to the U.S. Coastal Energy Impact Program (which provides financial assistance in the form of grants and loans to communities affected by offshore energy development) the Shetland Islands approach to securing front-end financing and compensation must appear particularly attractive to local authorities.[25]

IMPACT ASSESSMENT AND MITIGATION

As the emphasis has shifted from oil exploration to development and production, the magnitude and "visibility" of oil-related impacts have markedly increased. Offshore, for example, an immediate threat to the long-term vitality of the fishing industry has been the disturbance of inshore shell fisheries by oil vessels engaged in laying the pipeline from the Brent field to Sullom Voe. "Such conflicts are inevitable, as projected pipeline routes crossing inshore waters are liable to result in partial destruction of the major shell fish beds . . . These circumstances have already resulted in refusal by the Council of planning permission for a temporary floating base to service the Brent pipeline project, on the grounds that pollution from the base might harm shell fish grounds" (H. D. Smith, 1975, p. 70).[26] Many fishermen appear concerned about the potential "sterilization" of fishing grounds as a result of the high risk of losing gear through fouling unburied pipelines. It should be stressed, however, that most potential conflicts have been resolved through close consultation between the oil companies and the extremely active Shetland Fisherman's Association—thus the Brent pipeline was rerouted to avoid important fishing grounds east of Unst.

The intensification of environmental and social impacts initially strengthened local support for the council's policy of strictly regulating the activities of the oil companies. Thus an early issue, particularly in the crofting communities around Sullom Voe, was the compulsory purchase of land, "which led in the first instance to the majority of local people being opposed to County Council policy . . . With increasing realisation of the implications of development, this attitude has been largely replaced by one of support for County policy" (H. D. Smith, 1975, p. 70). As the consequences of growth have become more apparent, however, there has been increased criticism of the council's policies and priorities. The spill of Bunker C fuel oil from the *Esso Bernica* on December 31, 1978, did little to allay concern about the council's ability to manage growth. As already noted, the spill followed the resignation of the Island's Chief Pollution Officer and a warning by the Director of Ports and

Harbours that the efficient and safe operation of the port was being jeopardized by a shortage of qualified master pilots, by delays in providing navigational aides, and by a general tardiness in completing surveys of alternative approaches to Sullom Voe through Yell Sound. In a candid appraisal of the incident the Shetland Oil Terminal Environment Advisory Group[27] concluded that

> it was not satisfied with the present capability at the terminal for the containment and recovery of spilt oil, and recognized that urgent action is necessary for safeguarding the Shetland environment. Its main recommendation to SVA [Sullom Voe Authority] Limited is that steps be taken immediately to provide, and if necessary develop, technology which will ensure that any future oil spill at the terminal is contained within Sullom Voe and prevented from escaping into Yell Sound. Taking due note of the fact that this spill involved fuel oil which presents special problems with regard to its recovery, the Group was nonetheless very concerned that the equipment designed to contain spilt oil proved to be unreliable, and recommends immediate action to remedy this. It is essential that techniques be developed for the recovery of spilt fuel oil and this matter must be brought to the attention of relevant authorities as a matter of top priority. (*Marine Pollution Bulletin*, vol. 10, no. 4, 1979, p. 96)

At the same time the council's working relationship with the users of the terminal has been strained by disagreements over the design, construction, and operation of the terminal facilities. In approving the initial terminal design submitted by Shell in 1974, for example, the council accepted plans for four surface storage tanks but requested that for visual and safety reasons future storage capacity should be in the form of underground caverns. Shortly after taking over responsibility for project construction in November 1975, British Petroleum informed the council that the original timetable (which called for completion of the terminal in two years with a peak construction force of 1,200 workers) significantly underestimated the complexity and magnitude of the task, and that underground caverns were not economically feasible. Further disagreements arose with respect to the design of the gas separation facilities, while in late 1978, with the first oil tanker entering Sullom Voe to load oil from the Dunlin field, difficulties arose over the terms of the lease for the Sullom Voe site, with the council refusing to allow the tanker to load until the users' association accepted responsibility for cleaning up any spilled oil.

As initially conceived, Sullom Voe was to be an oil storage and transshipment complex. A key question, however, has always been whether this initial development would not attract some form of downstream activity.

> The long-term economic impact of jetty and storage facilities is limited, since few men are needed, but the visual intrusion may be considerable in remote and otherwise undeveloped anchorages. Although careful design and landscaping may minimise such intrusions, the scope for this kind of cosmetic operation is limited where substantial quantities of oil are to be landed from several fields, as is proposed for Sullom Voe . . . [moreover] an oil refinery there would have an economic, social and environmental impact of an entirely different order of magnitude. (Chapman, 1976, p. 162)

In this situation, the council took the precaution of setting aside land within the Sullom Voe complex for both a refinery and an NGL plant. "The concentration of major developments and maximising use of land at Sullom Voe is crucial to the Council's objective of reducing the general environmental impact" (Shetland Islands Council, 1976, vol. 1, p. 42). Yet is is clear that such downstream developments will not necessarily be welcomed by local planners. Thus the structure plan emphasizes that the land has not been allocated for a refinery or NGL plant in the normally accepted planning sense but reserved against the possibility of such development (ibid., vol. 2, p. 82).

Moreover the plan specifically identifies refineries, NGL plants, and petrochemical complexes as examples of development that will not be considered until terminal facilities are completed, by which time the impact of first phase oil development will have become apparent. The criteria established by the council for any future decision to allow significant expansion of oil-related development include a local referendum on any proposal to build an oil refinery in the Shetlands (Shetland Islands Council, 1976, vol. 3, p. 44).

In regulating onshore development outside the terminal the Shetland Islands Council has made full use of its planning powers. Following selection of the Sullom Voe site, the council began work on a county development plan and engaged a consultant (Livesey and Henderson) to prepare a land use plan for the terminal area. A major policy decision was to disperse the permanent work force for the Sullom Voe development (initially estimated by the consultant to be on the order of 550 persons by the mid-1980s) among four existing communities—Brae, Voe, Mossbank, and Firth. This choice was considered preferable to the construction of a single new town, an action which might have encouraged a sense of exclusion. "The desirable objective . . . of all policies relating to the assimilation of newcomers . . . must be that of encouraging people, irrespective of social differences, to feel a positive sense of identity with and responsibility within the community" (Shetland Islands Council, 1976, vol. 1, p. 11).

Nevertheless, as Baldwin and Baldwin emphasize, the settlements involved are little more than hamlets that, with an influx of around 150 workers and their families, could easily be overwhelmed. "Brae will also become the largest centre of growth for community facilities. Among those planned are: an education centre for nursery through secondary school, a health clinic, a community hall, a church, a hotel, a sports complex, a shopping centre, a public house, and a fire and police station" (1975, pp. 114–115).

The structure plan anticipates a total of 1,585 oil-related jobs by 1981, increasing to 1,850 by 1986 (Shetland Islands Council, 1976, vol. 2, pp. 89–100). Other estimates are considerably higher. Lewis and McNicoll, for example, conclude that the total local employment generated by oil could range from 2,500 to 3,500 for most of the eighties (Lewis and McNicoll, 1978, p. 121). To place these figures in an appropriate context, it should be noted that the total labor force in 1971 was estimated to be about 6,900 (although this figure should be treated with some caution, as there clearly existed a considerable amount of part-time and multioccupation employment). In this regard it is hardly surprising that the structure plan identifies housing as a "key need" (Shetland Islands Council, 1976, vol. 1, p. 25).

The estimated requirement (based on committed development) is for 1,300 new housing units by 1981 for oil-related development, in addition to 750 new units and 600 "improved" units for local needs. "The needs for the local population for improved housing accommodation must be met alongside programmes for the incoming population . . . If this fails, the danger is that a dual standard of accommodation may develop" (Shetland Islands Council, 1976, vol. 1, p. 28). However, the magnitude of this undertaking becomes apparent when the rate of construction is compared with the figure of approximately twenty-four houses constructed annually by the public and private sectors combined during the 1960s.

With respect to the construction work force, the structure plan notes that "the existence of this temporary, fluid, male population could cause much greater social disruption than the build-up of permanent population" (Shetland Islands Council, 1976, vol. 3, p. 9). The plan calls for the provision of high-standard accommodation, recreation, and entertainment facilities to "deflect potential pressures on existing communities" (ibid., vol. 1, p. 11). The plan expresses a strong preference for keeping the work force (and disruption) to a minimum by avoiding peaks in construction activity. Nevertheless, difficulties have already emerged with respect to the size of the construction force. "Having assured the Council that a maximum of 1200 workers would be needed for the terminal developments, the oil industry are now indicating a figure of at least 2000 and possibly as much as 3000 at peak (1977/78)" (Shetland Islands Council, 1976, vol. 1, p. 19). As noted earlier, the present construction force exceeds even this figure and now totals more than 5,000 persons. In 1978 British Petroleum was forced to hire a converted car ferry to provide additional housing for construction workers.

Clearly, accommodating growth on this scale is severely testing the Shetland planners. In large measure the success of Shetland's oil policies will depend upon accurately identifying the numbers and needs of the newcomer population, upon achieving a balanced and phased development that will both minimize major fluctuations in the construction labor force and protect indigenous industries from job competition, and upon a sensitive response to the concern of the resident population. In this context, while the Shetland Islands Council has taken a number of innovative measures with respect to the siting of major industrial developments and the financing of required infrastructure, it still remains to be seen whether the council can deal with the social consequences of rapid energy development.

Conclusion

This study has focused on the relationship of British regulatory policies to the pace and character of North Sea oil development. A major objective has been to describe and evaluate the framework for land use planning in the United Kingdom, with particular reference to the opportunities for mediating national/local, development/conservation interests and priorities. A key question for U.S. policymakers is whether British planners have been able to expedite the development of North Sea oil while simultaneously absorbing the major onshore impacts without significant social disruption or environmental degradation. At a time when energy needs and environmental priorities are a subject of intense debate, the overall goal has been to improve the factual basis for determining U.S. offshore policy. However, the insights derived from the North Sea experience of the United Kingdom are clearly applicable elsewhere as other nations seek to resolve the conflicts and mitigate the impacts that arise in the course of energy development.

Yet, as this report has documented, the complexity of the issues and the range of impacts associated with offshore exploration and development make easy generalizations and simple judgments impossible. At best, a number of observations can perhaps be made with respect to particular aspects of the U.K. experience with offshore oil, notably the attitude toward oil development, the quality of planning initiatives, and the relationships between energy development and political devolution.

Attitudes toward Oil Development

Even a brief exposure to the U.K. (and more particularly Scottish) media is sufficient to reveal markedly different interpretations of the North Sea experience. One view is that North Sea oil development has been unnecessarily delayed as a result of government intervention and regulation: that if the government had been less restrictive in terms of its leasing requirements and more willing to expedite onshore planning decisions, oil could have been flowing earlier and in sufficient quantity to ensure that the United Kingdom would by now be a major exporter of oil.

In this assessment there is widespread criticism of the British National Oil Corporation and frustration at the delays experienced in obtaining planning permission for onshore support facilities. This view is exemplified in a 1975 editorial in the *Oil and Gas Journal*. Criticizing the rules and regulations that make offshore regions less attractive for private companies, the editorial noted that the "preoccupation [in the United Kingdom] with increasing the government's financial share in and control of operations and development is casting a pall over North Sea prospects. Government interference and rapidly rising costs are making some very promising fields economically marginal and are proving a disincentive for further exploration in the U.K. area" (*Oil and Gas Journal*, vol. 73, no. 48, December 1, 1975). A more recent editorial complains of the favoritism shown by the U.K. government. "The government is giving the national company [BNOC] its choice of tracts outside licensing, 51% of new licenses issued to others . . . and a monitoring role which merely slows down action by private operators. No wonder one North Sea licensee was moved to describe BNOC as an albatross that is contributing nothing to the British economy" (*Oil and Gas Journal*, vol. 76, no. 19, May 8, 1978). From this perspective there is still strong criticism of the restrictive nature of central government offshore policies (particularly as these relate to the tax regime and to depletion controls) despite the election of a Conservative government in 1979 committed to encouraging private sector interests in North Sea development.

> It would appear that the dampening impact on development [of the new supplementary petroleum duty introduced in the 1981 budget and the proposed tightening of petroleum revenue tax reliefs] is seen by the Government as a happy by-product contributing to smoothing out the UK production plateau towards 1990. The danger is that the long-term effect will be a smothering of production potential through the 1990s. . . . While the tax changes are not having an immediate impact on exploration, it is likely that 1982 exploration budgets will be smaller, and plans for deeper wells in more hostile waters such as north and west of Shetland will be dropped. The industry

> view is that UK cannot afford this, as its only hope for large finds in the future are in these virgin territories. (Steven, 1981, p. 111)

In marked contrast is the view that North Sea development has proceeded too rapidly, that the central government in Whitehall, prompted by the immediate balance-of-payments crisis, promoted the exploitation of North Sea oil without sufficient attention to the social and environmental costs of rapid development. From this perspective national planners are widely criticized for their insensitivity to the concerns and priorities of impacted communities and for their willingness to accept longer-term costs in the interest of short-term economic gains. Other commentators question whether the benefits of oil development are being utilized and distributed in such a way as to create a permanent basis for stability and prosperity in Scotland's post-oil era. John Prebbel, a historian whose research has covered the destruction of the Scottish clan system after Culloden and the massive depopulation that accompanied the "Highland Clearances," offers a particularly pessimistic appraisal. "Oil is just a Band-Aid. While it is, of course, encouraging that people are going back to the Highlands, it is surely a short-term thing that will not last more than about 25 years at the most. After that, it will revert to being a destitute area. I doubt whether oil is a solution to the long-term problem of depopulation" (quoted in the *Times* [London], Friday, October 6, 1978).

A third interpretation is that North Sea development has proceeded at about the right pace, balancing the national need to achieve energy self-sufficiency with the equally pressing need to protect the environment and minimize community disruption, that the planning strategies implemented have been effective in reconciling social and environmental objectives with offshore development. From this perspective, the planning process is viewed as having been responsive to the needs and priorities of both the nation and locally impacted communities. This view is reflected in a Scottish Development Department paper (1975a, p. 1) which concludes that "the planning system in Scotland has been able to cope remarkably well with the extraordinary pressures that have arisen from the oil developments. In general terms it does appear that the procedures for forward planning, development control and infrastructure provisions have proved to be adaptable to most of the new demands."

These, it must be emphasized, are all highly generalized statements that treat the development process as a single, monolithic entity. Moreover, it is necessary to recognize the highly subjective nature of any verdict on the relative "success" of British planning. In this respect, any value judgment as to whether offshore oil policies and planning strategies have been "good" or "bad," "successful" or "unsuccessful," is likely to incorporate a subjective element.

It should also be stressed that the way in which the costs and benefits of oil development are distributed (or at least their perceived distribution) may well alter both individual and community attitudes. As oil and gas move ashore in large quantities, many communities now confront the consequences of early planning decisions. The economic benefits of North Sea oil have already been translated at the national level into a resurgent pound and an improved balance-of-payments situation and at the community level into increased employment opportunities. Yet the disamenities—the visual intrusion of support facilities, the cost of maintaining community infrastructure, the presence of larger-than-anticipated construction forces, the risks associated with handling liquefied natural gas, the incidence of oil spills in previously unpolluted waters—are now facts of life and can no longer be discounted. Moreover, there is a growing recognition that many of the initial planning decisions, made on a pro forma basis, have constrained subsequent options. Thus, decisions about supply bases and pipeline landfalls have often led inevitably to proposals for downstream refining and processing facilities that raise more complex and contentious issues of safety and acceptable risk.

At Nigg Bay, the site of the platform fabrication yard and controversial oil terminal and refinery discussed in Chapter 4, uncertainty over the timing and level of demand for production platforms undoubtedly solidified community support for the establishment of additional oil-related development. A major concern of the local authority was to justify the high level of social capital already invested in the Cromarty Firth area. In this situation, the force of environmentalists' arguments against transporting oil from the nearby Beatrice field to the Cromarty Firth was considerably diminished. Elsewhere local support for oil development has declined. Early in 1978, Bourne suggested that "the Shetlanders are becoming accustomed to their role as partners in the development of the Sullom Voe base serving the northern oilfields, and few grumbles are now heard" (Bourne, 1978, p. 29). This evolving partnership, derived from the Shetlanders' determination to have oil only on their own terms, now appears threatened by repeated minor spills at the terminal and particularly by the failure of the much-publicized pollution control measures to cope with the spill of fuel oil from the *Esso Bernica*.

The spills have articulated the continued risk that the community must deal with in the years to come.

It is possible to detect a certain ambivalence in Scottish attitudes toward oil development. After a century or more of out-migration and economic neglect, it is hardly surprising that most communities have been happy enough to receive the benefits of oil-related development and expanded employment opportunities (J. K. Mitchell, 1976, p. 396). Opposition to the location of onshore support facilities has tended to come from conservation groups and from the so-called "white settler" community. Bourne (1978, p. 30) suggests that as the rising tide of prosperity engulfs the countryside, even the conservationists are ceasing to complain.

In these circumstances, "it seems safe to say that no sense of local outrage or deep-seated discontent will lead to any immediate local challenge to the oil decisions of the Scottish office or of Whitehall" (Baldwin and Baldwin, 1975, p. 143). Nevertheless, in numerous conversations an undercurrent of dissatisfaction can be detected that is perhaps related less to the presence of oil-related development per se than to the way in which the costs and benefits have been distributed and the manner in which decisions have been made.

The Quality of Planning

The early phases of offshore activity exposed major weaknesses in government planning at both the local and the national level. Despite the intensity of the exploration effort in the North Sea during the latter half of the 1960s, only the most cursory consideration appears to have been given to the likely onshore impacts of a major oil strike. It has been suggested that this complacency was due to the deceptively low level of onshore development during the early exploration phase of offshore activity and to the ease with which the gas fields off the East Anglian coast had been brought into production.

Nevertheless, planners were clearly ill prepared for the number, size, and complexity of the oil-related planning applications that swamped local authorities in the early 1970s. Lacking the necessary staff and expertise to adequately evaluate these applications, local planning authorities made many decisions on a pro forma, site-by-site basis. They lacked the time and perhaps the inclination to undertake detailed project appraisal, to obtain independent verification of the data provided by the applicant, to evaluate alternative sites, to coordinate land use policies with other planning authorities, and to assess the application in the broader context of long-term market forces and national needs. With few exceptions, long-term planning goals were compromised as planners reacted to the immediate problem of coping with the short-term demands of offshore developers (J. K. Mitchell, 1976, p. 396). The lesson, as James K. Mitchell observes, is that although offshore oil may appear to local communities to be "a heaven-sent solution to their problems, the ecological, economic, and aesthetic values of coastal areas counsel caution and comprehensive planning preparations. With such an outlook, we can reap the undoubted benefits of oil secure in the knowledge that they are not the fruits of a Pyrrhic victory" (1976, p. 397).

During this critical phase of offshore development, little guidance was forthcoming from the Scottish Development Department. Briefing papers on supply bases and pipeline landfalls appeared some time after they were really needed. The *Coastal Planning Guidelines*, appearing in final form in 1974, were in many respects a post-factum rationalization for what had already occurred rather than a framework for forward planning. Moreover, the stated policy of concentrating development in those areas of greatest economic need (highest priority being given to Scotland's Central Belt) with existing communications and infrastructure has not been forcefully implemented. Thus the most significant developments have been concentrated along the coastline of the Grampian and Highland regions, with comparatively fewer benefits flowing to the more depressed areas around Clydeside. To this extent, private industrial siting criteria have continued to dominate the location of oil-related facilities (J. K. Mitchell, 1976, p. 396).

Nevertheless the principle underlying the *Coastal Planning Guidelines* is a sensible one. By seeking to identify ahead of time preferred development zones where oil-related activity will not compromise coastal amenities or impose significant social costs, planners can minimize locational conflicts and expedite offshore development. Time is a crucial element in any developer's calculations about project feasibility. Knowledge that applications for planning permission within a designated zone will be more favorably reviewed would go some way toward compensating for the selection of what might be a less than optimum site on strictly economic criteria. From the public perspective, provided that there has been active involvement at an early stage in identifying the preferred development and conservation zones, many of the suspicions which presently surround the activities of oil developers would be allayed. The net result would be a more orderly pattern of development, without many of the present frustrations for conservationists and developers alike, in accordance with clearly defined and publicly supported guidelines.

The experiences of the last decade have undoubtedly contributed to a marked improvement in the quality of planning initiatives at both the local and national level. The *National Planning Guidelines*, for example, provide local authorities with a much improved information base with which to evaluate a wide range of downstream petrochemical plants. The utility of positive, forward planning has been demonstrated by several local authorities. Thus the Shetland approach of identifying the most appropriate area for oil-related development and directing all oil activity to that location is an example of accommodating development by ensuring that coastal amenities are only minimally impaired.

In a similar vein, the identification of suitable sites for a range of possible petrochemical developments, as undertaken in Grampian and Strathclyde, represents a sensible way of combining forward planning initiatives with environmental assessment. Such an approach can be applied equally well to the siting of supply bases, pipeline landfalls, storage farms, or refineries. Thus the initial appraisal eliminates unsuitable sites (on grounds of environmental amenities, social cost, safety, or infrastructure needs) without prejudging individual proposals, which would continue to be subject to a detailed impact statement.

In this context, everything in the Scottish experience reinforces the need for a formal appraisal of projected impacts. While there is clearly little likelihood that such an assessment will be statutorily required in the United Kingdom, the trend toward a more formal impact assessment routine and the use of independent consultants is likely to continue. Recent impact statements reveal a marked improvement in methodology and sophistication, yet the Scottish experience with project appraisal reinforces the need for caution with respect to interpretation in view of the difficulties inherent in ranking and weighting potential impacts.

The issue of the adequacy, timing, and quality of information available to both planners and communities is clearly a crucial one. As noted by John Francis and Norman Swan, it is unreasonable to expect planners to cope effectively with development applications when they "are obliged to make major assumptions concerning the rapidly developing and changing situation. They are required to work from the job specifications of the incoming contractors, through regional multiplier effects to define housing targets for the short-term view. The long-term parameters are even more uncertain" (1973, p. 103).

It must be recognized that in the absence of a firm stance by planning authorities, companies involved in offshore development are unlikely to give high priority to community needs.

> It is for citizens and governments at all levels . . . to protect community, environmental, and other public interests in the face of oil development. But the challenge of matching wits and forecasts with the oil industry is great. It requires sufficient money and expertise to ensure—as the Shetland Islands have done—that oil companies will meet citizens on the citizens' terms. This must be a cooperative effort with the industry if possible at all levels of government and must include a great amount of public involvement. (Baldwin and Baldwin, 1975, pp. 168–169)

The issue is a good deal more complex than a simple question of accurate and timely information about developers' intentions; the issue includes the way in which such information is obtained, the extent to which it is diffused, and the manner in which it is evaluated in the context of long-term community impacts.

The need to deal with oil companies from a position of strength is sometimes advanced as a justification for centralized decision-making. Yet there is little in the U.K. experience to support the view that centralized decision-making (at least in the context of land use planning) is necessarily better informed, more decisive, or more efficient than local decision-making. The underutilization and closure of platform fabrication yards—many in small, rural communities least well equipped to deal with either the social costs or the economic uncertainties that characterize such developments—are the most glaring examples of inept planning at the national level. Such criticisms, however, must be seen in the context of the very real difficulty of projecting the nature and scale of oil-related needs when economic circumstances and technological appraisals are subject to rapid change.

Whether, on the other hand, the Shetlanders' firm approach to oil development could have been replicated in other less independent or culturally homogenous communities is uncertain. Yet, in the light of the Shetlanders' experience, there would appear to be a strong case for strengthening the resources and supporting the judgments of local authorities. This attitude is well summarized by MacKay and Mackay:

> . . . much more executive power should be given to local planning authorities . . . once they are provided with reasonable advice on the probable outcomes of alternative policies, they

> should be free to decide in accordance with their interpretation of local interests. In doing so they should be able to insist on detailed information being produced by prospective developers which should reduce the number of speculative projects. Only when there is an evident and overriding case for the imposition of the national interest should central government intervene. (MacKay and Mackay, 1975, p. 149)

Oil and Political Devolution

As noted in Chapter 4, the early phases of North Sea activity coincided with a major restructuring of local government entities and planning responsibilities. The reforms were intended to reduce administrative delays in the preparation of development plans and the processing of planning applications, to emphasize positive forward planning rather than negative control of development, and to provide greater opportunity for public involvement in planning decisions. The guiding philosophy was to make central government less intrusive in planning decisions that were essentially a local concern and to assure that local responsibility in local matters should become a reality. The emphasis on political devolution and decentralization of planning powers was not entirely an exercise in democratic altruism but reflected the political realities of the moment and the growing strength, particularly in Scotland, of regional and nationalist sentiments and aspirations. As noted, these reforms (in terms of both the reorganization of administrative units and the flexibility enjoyed by regional planners) were a good deal more radical in Scotland than elsewhere. As a result, there might have appeared to be, as J. W. House suggests, "hopefully the basis upon which more comprehensive regional planning and even a measure of political devolution will develop during the next decade" (1977, p. 1).

In practice, it is clear that serious weaknesses remain, not least between the theory and the reality of land use planning. A coherent, fully integrated regional planning policy does not yet exist, nor has a satisfactory balance been achieved between national, regional, and local interests and priorities. Serious deficiencies and contradictions exist in the preparation and integration of development strategies and plans, in the continued parochialism of many authorities that inhibits effective dialogue, in the extent to which alternative courses of action are adequately appraised and evaluated, and in the level of public involvement.

Many of these weaknesses reflect the tensions existing within the planning hierarchy as a result of ill-defined functions and responsibilities. In such circumstances, planning goals are unlikely to be clearly formulated through rational debate involving all affected groups; rather there are likely to be multiple, ambiguous, and conflicting goals, differing between nation and region and from agency to agency, with actual planning measures evolving through the complex interaction of bureaucratic inertia, opportunism, and vested interests (O'Riordan, 1971, pp. 110–111).

The extent to which any real transfer of power has occurred is also a matter for speculation. As noted by Bourne, "a first attempt to appease the restless natives by inserting an extra tier in their local government which provided a good deal of employment has merely whetted their appetite for change" (1978, p. 29). Moreover, the fear has been expressed (most notably in the Shetlands) that the net effect of political devolution would be to substitute Edinburgh for Whitehall and even, in the long run, to curtail those planning and fiscal powers presently enjoyed by local authorities.

A key issue is the extent to which this slow move toward political devolution and greater sensitivity to regional feelings and concerns has been threatened by the overriding need (at least as perceived by the national government) to develop North Sea oil as rapidly as possible. Whereas other development issues (as for example evidenced in public inquiries over the routing of motorways) have increasingly been influenced by local sensitivities and concern for "quality of life" in impacted communities, energy development policies bear the imprint of Whitehall (and more particularly the Ministry of Energy). Baldwin and Baldwin note that the consequence of the pressure for rapid offshore development "is a limiting policy for all subsequent decisions, large and small" (1975, p. 169). Everything is subordinated to that single goal. "To pause for reflection, or to reject a proposal as less than the ideal, is a difficult political decision for local communities or even for the Scottish Secretary. It is clear that the marching orders come primarily from London" (ibid.).

The most intrusive central government actions have involved the exercise of ministerial discretionary powers to "call-in" planning decisions "involving the national interest" and the enactment by Parliament of special legislation exempting oil development from normal planning procedures. The Harbours Development (Scotland) Act of 1972 and the Offshore Petroleum Development (Scotland) Act of 1975 are symptomatic of the way in which energy development has resulted in increased involvement of the central government in land use decisions.

Equally disturbing for those Scottish communities and regions that must bear the full social and environmental impact of rapid North Sea oil development has been the way in which community concerns and apprehensions, even when legitimized by the findings of a public inquiry, have been so frequently overturned by the Secretary of State. In the context of the controversy over the NGL separation plant and ethane cracker at Mossmorran, for example, Bourne suggested that "the results of a public enquiry are being discounted before they are known on the grounds that the Secretary of State has already overruled those of two others into the location of Edinburgh airport and the Cromarty Firth oil refinery, and dealt with the decision at a third over the development of a platform construction site at Drumbuie by choosing a dubious alternative at Kishorn instead. It will be interesting to see what sort of explosion eventually occurs in Fife" (Bourne, 1978, p. 30). Such actions run counter to the often-expressed desire of the government for a more localized and truly participatory approach to policy formulation and decision-making.

Bourne's 1978 editorial in the *Marine Pollution Bulletin* catches the sense of uncertainty with regard to the U.K. experience in the development of North Sea oil. "Macbeth's question, 'Stands Scotland where it did?' could have received many complicated answers over the years, but rarely more so than at the moment" (Bourne, 1978, p. 29). Implicit in many of the issues now being debated, particularly in Scotland, is concern over the uneven distribution of the costs and benefits of rapid oil development. And while it is impossible to provide a neutral answer to the question of equity, the belief appears to be growing that too many of the benefits have accrued at the national level and that Scottish communities have been forced to bear a disproportionate share of the costs.

Notes

Introduction

1. Since the first federal sale of outer continental shelf (OCS) leasing rights in 1954, a total of 4,579 wells have been drilled and 254 production platforms installed on federally leased OCS land in the Gulf of Mexico (*Offshore*, vol. 39, 1979, p. 45). Ship Shoal 32 was still producing oil in 1979, by which time Gulf OCS fields had produced some 4 billion (i.e., 4×10^9) barrels of oil.
2. The present ability to explore for oil in deep-water locations is an indication of the enormous advance in offshore technology in the course of the 1970s. The record for deep-water drilling has been repeatedly broken, almost on a year-to-year basis. In 1979, for example, the drillship *Discoverer Seven Seas* achieved a new drilling record of 4,441 feet off the coast of Spain in March, and the following month completed an exploration well in 4,876 feet of water some 200 miles northeast of St. Johns, Newfoundland. Drilling technologies of course have always been far ahead of development and production capabilities. As of early 1981, the deepest offshore development was Shell's Cognac field, located in just over 1,000 feet of water off the Louisiana coastline, which was due to start production in 1982, although a decision was still pending on development plans for Exxon's Lena prospect, located in 1,200 feet of water immediately south of the Cognac field. There appears to be some consensus amongst offshore operators that production from depths of 2,000 feet will have become technically and financially viable by the mid-1980s. Much will depend upon the success of new platform concepts (such as Exxon's guyed tower, a flexible structure designed to move with the waves rather than rigidly resist wave stresses as in the case of conventional platforms) and subsea production systems (where the wellhead is placed on the ocean floor rather than on a platform) (Manners, 1980a).
3. See, for example, "The Public Speaks Again: A New Environmental Survey," *Resources*, no. 60, 1978, pp. 1–6.

1. The Development of North Sea Oil and Gas

1. In contrast to the traditional jack-up rigs which are essentially restricted to drilling in depths of less than 350 feet, semisubmersibles have been developed for drilling in deeper offshore waters such as those encountered in the northern portions of the North Sea. The jack-up rig is floated (or on occasion "piggy-backed") into position, its elevator legs lowered to the seabed, and the platform "jacked-up" to a safe level above the ocean surface. A semisubmersible rig, as its name implies, is partly submerged and anchored (rather than fixed) to the seabed. The rig is then maintained in position through use of a sophisticated acoustical system. Semisubmersibles may displace as much as 40,000 tons and cost more than $50 million to construct. Rigs of this type have drilled wildcats in water depths in excess of 1,000 feet, and a number have been designed to operate in depths to 3,500 feet. In some instances, semisubmersible rigs intended for use in deep-water environments maintain their drilling position through the use of propellors rather than a mooring system. In this respect they are similar to the drillships designed for exploration in the deepest offshore waters.
2. C. Robinson (1976, p. 15), for example, argued that even a very slight change in the global pattern of supply and demand, resulting from the emergence of a non-OPEC source with a production capacity of perhaps as much as six million barrels per day oil equivalent, would probably have a significant effect on oil prices. He argued that "producers' price expectations will change as they see non-OPEC energy supplies appearing on the market, even in comparatively small volume. Instead of expecting substantial future price increases, which have led them in the last two years to hold back output, they may well conclude that the outlook is for a rate of increase in prices below the going rate of interest and, as a consequence, the incentive to hold oil in the ground may disappear." By 1979, following the decline in Iranian output (and the loss of all exports in January and February), this scenario looked increasingly implausible. Spot market prices, particularly for light crude oil, rose sharply, triggering further price increases and surcharges by OPEC producers. As the new decade began, however, the combined effect of higher prices and a deepening recession had brought about a small but significant reduction in oil consumption, particularly in the United States. The long-term effect on OPEC pricing and production policies remains unclear at this time.
3. It should be noted that not all North Sea oil is used to meet U.K. energy demand. Imports and exports of different qualities of crude oil and some refined products are necessary to balance supply with refinery and downstream demands. In 1979, total disposals of North Sea crude (i.e., total production less changes in stocks held by producers) amounted to 77 million tonnes, of which 38 million tonnes were refined in the United Kingdom and the remainder exported. The major markets for

North Sea crude were the European Economic Community countries (22 million tonnes) and the United States (7.2 million tonnes). The remaining exports were to Canada, Denmark, Finland, Norway, and Sweden.

4. D. I. MacKay and G. A. Mackay, *The Political Economy of North Sea Oil*, p. 5. As the authors themselves point out, this figure does overstate the magnitude of the deficit. In particular, it omits such factors as the use of British tankers, investment of foreign oil companies in the United Kingdom, and the profits of British oil companies remitted to the United Kingdom—all of which would serve to reduce the net deficit in the nation's "oil account."
5. The corresponding figures for 1978 were 88 percent and 16 percent. This decline in the proportion of gas requirements derived from the U.K. continental shelf is attributed to an increase in imports from the Norwegian part of the Frigg gas field (Department of Energy, 1980, p. 21). Production increases from Frigg have permitted a planned reduction in the amount of gas taken from fields in the southern North Sea.
6. "*Dry gas*" is predominantly methane but can contain small amounts of other gases. The Frigg field, a "dry gas" field, for example, produces a small amount (equivalent to 2 percent of total output or ca. 100,000 tonnes per year) of natural gasoline. "*Wet gas*" is predominantly methane but will contain a significant amount of other gases (ethane, propane, and butane) and liquid hydrocarbons (recovered as natural gasoline) in varying proportions. When the term *natural gas* is used it usually refers to methane only. *Natural gas liquids* (NGL) are the hydrocarbon liquids obtained from wet gas, comprising ethane, propane, butane and natural gasoline and are also referred to as condensate.
7. Government approval in principle for a new gas-gathering network (the U.K. Gas Gathering Pipeline or GGP) was announced by David Howell, the Secretary of State for Energy, in July 1980. As originally conceived the network was to consist of a thirty-six-inch trunkline from the St. Fergus terminal to a T-junction close to the Phillips T block (the oil-bearing complex formed by the Thelma, Tiffany, and Toni fields). From there the spine of the network would run northward to Magnus with laterals to other fields in the East Shetland Basin and southward as far as Fulmar (see Figure 2). Under this proposal a total of fifteen fields would be integrated into the network, several of which have yet to receive final development approval. A complete range of gas from methane to condensate would be gathered for separation at the St. Fergus terminal. The cost of the 572-mile pipeline network (at 1980 prices) was put at £1,100 million. In December 1980 it was announced that the British National Oil Corporation (BNOC) would exercise its option under existing participation agreements to purchase 51 percent of the natural gas liquids. Those petrochemical companies seeking the gas as feedstock would be required to negotiate supply contracts with the state-owned oil company.

   It should be noted that the future of the GGP network still remains uncertain. There have been unanticipated delays in securing the necessary financing for the pipeline. According to *Offshore* (vol. 41, no. 7, 1981, p. 115), financial institutions have not come forward as anticipated, due to belief that recoverable reserves do not justify the required level of investment. At the same time several of the offshore operators who would be involved as joint users of the pipeline appear to be having second thoughts. Thus plans for the southernmost fifty-mile section between the Lomond condensate discovery and the Fulmar field appear to have been shelved. This appears to be related to hopes that the Fulmar area will ultimately justify a separate pipeline, but such a decision would also ease the immediate financing requirement for the Gas Gathering Pipeline. Similarly, Shell is reported to be reconsidering its support for the northern section of the pipeline in the belief that there will be sufficient surplus capacity in the FLAGS trunkline to transport gas from Magnus and other northerly fields. This could result in the GGP network terminating in the region of the Beryl field, considerably farther south than originally planned. Clearly, with these uncertainties there is little likelihood of being able to adhere to the original schedule of bringing gas ashore by the end of 1984.

8. More recently, in a reversal of the move offshore, exploration drilling by the British Gas Corporation in southern England has identified Britain's largest onshore oil field. The Wytch Farm field near Wareham, Dorset, is still small by North Sea standards but its discovery has implications for oil prospects in the English Channel.
9. Keith Chapman points out that earlier exploration efforts had concentrated upon younger geological formations (Cretaceous and Jurassic series) which had provided the reservoir rocks for the small onshore oilfields, whereas the Slochteren discovery was made in the Lower Permian Sandstone formation. "After the Slochteren discovery, the basal Permian became the primary target, with the more recent Bunter, Jurassic, Cretaceous and Tertiary acting as subsidiary objectives of progressively diminishing importance. . . . The three essential factors of the occurrence of gas deposits were recognised as a combination of underlying Carboniferous coal measures, which represented the probable source of the gas; an overlying impervious layer of salt to prevent its upward migration; and an intermediate reservoir rock of high porosity. It remained to prove the existence of this combination under the North Sea" (Chapman, 1976, p. 41).
10. This figure for the Southern Basin is based on the current estimate of proven recoverable reserves (405 billion cubic meters) plus gas already extracted (309 billion cubic meters) (Department of Energy, 1980, pp. 5, 33).
11. "Significant discovery," as defined by the Department of Energy, relates to flow rates achieved in well tests and is not necessarily an indicator of the commercial viability of the find.
12. The record in this area is in marked contrast to the success rate in the North Sea. Exploration drilling in the West Shetland Basin began in

1972; British Petroleum's strike followed a string of fifteen dry wells. Indeed the flurry of activity in 1977 leading to BP's strike was primarily a result of the licensee's obligations to drill on acreage due for surrender early in 1978 rather than any great expectation of discovery.

13. Current estimates indicate the oil in place may amount to four billion barrels but that the recovery rate is likely to be less than 20 percent, considerably below the 40 percent recovery rate achieved on most North Sea fields. For an excellent discussion of the technical challenge to be overcome, see Steven, 1979. Technical problems include the low permeability and the shallow, widespread nature of the oil-bearing structure. In these circumstances a single production platform would be inadequate (since even wells "deviated" outward would still tap only a small area), a multiplatform development would be too expensive, and a subsea production system would be too problematic in view of the low productivity of the reservoir. Quite apart from geological problems are the extremely adverse weather conditions. These include storms of greater intensity and frequency than encountered even in the North Sea as well as exposure to the full vigor of the Atlantic Ocean swell. "Gale force winds of force 8 and greater occur on average one day in four during the winter. Wind gusts of up to 120 mph and 100-foot waves are not uncommon" (Steven, 1979, p. 114). In 1978 BP abandoned its attempt to drill a third appraisal well when the wellhead was damaged in a severe storm.
14. The average rate of flaring in 1979 was approximately 18 million cubic meters a day, equivalent to around 7 percent of oil production or 14 percent of total gas supplies to the British Gas Corporation (Department of Energy, 1980, p. 11). This level of flaring should be considerably reduced as gas reinjection and collection schemes come into operation. In order to monitor progress in gas conservation, flaring consents were only granted for three months at a time during 1979. At Brent (where the ratio of gas to oil is particularly high) restrictions on gas flaring were further tightened in November 1979 following an economic analysis that indicated significant net benefits from conservation.
15. It should be noted that the rate of flow in many North Sea fields has exceeded expectations. The performance of the Argyll field, however, has been disappointing. Water levels have advanced rapidly in the Zechstein reservoir and the field is now projected to have a shorter life span than originally anticipated. The Buchan field has provided major headaches for British Petroleum including costly delays and difficulties in converting a semisubmersible drilling rig into a floating production platform and uncertainty over the ability of the Devonian sandstone reservoir formation to sustain production. *Offshore* (vol. 41, no. 2, 1981, p. 90) reported that "after struggling against adversity for so long to bring the field onstream, wells could start drying up almost immediately."
16. Under current development schedules, Argyll and Auk will be depleted in the early 1980s, while production from Beryl, Claymore, Forties, Montrose, Piper, and Thistle will have peaked by 1983.

17. These figures represent the official government estimates; the uncertainties involved in any such calculations are immense. Future exploration activity as well as recovery rates from fields already discovered will be highly sensitive to changes in prices, costs, and tax structure, and to improvements in offshore production technology. In these circumstances, it is hardly surprising that estimates of reserves are highly controversial. Peter R. Odell and K. Rosing in particular have seriously questioned official estimates of North Sea discoveries, suggesting that they should be increased by a factor of between two and three (see Odell and Rosing, 1974a and 1974b). Odell and Rosing's conclusions are essentially based on past experience of the "appreciation factor"—this is essentially the difference between the reserves as announced at the time of the discovery and the actual reserves as proven during operation when the reservoir characteristics are better understood. Others (see, for example, MacKay and Mackay, 1975, p. 62) take an intermediate position, arguing that reserves will be revised upward but not by the amount suggested by Odell and Rosing, since seismic surveying and reservoir estimation have improved significantly in recent years. While such disagreements may seem a somewhat academic exercise in futurology, they are crucial to the formation and implementation of any rational depletion policy.
18. The original Brae strike was made in 1975. The rate of flow in this first well was so encouraging that some analysts spoke in terms of another Forties discoveries with possible recoverable reserves of 250 million tonnes. Subsequent appraisal drilling revealed an extremely complex geological structure with reservoir characteristics varying markedly over quite small areas. Development plans were finally approved by the Secretary of State for Energy early in 1980 and current estimates are for recoverable reserves of around 36 million tonnes.
19. See Robinson and Morgan, 1978, pp. 40–72, 186–201, for development of a "surprise-free" projection of North Sea oil supplies up to the end of the century and excellent discussion of geologic, technological, economic, and political uncertainties.
20. In 1979–1980 direct government revenues attributable to North Sea oil and gas amounted to £2,230 million. For further discussion of the tax structure, see Chapter 2.
21. In contrast to the conventional fixed production platforms that rest on the sea floor, the tension leg platform (TLP) is basically a floating platform anchored to the seabed by clusters of heavy steel cables. The platform for the Hutton field is described as a six-column structure (modeled on the lines of a semisubmersible drilling rig) that will be attached to the seabed by sixteen vertical tension tubes, four at each corner. These will allow a degree of lateral motion but not the vertical motion associated with surface vessels (see *Oil and Gas Journal*, vol. 79, no. 27, 1981, p. 90).

2. Offshore Concessionary Terms

1. Such criticisms are easy to make in retrospect and are noticeably less strident in those local communitites that moved quickly and decisively as soon as it became apparent that they would be affected by offshore activities. Thus during the third round of production licensing in the U.K. sector, record bids were submitted for those blocks lying immediately to the east of the Shetlands, suggesting a considerable degree of optimism about the chances of striking oil. The local community moved immediately to establish a planning council and to obtain sweeping planning powers that were ultimately conferred under the Zetland County Council Act of 1974 (see Chapter 4).
2. The election of a Conservative government in 1979 clearly presaged significant changes in offshore licensing. In this respect the terms for the seventh round of production licensing (as announced in May 1980) contained a number of innovative features intended to offer greater encouragement to private sector investment in the North Sea. Licensing of approximately ninety blocks was anticipated, reflecting the Conservative government's concern that the rate of exploration drilling had declined to a level that seriously jeopardized the United Kingdom's continued self-sufficiency in oil in the second half of the 1980s.
3. As noted by Chapman, "with uncertain prospects of success and with an absence of detailed information on both sides . . . governments are in no position to direct company activity to one part of their offshore territory rather than to another" (1976, p. 88). Subsequently, however, a more restrictive approach to the issuance of licenses may be used to slow down activity or to redirect exploration. The dilemma confronting governments is to strike a balance between offering blocks in numbers, locations, and on terms sufficient to attract interest and promote exploration, and safeguarding their long-term interests in royalties, returns, and participation should the discovery of oil and gas exceed initial expectations.
4. The formal announcement of the third round of licensing, with details of the blocks on offer and the relevant financial terms, was made in the *Official Gazette* on September 23 and 26, 1969. However, production licenses were not actually granted until June 1970, sometime after the announcements of the Montrose and Ekofisk oil strikes.
5. The Department of Trade and Industry felt unable to provide the Committee of Public Accounts with reliable estimates of the costs, revenues, or profits associated with North Sea operations. The committee's estimates were that a large field, such as Forties, would provide a company with an £8-per-ton profit in 1975 (compared with a return to the government of £1 per ton), declining to a £4-per-ton profit in 1980 (when the government's "take" would be £5 per ton, of which £4 would be tax). As a result of double tax relief provisions, however, most companies were

not liable for any tax. As the Committee of Public Accounts concluded, "under the present arrangements the U.K. will not obtain either for the Exchequer or the balance of payments anything like the share of the 'take' of oil operations on the continental shelf that other countries are obtaining for oil operations within their territories" (1973, p. xxxii).

6. The Department of Trade and Industry estimate in 1972 was for an annual production rate of 75 million tons by 1980. However, as the Committee of Public Accounts noted (1973, pp. xvii–xix), the Forties field alone was believed to have a production potential of 20 million tons a year by 1980; the recent drilling program had been extremely successful, revealing the potentially productive East Shetland Basin; and independent estimates prepared for the department (International Management and Engineering Group, 1972) indicated a production range of 105 to 130 million tons for 1980. "The Department considered that the difference between their estimates and those in the IMEG Report . . . was that they took a more cautious view of the rate of build up. . . . The Accounting Officer, who wished to dismiss any suggestion that the Department were not being provided with adequate information or that they had any reason to play down the estimates, summed up by saying that the 75 million figure remained the best estimate of the middle of range but that he would hope that the results would be rather better than that and that the amounts would rise subsequently" (Committee of Public Accounts, 1973, p. xix).
7. Since the introduction of PRT, there have been three sources of state oil revenue: (1) *royalties* paid at a rate of 12.5 percent on gross revenues (essentially equivalent to the market price of the oil produced); (2) the *petroleum revenue tax,* assessed since 1980 at a rate of 70 percent on the post-royalty revenue of each field less operating costs and capital allowances; and (3) *corporation tax,* levied on gross revenues at a rate of 52 percent less royalty payments, operating costs, capital allowances, and PRT. In August 1978 the government announced that it would begin to exercise its option under the Petroleum and Submarine Pipe-lines Act of taking its royalty payments in the form of oil to be delivered to BNOC for marketing. In March 1981 the Conservative government announced a new supplementary tax (Supplementary Petroleum Duty) to be charged at a rate of 20 percent on gross revenue after a tax-free production allowance of 20,000 barrels per day. Effective January 1, 1981, the supplementary tax is to be reviewed after eighteen months.
8. In the course of the negotiations, the Department of Energy and BNOC signed participation agreements with a total of sixty-two companies. The final agreement (with Imperial Chemical Industries) was signed in July 1978. These agreements cover all proven or prospective commercial discoveries on pre–fifth round production licenses. As other discoveries are made or prove commercial under these production licenses, participation agreements will be negotiated with the companies involved.
9. The companies involved were perhaps even more successful in forcing the government to modify its North Sea policy than the buy-back pro-

vision alone would suggest. Various reports circulating at the time suggested that the confrontation between the oil companies and the Department of Energy and BNOC extended not only to the terms of participation but also to the selection criteria for the fifth round of licensing, with companies threatening to opt out unless the criteria were modified. Of particular concern was the government's stipulation that BNOC would have the option of deferring its share of the development costs for any field discovered under fifth round licenses. The government's subsequent decision not to pursue this option (i.e., that BNOC would contribute to development costs as incurred) was viewed in some quarters as a calculated step to persuade oil companies into signing agreements and participating in the fifth round of licensing. (See for example, B. Smith, 1976, pp. 55–57.)

10. *Oil and Gas Journal*, vol. 77, no. 12 (1979), pp. 48–49. In fact, as things stood in 1980, BNOC would have access to oil from four sources: (1) crude sold to BNOC at market prices under the participation agreements; (2) crude from fields where BNOC has an equity interest; (3) agreements with the British Gas Corporation and Burmah Oil to handle their North Sea entitlements; and (4) royalty payments in kind. Participation agreements, however, would account for at least 80 percent of BNOC's supply.
11. The British Gas Corporation retained its existing license interests.
12. Although BNOC had sunk successful wells in the Thistle field and had identified two other promising structures to the north of the Thistle field, these were all on licenses acquired as a result of the takeover of Burmah Oil's offshore interests. The discovery on block 30/17b was the first where BNOC had received the original license as operator. Although development plans for this field were never formally submitted to the Secretary of State for Energy, it would appear that BNOC had hoped to have the field in production by 1985. As already noted, the election of the Conservative government in 1979 resulted in a broad review of North Sea production policies, and in December 1980 it was announced that development of the Clyde field would be deferred for two years "to prolong higher levels of U.K. production to the end of the century." While represented as part of the government's depletion policy, the decision also conformed to the Treasury's desire to reduce future public sector borrowing requirements. Deferring development (BNOC's share of the development costs were estimated at around £250 million) will significantly increase the state corporations' earnings and contributions to the Treasury over the medium term.
13. The reference is to the fact that the strike occurred on half of a block relinquished by Philips Petroleum from a previous round.
14. For discussion of the background to government policies toward the depletion of North Sea oil and gas, see Robinson and Morgan, 1978, pp. 18–50.
15. Under the terms of the fifth and sixth rounds of offshore licensing, all awards were subject to two major conditions: (1) an agreement between the Department of Energy and the prospective licensees on an obliga-

tory program of exploration for each block, and (2) the conclusion by BNOC and each group of co-licensees, with the approval of the Secretary of State for Energy, of a joint operating agreement. Subsequent development plans must also be approved by the Department of Energy. Companies with U.K. continental shelf oil production are also required to consult with the Department of Energy on their plans for refining and disposing of their crude supply. "The objective in these consultations has been to ensure that the national interest is properly taken into account in the companies' plans" (Department of Energy, 1978, p. 23).

16. Consents to flare gas were issued for only limited periods of time—three months or less—as a way of monitoring companies' progress toward conserving gas, but renewals were usually routine. In the case of the Argyll and Auk fields, the Department of Energy agreed to long-term flaring consents, since the quantities of associated gas were too small to permit economic recovery.
17. With the increase in oil prices during 1979, the current projections are for an exportable surplus of up to one million barrels per day in 1985, the year of peak production from existing fields. According to Wood Mackenzie, the Edinburgh stockbrokers, full application of the Varley guarantees could reduce production by 300,000 to 400,000 barrels per day from 1983. "If Britain decided to postpone royalty oil (i.e., one barrel in every eight produced—12 1/2 percent—goes to the government as royalty in cash or kind) that would reduce the surplus more dramatically. It would keep the surplus (assuming 1.8 m b/d domestic consumption) to a maximum of 400,000 in 1985 and closer to 200,000 b/d in most other years. Would the Treasury be willing to forego revenue of £600 m a year at today's prices in 1985?" (*Economist*, May 17, 1980,p. 59).
18. Offshore operators argued that the latter change in particular would inhibit the development of marginal fields. According to Robinson and Morgan, "the two cardinal principles of PRT were that it should be levied on a field-by-field basis and that the normal definition of profits for tax purposes should not apply. So rigid was the Government's adherence to these principles that, when persuaded of the need to reduce the tax burden on so-called 'marginal' prospects, rather than relaxing either of its two principles it chose instead to accommodate such fields by granting a fixed tax-free allowance to participants related to production from each field and by placing a limit on the amount of PRT payable in any year from each field" (1978, p. 108).
19. One major change under the Conservative legislation was that the state corporation, BNOC, lost its exemption from the petroleum revenue tax. Any reluctance to enact the tax changes in full was presumably dispelled by the sharp increase in oil prices during the first half of 1979 following the disruption of supplies from Iran. In this respect the "disappointment" expressed by the oil companies is noteworthy. In arguing against the Labour government's proposals, the companies had claimed that changing the tax regime would be justified only if world oil prices rose sharply. In 1980 the government announced a further increase in

the basic rate of PRT from 60 percent to 70 percent, retroactive to the beginning of the year. In addition 15 percent of PRT liability was to be paid six months in advance.

The introduction of a new supplementary petroleum tax by the Conservative government in March 1981 (see note 7), combined with the announcement that the Chancellor of the Exchequer, Sir Geoffrey Howe, had asked the Inland Revenue to review existing PRT reliefs in the light of changed circumstances since the introduction of PRT in 1975, was a bitter disappointment to offshore operators. Companies were quick to point out that the imposition of a new tax and the prospects of future changes in PRT would make it more difficult to finance existing development programs and would result in a reexamination of future exploration and production schedules. An editorial in *Offshore* (vol. 41, no. 1, 1981, p. 49) noted that although the Conservative government had promised to create a new climate and to restore momentum and confidence to North Sea activity, it had "outdone itself" putting short-term financial expediency before potential long-term benefits.

20. The British National Oil Corporation's controversial discovery on block 30/17b closely followed the announcement of the proposed terms for the sixth round, which included the stipulation that BNOC would be the designated operator on six of the licensed blocks.
21. For discussion of the changing relationships between governments and companies in the Middle East during the 1960s, and particularly the role of the independents in breaking the traditional concession arrangements by offering more favorable terms with respect to profit-sharing and state participation, see Odell, 1968.
22. In one particularly controversial decision Conoco awarded a $70 million contract to McDermott of Ardersier to build the production platform for the Murchison field. It was later claimed that the French firm Union Industrielle et d'Enterprises had underbid McDermott by $26 million. "Under these circumstances, and with the threat of [an] inquiry by the European Economic Community, it may be small wonder that Texaco U.K. Ltd. last month announced its $42 million Tartan field jacket as a joint venture between British Steel's subsidiary, Redpath Dorman Long, and France's UIE" (*Offshore*, vol. 37, no. 12, 1977, p. 55).
23. In addition BNOC was reportedly offered more than the 51 percent minimum share in several of the more promising blocks. An article in *Offshore*, however, suggests that the Department of Energy figures are misleading in that there were thirty fewer companies involved in the sixth round than in the fifth round. Moreover "two factors seem likely in explanation of the surprising 100% application. Firstly the state companies BNOC and BGC have bid for many, if not all, of the blocks. Secondly, some companies may be using the tactic of offering to take unattractive blocks to gain consideration for more attractive acreage" (*Offshore*, vol. 39, no. 1, 1979, p. 81).
24. According to UKOOA estimates, U.K. production from proven fields would have declined to 400,000 barrels per day by 1985 against a pro-

jected need of 2.5 million barrels per day. To make up the deficit, UKOOA calculated that twelve to nineteen rigs would need to drill sixty to ninety-five wildcats each year. In addition it would be necessary to sustain an appraisal and development program sufficient to bring on stream between eighteen and thirty-two new fields during the next decade. This could only be accomplished with the cooperation of the international oil industry, since the government lacked the expertise and the risk capital to sustain such an effort (see *Offshore*, vol. 39, no. 1, 1979, p. 41).

25. A recent World Bank study draws particular attention to the difficulties many countries have had in promoting offshore exploration and development when they have been dependent upon foreign operators for technology and risk capital (see World Bank Group, 1979).
26. In December 1980 it was announced that 125 applications had been received for production licenses from consortia representing 204 companies. A total of 42 production licenses were to be allocated under the company-nominated procedure, including 3 blocks to consortia with the British National Oil Corporation as the designated operator. The award of 37 additional production licenses for blocks designated by the Secretary of State for Energy were announced in March 1981. The new round of awards remains conditional on each of the licensees agreeing to offer BNOC the option to take up to 51 percent of the petroleum produced at market prices.

3. Offshore Oil: The Risk of Marine Pollution

1. See, for example, E. P. Danenberger's analysis of OCS activity in the Gulf of Mexico. "No spill in excess of fifty barrels has been recorded during exploratory drilling either on the Federal OCS or, to our knowledge, in any other offshore area throughout the world. Nevertheless, exploratory drilling is often labelled as environmentally the most hazardous aspect of offshore operations. The record does not support such contentions" (Danenberger, 1976, pp. 9–10).
2. In this particular instance, offshore operators and the U.S. Department of the Interior found themselves opposed at various times by a powerful coalition consisting of the state of Massachusetts, the Conservation Law Foundation of New England, environmental groups, and fishing interests, primarily on the grounds that the environmental impact statement (required under the National Environmental Policy Act before any federal leasing sale may take place) had failed to adequately assess the threat to the Georges Bank fishing grounds that would result from acute and chronic discharges from platforms, pipelines, and tankers. The courts ultimately ruled in favor of the Department of the Interior and the first round of leasing sales were completed early in 1980 when the department accepted high bids totaling $816.5 million for sixty-three offshore tracts.

3. The U.S. data base includes information on outer continental shelf activity (i.e., the production and transportation of offshore oil and gas) compiled by the U.S. Geological Survey and the nationwide Pollution Incident Reporting System (PIRS) maintained by the U.S. Coast Guard. Neither source is without its limitations, but together they represent the most comprehensive pollution incident data base in the world.
4. The effluent from the oil-water separators mounted on production platforms contains small amounts of oil. For this reason offshore operators in the U.K. sector of the North Sea may be exempted from complying with the Prevention of Oil Pollution Act and may discharge effluent containing up to 100 parts per million (ppm) crude oil, although they must maintain the overall content of oil discharges below an average of 50 ppm. According to the Department of Energy, a total of 160 tonnes of oil was permitted to be discharged in this way during 1979 (Department of Energy, 1980, p. 17). This represents an exceptionally small amount when distributed between several platforms in the open ocean. However this type of continuous input may become significant in shallower, more confined coastal waters where conditions do not favor dispersal of the effluent or where the water is already a receiving body for other chronic inputs.
5. While avoiding the issue of whether or not such effects are "significant," marine pollution has been defined by the United Nations as "the introduction by man, directly or indirectly, of substances or energy into the marine environment (including estuaries) resulting in such deleterious effects as harm to living resources, hazards to human health, hindrance to marine activities including fishing, impairment of quality for use of seawater and reduction of amenities" (United Nations, 1972, p. 1).
6. Molecular weight is simply the weight of a molecule in a substance expressed in atomic mass units (amu). A molecule of water, for example, weighs 18.015 amu, i.e., 2 atoms of hydrogen each with an atomic weight of 1.008 amu plus 1 atom of oxygen weighing 15.999 amu ($[2 \times 1.008] + [1 \times 15.999]$). These represent average values. The structure of hydrocarbons (and hence molecular weight range) varies from methane (which is composed of 1 carbon atom and 4 hydrogen atoms) to hydrocarbons containing more than 60 carbon atoms and 120 hydrogen atoms (Ryan, 1977; Farrington, 1977).
7. On the basis of molecular type there are three main categories of hydrocarbons found in crude oil: (1) the *alkanes* (paraffins) such as methane, ethane, propane, and butane; (2) the *cycloalkanes* (naphthenes), and (3) the *aromatics* (compounds whose structure contains the benzene ring). Some of the polynuclear aromatic compounds and their metabolites are known to be potential carcinogens (Ryan, 1977). Other classes of hydrocarbon compounds (for example the *alkenes* (olefins) are usually present only in small amounts in crude oil but may be produced in the refining process.
8. Vaporization takes place at different temperatures for different hydrocar-

bons. A fraction refers to that portion of the crude oil that vaporizes between particular temperature limits. In general fractions vaporize in this sequence: dissolved methane, ethane, and other natural gases, gasoline, benzene, naptha, kerosene, diesel fuel, heavy heating oil, tar (Ryan, 1977).

9. As noted by Farrington (1977, p. 8), it is frequently stated that the *Florida* spill was unique. And indeed it would be exceedingly unlikely that the external circumstances of any spill would be exactly duplicated. However, a second spill of No. 2 fuel oil occurred in Buzzards Bay in 1974 following the grounding of the barge *Bouchard 65*. The spilled oil again penetrated the coastal marsh sediments with results (in terms of weathering and microbial degradation) that closely paralleled the *Florida* spill. "These two spills are an example of chronic spillage of oil into coastal areas. During the winter of 1977, the same barge—the *Bouchard 65*—again spilled Number 2 fuel oil into Buzzards Bay. . . . A critical question is whether or not areas receiving repeated small spillages will be able to cleanse themselves of oil prior to the next spill, and the next" (Farrington, 1977, p. 8).
10. The most widely publicized description of tar balls in the open oceans followed Thor Heyerdahl's *Ra* expedition. Heyerdahl reported that on occasions the *Ra* had sailed for several hours through waters in the mid-Atlantic covered with lumps of tar and asphalt (Heyerdahl, 1971).
11. According to NAS estimates, the total amount of petroleum floating on the surface of the ocean in the form of lumps and tar balls is on the order of a year's input. Clearly there will be a strong correlation between levels of pelagic tar and the major tanker routes. In enclosed seas such as the Mediterranean where there is less chance for pelagic tar to be degraded naturally on the ocean surface, it has been estimated that as much as 30 percent of the oil spilled will become stranded on beaches (National Academy of Sciences, 1975, pp. 53–57). "Modification of tanker practices to emulsify wastes before dumping would decrease the amount of pelagic tar but would increase amounts in the subsurface water column. Whether degradation would be enhanced or would result in sublethal effects on the ocean ecosystem is open to question" (ibid., p. 55).
12. Several authors have drawn attention to the dearth of information on the likely behavior of oil spilled in ice-covered Alaskan waters. One study postulated that any oil spilled during offshore drilling (for example in the Beaufort Sea) would spread under the ice, where it would be trapped within underwater pockets. This oil would remain largely unweathered, coming to the surface in the following summer if under first-year ice or in subsequent summers if the cover consisted of multiyear ice floes (see Harrald, Boyd, and Bates, 1977, for discussion of these studies).
13. On the basis of field experiments involving ocean spills of crude oil, McAuliffe (1977b) concludes that solution is a minor process when compared with evaporation even for aromatic hydrocarbons. McAuliffe's data

confirm previous predictions and laboratory evidence that the ratio of evaporation to solution may be 100 to 1 for aromatic hydrocarbons and as much as 10,000 to 1 for alkanes. Even those hydrocarbons that do dissolve in water tend to evaporate subsequently.

14. For discussion of the toxic effects of oil on a wide range of organisms, see Nelson-Smith, 1977, pp. 54–57. Physical coating affects flora and fauna in various ways. Temperatures in tide pools and across coated beach and rock surfaces may rise rapidly due to increased absorption of solar energy with fatal effects on flora; in other cases the coating of oil may effectively prevent light penetration. In the case of seabirds the microstructure and waxy covering of the feathers provide insulation and bouyancy. Oil contamination permits water and cold to penetrate the plumage, while thicker residues weigh a bird down. General stress plus efforts to clean leading to ingestion of the oil may well accelerate death. Marine mammals (such as seals, otters, sea lions) will be affected in much the same way as birds when oil penetrates the fur.
15. According to John M. Teal (1977) there remain considerable gaps in our understanding of the processes whereby hydrocarbons are taken up by marine organisms and transferred through the food web. However, earlier fears that petroleum hydrocarbons might be concentrated along the food chain (biomagnification) appear to have been unfounded as far as aquatic organisms are concerned.
16. There is a tendency to assume that *all* biological consequences of an oil spill are negative, but this need not be the case. In the process of recolonization following a spill, for example, opportunistic species (such as polychaete worms) can take advantage of the disturbed condition and multiply very rapidly. For a brief period of time they will be the dominant species, providing a food bonus to other organisms (Michael, 1977, p. 133). Cowell, Cox, and Dunnet (1979, p. 5) note that in several well-documented cases commercial fish catches have actually increased in the years following a spill.
17. Calculating the direct and indirect costs of a spill is impossible. Many claims, for example in terms of lost revenue from tourism, are difficult to verify. The direct cost of clean-up operations following the Santa Barbara blow-out is estimated to have been in excess of $10 million (Organization for Economic Co-operation and Development, 1977a, p. 25). According to one estimate the direct cost of cleaning up a North Sea spill of a quarter of a million barrels of crude oil might approach $20 million, a figure that does not include the capital costs of equipment and warehousing (ibid.). The Council on Environmental Quality reported in 1978 that the costs of cleaning up the *Amoco Cadiz* spill were in the order of $50 million. Clean-up operations along the Texas coastline following the *Ixtoc 1* blow-out are estimated to have cost the U.S. government $6 million. In this incident, damage claims filed in the U.S. District Court so far total $365 million, including three class action suits brought by shrimpers, fishers, and property owners and a claim by

the state of Texas for damages to natural resources, loss of tax revenues, clean-up costs, and civil penalties. (See Texas House of Representatives, 1980.)

18. Succession involves the replacement of one community by another. It is reasonably directional and therefore predictable. Succession results from modification of the physical environment by one community for colonization by a new community; i.e., succession is community controlled although the physical environment determines the rate and pattern of change and often sets limits on how far development will proceed. Succession culminates in a stabilized ecosystem in which maximum biomass and symbiotic functions between organisms are maintained per unit of available energy flow (Odum, 1971, p. 251).
19. Whether an ecosystem ever *fully* recovers and the time frame for recovery are controversial issues. In an exchange of letters in *Science*, for example, T. S. Wyman's claim (1978, p. 1218) that areas affected by the *Torrey Canyon* and Santa Barbara spills "fully recovered in a surprisingly short time" was disputed by John H. Vandermeulen et al. (1978, p. 7), who argued that physiological and community disruption persisted for at least a decade. Moreover, Louis G. Williams (1977, p. 636) suggested that some of the most productive areas of eelgrass and clams along the North Carolina coastline still had not recovered from the effect of World War II spills resulting from the sinking of oil tankers in coastal waters by German submarines.
20. For futher discussion see Mix et al., 1977, pp. 421–431. What is most striking perhaps in this review is the considerable ignorance that exists with respect to sources, possible uptake through the food chain, incorporation into shellfish tissues, and metabolic rates—all vital to evaluating the public health risk.
21. Discussion of environmental decision-making in terms of the actual choice as contrasted with the theoretical range of choices owes a great deal to the work of the geographer Gilbert F. White (1961, 1963, 1969). White's model of decision-making involves appraisal of the quality and quantity of the resource base, as well as consideration of the level of technology available, prospective returns from alternative courses of action, the impact of resource use on functionally related areas, and the social-political norms that guide behavior. The most widespread application of this model has been in the area of natural hazards research. This research has addressed a general problem defined by White as "How does man adjust to risk and uncertainty in natural systems, and what does understanding of that process imply for public policy?" (1974, p. 194). More specifically, hazards researchers have addressed five major issues: (1) the extent of human occupancy of hazard-prone areas; (2) the range of possible adjustments to extreme events; (3) the way in which individuals and communities perceive and appraise the extent of the hazard; (4) the process by which specific damage-reducing adjustments are selected; and (5) the need for policies that ensure selection of the optimal set of adjustments (J. K. Mitchell, 1974, pp. 312–313).

These issues have been primarily investigated in the context of natural phenomena. More recently the hazards paradigm has been extended to other forms of environmental stress, notably those arising from technological hazards (Kates, 1977). As noted by Roger E. Kasperson, "The continuing widespread adverse side effects of technological change, in the face of new control institutions and mechanisms, raise a number of basic questions over society's management of technological hazards. Is technological growth producing a flood of hazards which threaten to outstrip society's capability to respond? Are the hazards themselves becoming more resistant to solution? Is the widespread concern over these hazards a product of managerial failure or the result of increased societal awareness and changes in public values? Are we failing to identify hazards, or do managerial breakdowns lie in the decision-making process?" (1977, p. 52). The evolution of hazards research reflects Robert W. Kates' contention that attempts to survive and prosper under conditions of environmental uncertainty are "symptomatic of a variety of resource problems accompanying the increase of man's numbers and the spread of his works" (1972, p. 520). Whether natural hazards research will aid, as its proponents suggest, in developing general theories of human-environmental relations remains unclear. However an approach that emphasizes the interaction of society and nature, based upon changes in the natural system and responses in the human system, is particularly helpful in clarifying both the nature of marine oil pollution and the ways in which society evaluates and deals with the problem.

22. Even efforts to reduce the flow of oil are likely to provoke controversy. Following the collision of the *Eleni V* off the Norfolk coast on May 6, 1978, discussions involving the Department of Trade and local authorities as to how best to dispose of the fuel oil remaining in the hulk were protracted and acrimonious. Ultimately the decision was made to tow the wreck to another location for sinking; however, the first two disposal sites selected proved to be unsuitable (because of soft sands and strong tides). The wreck was finally towed out to sea on May 30 and sunk with explosives off Lowestoft.
23. The limitations of existing recovery equipment with respect to fuel oil were evident in both the *Eleni V* and the *Esso Bernica* spills. Such oils tend to be highly viscous, with pour point temperature that is usually at or just above the temperature of the North Sea; as a result heavy fuel oils tend to quickly congeal. At Sullom Voe there were no pumps available capable of pumping spilled oil from the *Esso Bernica* after it became viscous at low temperatures (Bourne, 1979, p. 93). In the *Eleni V* incident, towing equipment, booms, and breaker bars (designed to agitate the slick to increase the effectiveness of chemical dispersants) were badly damaged following contact with the slick (Select Committee on Science and Technology, 1978, p. 123).
24. The reference is to the U.K. government's submission to the sixth session of the Marine Environment Protection Committee (MEPC) of the Intergovernmental Maritime Consultative Organization (IMCO). Similar

views about the desirability of avoiding the use of chemical dispersants have been expressed by the U.K. Department of the Environment (1974c). However this has yet to be matched by any real commitment, in terms of support for research and development, to enhanced recovery techniques. In 1979, for example, the government announced that as an economy measure there would be a "phased run-down of personnel" at the Warren Springs Laboratory "to a degree which will permit some continuing work on oil pollution and allow the development of chemical pollution clean-up techniques" (*Hansard*, December 11, 1979, pp. 556–557).

25. This of course applies only to oil deposited from one isolated incident. Chronic pollution involving repeated or continuous applications of oil to the coastline requires a very different response.
26. Cowell draws attention to examples of local authority contingency plans that have operated successfully. At Milford Haven, the second largest oil terminal in the U.K., "should oil reach the shore, then cleaning is carried out using the minimal amount of dispersant and in consultation with conservation societies. The system shows just how well the apparently conflicting interests of the oil industry, the harbour authority, local authorities and biologists can be coordinated. Only amenity beaches are automatically cleaned, some beaches only after consultation with biologists while others are not cleaned at all" (Cowell, 1976, p. 378).
27. By way of contrast it should be noted that a compensation program has been established for oil pollution damages caused as a result of offshore exploration and production. Under the terms of an international convention (Convention on Civil Liability for Oil Pollution Damage Resulting from Exploration for and Exploitation of Seabed Mineral Resources, as adopted in London in 1976) compensation will be payable by offshore companies up to a maximum of $45 million per incident. Operators are required to insure against the bulk of their liability, which will be on a "strict" basis, i.e., claims are not dependent upon proof of negligence or fault by the operator (Department of Energy, 1978, p. 16). The convention must be ratified by at least four signatory nations (United Kingdom, France, Belgium, Germany, Netherlands, Norway, Sweden, Denmark, and Ireland) before coming into effect. The Advisory Committee on Oil Pollution of the Sea (1977, p. 9) has expressed reservations over the legal difficulties involved in obtaining compensation under different conditions in different countries (claims may be filed in the country where the damage occurred or in that which licensed the installation), and concern that British policy on limited liability was decided on the basis of consultations restricted to representatives of the oil and insurance industries. Hitherto clean-up expenses incurred by local authorities in the United Kingdom as a result of spills from offshore installations (for example from the Beatrice field in 1976 and 1977) have been recovered under a voluntary industry agreement.
28. The references in the Secretary of State's letter are to the International

Convention on Civil Liability for Oil Pollution Damages (1969) and a 1971 supplement to that convention known as the Contract Regarding an Interim Supplement to Tanker Liability for Oil Pollution (CRISTAL). Under the 1969 convention tanker owners may limit their total liability to 210 million gold francs (equivalent to around $18 million at the time the convention was approved) except where damages occurred as a result of the owner's fault or negligence, in which case liability is unlimited. CRISTAL is a scheme set up by the oil industry to voluntarily supplement the 1969 convention. It provides for up to $36 million per incident for expenditures incurred in avoiding or minimizing oil pollution damages. Compensation is available to private parties as well as government entities and is not dependent on the tanker owner's fault or negligence. According to the U.K. Select Committee on Science and Technology (1978, p. 127) clean-up costs following the *Eleni V* incident might total some $6 million. It should be noted that an additional compensation fund, the International Oil Pollution Compensation Fund, has recently come into existence, covering incidents occurring after February 13, 1979. Like CRISTAL, the new fund is based on a supplement to the 1969 civil liability convention (the International Convention on the Establishment of an International Fund for Compensation for Oil Pollution Damage, 1971). As noted above, a shipowner's liability is limited under the 1969 convention and the new fund is intended to provide additional compensation for environmental losses incurred by signatory nations.

29. The interdepartmental group's report, entitled *Liability and Compensation for Marine Oil Pollution Damage,* was released in February 1979. Its stand against a national fund to cover costs or losses incurred as a result of pollution from unidentified sources was sharply criticized by local authorities. Disappointment was also expressed with regard to the report's view that any future system of compensation should be restricted to damages occurring within the three-mile territorial sea and that little could be done to speed up the processing of claims for compensation under existing programs. The Advisory Committee on Oil Pollution of the Sea described the tenor of the report as disquieting, contrasting its "moderate" proposals with the more imaginative and innovative approaches being pursued in other countries. Canada, Australia, and New Zealand all have established funds to cover clean-up costs and environmental losses regardless of whether the source can be identified. Similarly the "super fund" legislation enacted by the U.S. Congress broadens the scope of compensation, establishes time limits for settlement of claims, and allows for interim payments and payment of interest (ACOPS, 1979, p. 30).
30. Six months before the *Esso Bernica* spill at Sullom Voe the oil pollution officer and the Director of Ports and Harbours had advised the Shetland Islands Council that local resources and personnel were not sufficient to operate the terminal safely and efficiently. The council's chief executive dismissed the report as "opinionated and unbalanced," whereupon the

port director resigned. (For further details see Bourne, 1979.) The controversy however, also illustrates the problems confronting local authorities in acquiring trained personnel; restricted to local government salary scales, the council had been able to recruit only thirteen of the twenty pilots required to operate the harbor efficiently.

31. It should be noted that despite this requirement very little mechanical recovery equipment had actually been designed for use in the open sea and only prototypes were available (Norges Offentlige Utredninger, 1977, p. 53).

32. The 1954 Convention for the Prevention of Pollution of the Sea by Oil became international law in 1958 and has subsequently been amended three times, in 1962, 1969, and 1971. The 1969 amendments, which finally came into force in 1978, represented the first important effort to control operational discharges of oil from all vessels. The regulations are restricted in scope and extremely complex. Tankers, for example, may continue to discharge oily residues from cargo tank washing while en route provided that they are at least fifty miles from land, that the total discharge is limited to 1/15,000 of their cargo capacity, and that the oil content of the effluent does not exceed 100 parts per million (ppm). Other operational discharges from ordinary ships and tankers (including bilge waters) are permitted while proceeding en route, i.e., there is no distance from shoreline restriction, but vessels are restricted to discharge rates of less than sixty litres per mile and maximum effluent concentration of 100 ppm (Portmann, 1977).

33. The MARPOL convention has the unusual distinction of having been amended (in 1978) before coming into force. The 1978 amendments provided for stricter standards and controls on all maritime sources. For new tankers, permitted discharge levels from cargo tank washing will be reduced to 1/30,000 of total cargo capacity. In addition, tankers must be equipped to operate discharge monitoring and control systems. Retention facilities are to be required on all new tankers in excess of twenty thousand tons (MARPOL 1973 agreement affected only tankers of seventy thousand tons) and on existing carriers in excess of forty thousand tons (either segregated ballast tanks or crude oil washing system).

As the ACOPS *Annual Report* noted, such provisions "are to be welcomed as is the fact that the participating States agreed on a target date 1981 for the proposed entry into force of the two Protocols. States which do not comply with this [will be] invited to explain their intentions to the Secretary General of IMCO [Intergovernmental Maritime Consultative Organization] and this offers an important public relations exercise to accept target dates" (ACOPS, 1978, p. 31). Nevertheless, such progress was not achieved without a high price. In particular the 1978 protocol amending MARPOL includes less than desirable compromises with respect to those pollution control systems recommended for existing crude carriers, while Annex II (covering the discharge of chemical waste materials) was temporarily shelved. "The old cliche that the best may be the worst enemy of the good could well be true, but one cannot help concluding that MARPOL 1973 has now been not just cut down to size but rather had its wings clipped: it was widely believed that it was Annex II which made the convention so progressive" (Advisory Committee on Oil Pollution of the Sea, 1978, p. 32).

34. Crude oil washing is a recent practice whereby part of the crude cargo is recirculated during discharge, thereby dislodging some of the oil and sediment adhering to the inside of the tanks. The primary advantage is a more efficient recovery of crude residues, but the process also acts as a first step in tank cleaning. To be fully effective in this latter respect, additional tank washing with sea water is required, the oily waste being treated by LOT procedures en route or discharged to reception facilities onshore. Thus COW procedures have a rather different objective to segregated ballast tanks and as the 1978 protocol itself acknowledges, the best approach would be to combine the two systems. Moreover, as ACOPS has emphasized, the success of crude oil washing is highly dependent upon crew performance. "Past experience has taught us that most accidents are due to human error and that the load on top (LOT) system, much praised throughout the 1960s, largely failed to come up to expectations precisely because crew operations left much to be desired" (1978, p. 32).
35. Port state jurisdiction would allow a country to prosecute ships arriving at its ports or terminals for any pollution offense in violation of international law wherever committed. As UNCLOS III draws to a close it would appear that if the current Informal Composite Negotiating Text is incorporated into international law through treaty, convention, or customary usage, the powers of coastal states to detain and even prosecute vessels committing discharge violations within a twelve-mile territorial sea or a two-hundred-mile exclusive economic zone will be considerably increased. A port state will be able to investigate discharges which have taken place on the high seas only when a ship is voluntarily within one of its ports or terminals. Where a discharge occurs in the territorial water or economic zone of another state, a port state is able to take action only if so requested by that state or if the resultant spill enters the waters of the port state.
36. Under Scottish law an independent witness of the actual spillage is required if a prosecution is to be initiated. Additional difficulties arise in that a successful defense may be based on the concepts of "reasonable care" to prevent the discharge or "all reasonable steps" to stop the discharge as soon as possible. Neither meaning has been fully tested in the courts, but the RSPB report indicates that an action against Mobil was dismissed because a case for reasonable behavior was sustained.
37. Translation from Tamil by S.S. Iyer, in a letter to *Science*, vol. 185, no. 4149 (1974), p. 400, cited by Blumer (1975).
38. A greater concern with the environmental impact of offshore activity is evident in the proposals for the seventh round of offshore licensing announced in December 1979. At that time the Secretary of State for Energy invited any interested parties to express their views with respect

to those areas which might be licensed over the next few years and in particular those areas where they believed special care would be needed in the conduct of license activity for environmental, fishing, or other reasons. In a subsequent statement issued in May 1980, the Secretary of State indicated that he agreed with the views expressed to him "that the licensing of some areas, particularly in the English Channel and parts of the Moray Firth, is of major interest to the fishing industry and to organisations concerned with environmental matters, and that particular care and consideration are needed in the conduct of exploration and development in these areas" (Department of Energy, 1980, p. 49).

39. Although authorized under the OCS Lands Act amendments of 1978 (Public Law 95-372) as an addition to the Coastal Energy Impact Program, the OCS State Participation Grant Program (Section 503) has yet to be funded. The intention was to fund the program from revenues obtained through federal leasing.

4. Onshore Development: Land Use Planning in the United Kingdom

1. Although the impact of oil is most dramatically felt in the labor market, it should be noted that in Scotland oil-related employment (46,450 in 1978) accounted for only 2.2 percent of the total insured labor force. The regional impact was, of course, more pronounced, with the Grampian Region acquiring the largest share of oil jobs (more than 50 percent of all oil-related employment and 64 percent of employment in wholly involved companies). Between 1976 and 1978 only Grampian and Tayside experienced an increase in oil-related employment; all other regions lost jobs.
2. Natural gas liquids have become an increasingly important fuel and petrochemical feedstock source in the past decade. The risks associated with their use result from the fact that liquified gases require "a lower than normal temperature, or higher than normal pressure, if they are to be contained as liquids, which is the only economically feasible form for storage and transportation" (Fay, 1980, p. 89). Because of their low boiling points compared with ambient temperatures, if released accidentally into the environment the liquified gases quickly vaporize to the gaseous form. "They are then ripe for combustion by fire or, when further mixed with air, possibly by explosion. . . . two major accidents in the United States—a tank failure and fire in Cleveland in 1944 and an explosion and fire in an empty LNG tank in Staten Island in 1973, both involving substantial loss of life—have raised questions about the safety of LNG systems, which has affected both the rate of construction and the selection of sites for LNG terminals in the United States in recent years" (ibid., pp. 89–93).
3. The development at Mossmorran involves the use of associated petroleum gases from the Brent field, where Shell and Esso are the joint licensees. The gas will be piped to St. Fergus, where the methane will be

removed and sold to the British Gas Corporation. The remaining natural gas liquids (NGL) will be piped to Mossmorran for separation into ethane, propane, butane, and condensate. The propane and butane components will be liquefied for export by tanker from Braefoot Bay, while the ethane will provide the feedstock for an ethylene plant with a projected output of 500,000 tonnes per year.

4. Scotland has had a Secretary of State since 1885, although it is only since the late 1930s that the Scottish Office has acquired significant responsibilities. Within the Scottish Office, central government administrative functions are divided among five Scottish Departments based in Edinburgh. Town and country planning is the responsibility of the Scottish Development Department.
5. To some extent, the tension that exists between the Scottish Office and the Department of Energy is replicated within the Scottish Office, where two departments are directly involved in formulating and implementing oil-related development policy. The Scottish Economic Planning Department (SEPD), which is responsible for the formulation of regional economic development plans, advises the Secretary of State as to what will be required in terms of infrastructure and services if national objectives of rapid oil development and self-sufficiency by the early 1980s are to be realized. The SEPD is headed by a Minister of State who also chairs the Oil Development Council, an influential advisory group appointed by the secretary and consisting of prominent individuals with expertise and experience relating to oil development in Scotland. Its secretariat is provided by the SEPD. The Scottish Development Department (SDD) is responsible for social and environmental planning, for implementation of the Town and Country Planning (Scotland) Act, and for local government organization and finance.
6. Although local authorities were required to "have regard to the provisions of the development plan" when considering an application, the plan was not absolutely binding. Indeed local authorities were required to consider "any other material considerations" (see Cullingworth, 1976b, p. 78). As development plans became dated, it was almost inevitable that applications were approved or disapproved on the basis of factors not considered in the plan.
7. Cullingworth draws attention to a significant difference in the wording of the requirement for a survey in the new act. "Whereas the earlier Acts required a local authority to 'carry out' a survey, the new legislation refers to the duty to *institute* a survey. This change was deliberately designed to facilitate the employment of consultants, particularly by local planning authorities who were short of the necessary skilled staff" (1976b, p. 84). As is noted in a subsequent section, consultants have played an increasingly significant role in Scotland, where local authorities have felt ill equipped to cope with the number and complexity of oil-related planning applications.
8. Local plans are actually of three kinds. *District plans* are concerned with detailed planning for part of an area covered by a structure plan (for

example, a small town); *action area plans* deal with localities subject to "intensive change" as a result of improvement, redevelopment, or new development; *subject plans* deal with a particular issue (for example, mineral workings or housing).

9. The precise circumstances under which the regional authority may exercise its "call-in" power are defined in the act as: (1) when a local plan is urgently required to implement the provisions of an approved structure plan, and the district planning authority concerned has failed to adopt an appropriate local plan; or (2) when the district planning authority is likely to be affected by the local plan in question; or (3) when the local plan does not conform to a structure plan approved by the Secretary of State; or (4) when the implementation of the local plan will render unlikely the implementation of any other local plan relating to their district (Cullingworth, 1976b, p. 93).
10. In England the process was further complicated by the need to integrate structure plans with the regional economic strategy being formulated by Economic Planning Councils. Created in the mid-1960s, the eight economic planning regions extended across county boundaries. Although only consultative and advisory, "the Councils . . . have . . . come to represent effectively the regional interest upwards to central government [and] downwards to their constituent local authorities" (House, 1977, p. 27). Delays in strategic planning, however (particularly in the final stage of consultation between Whitehall, the Economic Planning Council, and the county and district authorities), meant that structure plans in England had either to be delayed or to be formulated independently and without the desired regional context (ibid., p. 43). This situation did not exist in Scotland, where responsibility for regional economic planning had been invested in the Scottish Office and where the subregions that had been identified for strategic planning corresponded more closely to the administrative entities that emerged in the local government reorganization.
11. In view of this requirement, local authorities may well ask an applicant to defer lodging a formal request for planning approval. In the case of Shell's application to construct a natural gas liquids separation plant in the Peterhead area, for example, the Banff and Buchan District Council asked the company not to advertise the application until further information was available on the implications of the proposal. As a result, although informal discussions between the council and Shell began in February 1975, it was not until early 1976 that the application was formally submitted.
12. It should, however, be noted that as oil-related developments have increased in numbers, complexity, and visibility, most Scottish local authorities have required developers to submit more detailed information before considering an application. In some instances, authorities have required developers to complete questionnaires; in other cases, developers have been requested to delay submitting a formal application (which statutorily must be acted upon within a specified time period) while the

information necessary to evaluate the application is gathered by the planning staff and/or consultants. Moreover, major planning applications now tend to be accompanied by considerable supporting material, including reports on environmental, social, and safety aspects.

13. The Cromarty planning application was submitted in December 1973. The site had already been the subject of a previous planning application in 1968 by Grampian Chemicals Limited for permission in principle for an oil tank farm at Nigg Point and a petrochemical complex across the bay near Invergordon. A public inquiry was held during 1969, but the Reporter's recommendations against the proposal were overruled by the Secretary of State. However, Grampian Chemicals did not proceed further with its application.
14. Areas designated as National Parks, Areas of Outstanding Natural Beauty (AONB), Nature Reserves, and Sites of Special Scientific Interest (SSSI) must be shown on development plans and considered in planning applications. The Countryside Commission is responsible for identifying areas suitable for designation as National Parks and AONB. The Nature Conservancy is the body responsible for overseeing Nature Reserves (which are actively managed) and for designating SSSI's (which are of less scientific importance than Nature Reserves). Local authorities are informed of these designations and are statutorily required to notify the Nature Conservancy of any proposed development affecting the site.
15. The first plans submitted by Mesa Petroleum, at that time the operator of the Beatrice field, involved a floating storage vessel at the field and the use of tankers to transport oil to the Nigg Point refinery. These plans were rejected as posing too great a risk of pollution in view of the field's proximity to the coastline and to major fishing grounds.
16. An interesting feature of the Ardyne Point site selection was that a hydrological survey subsequently had to be undertaken to ensure that completed platforms could be successfully navigated out of the Firth of Clyde (see Hutcheson and Hogg, 1975, p. 105).
17. Those areas included as preferred development zones partly on the basis of needing "environmental recovery and rejuvenation" are not identified. Nor does the report address itself to the intriguing question of which developments were thought to be capable of contributing to such a goal.
18. Constructed initially by French prisoners of war during the Napoleonic period, the Harbour of Refuge was greatly expanded in the late nineteenth century with the use of prison labor. Under an 1886 act, the outer harbor was designated as a shelter and protected from commercial development. Under the Harbours Development (Scotland) Act of 1972, the Secretary of State was empowered to undertake or to authorize others to undertake development of the Harbour of Refuge.
19. On the south side of the harbor, the ASC supply base is located on twenty-two acres of reclaimed land, the fill having been derived from offshore dredging. Additional construction has occurred on the north side of the bay to develop 4.5 acres for the BOC supply base. Further development

has occurred as a result of the decision to locate a 1,320-megawatt power station at Boddam, two miles south of Peterhead. "The North of Scotland Hydro-Electric Board has clearly been impressed by the harbour's capacity to take 35,000 ton oil tankers, and this is clearly a key factor in the decision to locate the £100 million development at the proposed site. This will, of course, necessitate an additional jetty in the south harbour to accommodate the large tankers visiting the port at a frequency of 6–7 days" (Francis and Swan, 1973, p. 35).

20. For an excellent review of the status of risk analysis for NGL and LPG (liquified petroleum gas) systems, see Fay, 1980.

21. At the time of the 1861 census, Shetland's population was 31,700. A long period of economic decline, broken only by a short-lived herring boom in the early years of the twentieth century, contributed to a persistent pattern of out-migration from the islands. The population in 1971, immediately prior to the discovery of oil, was 17,350 (Shetland Islands Council, 1976, vol. 1, p. 5).

22. David H. Rosen and Deborah Voorhees-Rosen have argued that "the change from an isolated, rural community to a more industrialized one, over a relatively short period of time, will adversely affect the mental health of the islanders and bring about an increase in such indices of social disorder as crime, divorce, and suicide. It is our hypothesis that industrialization and the consequent social changes will have a deleterious effect upon the islander's way of life and mental health that can be measured over time" (1978, pp. 49–50).

23. The major oil service bases are located at Lerwick (Norscot, Shell/BP, and Ocean Inchcape) and at Sandwick (Hudson's Offshore). As the Shetland Islands Council has noted, many bases (including those proposed at Basta Voe on Yell, Balta Sound on Unst, Ronas Voe on Northmavine, and several in Lerwick) have yet to materialize (Shetland Islands Council, 1976, vol. 1, p. 6).

24. Despite the financial resources available, formulating and implementing a suitable long-term development strategy presents a formidable challenge in view of the island's geographical remoteness and small indigenous market (McNicoll, 1976, 1977c).

25. There are four general types of assistance for states and local communities currently funded in the Coastal Energy Impact Program (CEIP). First, planning grants are intended to help state and local governments prepare for oncoming energy activity. They may be used, for example, to finance a community facilities inventory, to assess infrastructure needs, or to prepare a social impact statement. Second, loans and loan guarantees are available (to be repaid later from additional tax revenues and user fees generated by energy development) to upgrade a wide range of public facilities and services. Third, repayment assistance is provided for those local governments that cannot meet their loan or bond obligations because revenue from energy activity did not materialize as expected. Finally, grants are available to assist a state or local community

in preventing, reducing or ameliorating damage to valuable environmental and recreational resources (Manners, 1980b).

26. Under the Zetland County Council Act, the council has the power to license or refuse works within the sea or on the seabed (including pipelines) throughout a three-mile offshore zone.
27. The Shetland Islands Council initially set up a Sullom Voe Environmental Advisory Group to ensure that environmental considerations were taken into account in planning and operating the complex. This group was also to be responsible for establishing appropriate environmental monitoring procedures. Following charges that this group was unduly sympathetic to oil industry interests, a Shetland Oil Terminal Environmental Advisory Group (SOTEAG) was established, comprising representatives of the Shetland Islands Council, the oil industry, and a number of government departments and agencies, as well as several local islanders and independent university scientists. The responsibility of SOTEAG is to advise the Sullom Voe Association on the environmental implications of the development and operation of the terminal, to advise appropriate organizations as to potential threats to the Shetland environment, and to undertake monitoring studies to detect and assess environmental change. It is purely an advisory body and has no authority to implement its recommendations (*Marine Pollution Bulletin*, vol. 10, no. 4, 1979, p. 96).

Bibliography

Aalund, Leo R. 1979. "North Sea Crudes Are Examined." *Oil and Gas Journal*, vol. 77, no. 48, pp. 47–53.

Advisory Committee on Oil Pollution of the Sea (ACOPS). 1976. *Annual Report for 1975*. London.

———. 1977. *Annual Report for 1976*. London.

———. 1978. *Annual Report for 1977*. London.

———. 1979. *Annual Report for 1978*. London.

———. 1980. *Annual Report for 1979*. London.

Allison, L. 1975. *Environmental Planning: A Political and Philosophical Analysis*. London: George Allen and Unwin.

Ardill, J. 1974. *The New Citizen's Guide to Town and Country Planning*. Prepared for the Town and Country Planning Association. London: Charles Knight and Co.

Arnold, Guy. 1978. *Britain's Oil*. London: Hamish Hamilton.

Baldwin, Pamela L., and Malcolm F. Baldwin. 1975. *Onshore Planning for Offshore Oil: Lessons from Scotland*. Washington, D.C.: The Conservation Foundation.

Balogh, Lord. 1972. "A Socialist View of North Sea Developments." Paper presented at the Second North Sea Conference, Grosvenor House, London, December 12 and 13, 1972.

———. 1978. "Stormy Politics of the North Sea." *Observer*, July 9, p. 8.

Bates, C. C., and E. Pearson. 1975. "Influx of Petroleum Hydrocarbons into the Ocean." *Offshore Technology Conference Papers*, vol. 3, pp. 535–544.

Bedinger, C. A., Jr. 1979. "More Research Aimed at Oil Spill Woes." *Offshore*, vol. 39, no. 14, pp. 63–65.

Benn, Tony. 1978. "North Sea Investment Begins to Pay Off." *Offshore*, vol. 38, no. 7, pp. 80–81.

Blumer, Max. 1975. "Organic Compounds in Nature: Limits of Our Knowledge." *Angewandte Chemie* (International Edition), vol. 14, no. 8, pp. 507–514.

Blumer, Max, M. Ehrhardt, and J. H. Jones. 1973. "The Environmental Fate of Stranded Crude Oil." *Deep-Sea Research*, vol. 20, pp. 239–259.

Bourne, W. R. P. 1973. "Scotland, Development and Oil." *Marine Pollution Bulletin*, vol. 4, no. 8, pp. 113–114.

———. 1974. "Planning for Scottish Oil." *Marine Pollution Bulletin*, vol. 5, no. 6, pp. 81–82.

———. 1976. "Seabirds and Pollution." In *Marine Pollution*, edited by R. Johnston, pp. 403–504. London and New York: Academic Press.

———. 1978. "Where Scotland Stands." *Marine Pollution Bulletin*, vol. 9, no. 2, pp. 29–30.

———. 1979. "Sullom Voe Comes on Flow." *Marine Pollution Bulletin*, vol. 10, no. 4, pp. 93–94.

Cairns, W. J., and Associates. 1973. "Flotta, Orkney: Oil Handling Terminal."

Carter, Luther J. 1978. "*Amoco Cadiz* Incident Points Up the Elusive Goal of Tanker Safety." *Science*, vol. 200, no. 4341, pp. 514–516.

Catlow, J., and C. G. Thirthwall. 1976. *Environmental Impact Analysis*. Research Report No. 11. London: Department of Environment.

Central Office of Information. 1974. *British Industry Today: Energy*. Reference Pamphlet No. 124. London: Her Majesty's Stationery Office.

———. 1975. *Town and Country Planning in Britain*. Reference Pamphlet No. 9. London: Her Majesty's Stationery Office.

———. 1979. *Environmental Planning in Britain*. Reference Pamphlet No. 9. London: Her Majesty's Stationery Office.

Chaisson, Richard E., Lester B. Smith, Jr., and Jamie M. Fay. 1978. "Oil Spills and Offshore Drilling." *Science*, vol. 199, no. 4325, pp. 125–132.

Chapman, Keith. 1976. *North Sea Oil and Gas: A Geographical Perspective*. London: David and Charles.

Committee of Public Accounts (House of Commons). 1973. *North Sea Oil and Gas*. Cmnd. 122. First Report from the Committee of Public Accounts together with the proceedings of the Committee, Session 1972–1973. London: Her Majesty's Stationery Office.

Committee on Public Participation in Planning (Skeffington Report). 1969. *People and Planning*. London: Her Majesty's Stationery Office.

Commoner, Barry, 1971. *The Closing Circle*. New York: Knopf.

Conservation Foundation. 1977. *Source Book: Onshore Impacts of Outer Continental Shelf Oil and Gas Development*. Chicago: American Society of Planning Officials.

Cooper, B., and T. F. Gaskell. 1976. *The Adventure of North Sea Oil*. London: Heinemann.

Council on Environmental Quality. 1979. *Environmental Quality, 1978*. Ninth Annual Report of the Council. Washington, D.C.: U.S. Government Printing Office.

Cowell, Eric B. 1976. "Oil Pollution of the Sea." In *Marine Pollution*, edited by R. Johnston, pp. 353–402. New York: Academic Press.

Cowell, Eric B., Geraldine V. Cox, and George M. Dunnet. 1979. "Applications of Ecosystem Analysis to Oil Spill Impact." Paper presented to the 6th International Oil Spill Conference, Los Angeles, California, December 1979.

Cremer and Warner. 1976. "Report on the Environmental Impact of the Proposed Shell UK (Expro) NGL Plant at Peterhead." Prepared for Grampian Regional Council.

———. 1977a. "The Environmental Impact of the Natural Gas Terminal at St. Fergus." Report prepared for the Banff and Buchan District Council.

———. 1977b. "The Hazard and Environmental Impact of the Proposed Shell NGL Plant and Esso Ethylene Plant at Mossmorran, and Export Facilities at Braefoot Bay." Report prepared for Fife Regional Council, Dunfernline District Council, and Kirkaldy District Council.

———. 1978. "Guidelines for Layout and Safety Zones in Petrochemical Developments." Report prepared for the Highland Regional Council.

Cullingworth, J. B. 1976a. *Environmental Planning 1939–1969*. London: Her Majesty's Stationery Office.

———. 1976b. *Town and Country Planning in Britain*. Rev. 6th ed. London: George Allen and Unwin.

Dam, Kenneth W. 1976. *Oil Resources: Who Gets What How?* Chicago: University of Chicago Press.

Danenberger, E. P. 1976. *Oil Spills, 1971–75, Gulf of Mexico Outer Continental Shelf*. U.S. Geological Survey Circular No. 741. Washington, D.C.: U.S. Government Printing Office.

Dedera, Don. 1977. "The Disasters That Didn't." *Exxon*, vol. 16, no. 3, pp. 11–15.

Department of Energy. 1977a. *Energy Policy: A Consultative Document*. London: Her Majesty's Stationery Office.

———. 1977b. *Energy Policy Review*. Energy Paper No. 22. London: Her Majesty's Stationery Office.

———. 1977c. *Development of the Oil and Gas Resources of the United Kingdom, 1977*. London: Her Majesty's Stationery Office.

———. 1978. *Development of the Oil and Gas Resources of the United Kingdom, 1978*. London: Her Majesty's Stationery Office.

———. 1979. *Development of the Oil and Gas Resources of the United Kingdom, 1979*. London: Her Majesty's Stationery Office.

———. 1980. *Development of the Oil and Gas Resources of the United Kingdom, 1980*. London: Her Majesty's Stationery Office.

Department of the Environment. 1974a. *Local Government in England and Wales: A Guide to the New System*. London: Her Majesty's Stationery Office.

———. 1974b. *Structure Plans*. Circular 98/74. London: Her Majesty's Stationery Office.

———. 1974c. *Accidental Oil Pollution of the Sea*. Pollution Paper No. 8. London.

———. 1975. *Review of Development Control System*. Circular 96/75. London: Her Majesty's Stationery Office.

———. 1976a. *Assessment of Major Industrial Applications—A Manual*. Research Report No. 13. London: Her Majesty's Stationery Office.

———. 1976b. *The Separation of Oil from Water for North Sea Operations*. Pollution Paper No. 6. London: Her Majesty's Stationery Office.

———. 1978. *Accidental Oil Pollution of the Sea*. Pollution Paper No. 8. London: Her Majesty's Stationery Office.

Department of Trade and Industry. 1973. *North Sea Oil and Gas: A Report to Parliament*. London: Her Majesty's Stationery Office.

———, Warren Spring Laboratory. 1972. *Oil Pollution of the Sea and Shore: A Study of Remedial Measures*. London: Her Majesty's Stationery Office.

Economist Intelligence Unit Ltd. 1975. "Buchan Impact Study." Prepared for Aberdeen County Council and Scottish Development Department.

Exxon Chemical Company. 1980. *Oil Spill Chemicals Applications Guide*. 2d ed. Houston: Exxon Corp.

Farrington, John W. 1977. "The Biogeochemistry of Oil in the Ocean." *Oceanus*, vol. 20, no. 4, pp. 5–14.

Fay, James A. 1980. "Risks of LNG and LPG." *Annual Review of Energy*, vol. 5, pp. 89–105.

Francis, John, and Norman Swan. 1973. *Scotland in Turmoil: A Social and Environmental Assessment of the Impact of North Sea Oil and Gas on Communities in the North of Scotland*. Edinburgh: St. Andrew Press.

———. 1974. *Scotland's Pipedream: A Study of the Growth of Peterhead in Response to the Demands of the North Sea Oil and Gas Industry*. Edinburgh: Church of Scotland Home Board.

Fulleylove, R. J., and T. E. Lester. 1977. "Oil Spill Contingency Planning for the BP Forties Oilfield Production, Pipeline and Terminal Systems." *Proceedings*, 1977 Oil Spill Conference, March 8–10, New Orleans, pp. 87–90.

Gibson, David T. 1977. "Biodegradation of Aromatic Petroleum Hydrocarbons." In *Fate and Effects of Petroleum Hydrocarbons in Marine Ecosystems and Organisms*, edited by Douglas A. Wolfe, pp. 36–46. Oxford and New York: Pergamon Press.

Goldberg, Edward D. 1976. *The Health of the Oceans*. Paris: The Unesco Press.

Graham, Andrew. 1973. "Fuel Policy Sense and Nonsense." *New Statesman*, May 18, pp. 723–725.

Grampian Regional Council and Banff and Buchan District Council. 1980. *Contingency Plan for Petrochemical Industries*. Aberdeen.

Gregory, Roy. 1971. *The Price of Amenity*. London: Macmillan.

Grove-White, Robin. 1975. "The Framework of Law: Some Observations." In *The Politics of Physical Resources*, edited by Peter J. Smith, pp. 1–21. Harmondsworth, Middlesex: Penguin Books in association with the Open University Press.

Hamilton, Adrian. 1978. *North Sea Impact—Offshore Oil and the British Economy*. London: International Institute for Economic Research.

Hardin, Garret. 1968. "The Tragedy of the Commons." *Science*, vol. 162, no. 3859, pp. 1243–1248.

Harrald, J. R., B. D. Boyd, and C. C. Bates. 1977. "Oil Spills in the Alaskan Coastal Zone: The Statistical Picture." In *Fate and Effects of Petroleum Hydrocarbons in Marine Ecosystems and Organisms*, edited by Douglas A. Wolfe, pp. 1–7. Oxford and New York: Pergamon Press.

Heyerdahl, Thor. 1971. *The Ra Expeditions*. Garden City, N.Y.: Doubleday.

Hirst, Nicholas. 1978a. "Assuring a Future for UK's Offshore Oil." *Times*, July 13, p. 23.

———. 1978b. "North Sea Oil: How Fast a Pace for Exploration?" *Times*, November 11, p. 27.

Holcomb, Robert W. 1969. "Oil in the Ecosystem." *Science*, vol. 166, no. 3902, pp. 204–206.

House, J. W. 1977. (ed.) *The U.K. Space: Resources, Environment and the Future*. Rev. ed. London: Weidenfeld and Nicholson.

House of Commons. 1973. *See* Committee of Public Accounts, 1973.

———. 1974a. *United Kingdom Offshore Oil and Gas Policy*. Cmnd. 5696. July 11. London: Her Majesty's Stationery Office.

———. 1974b. *Parliamentary Debates*, vol. 186, no. 54. London: Her Majesty's Stationery Office.

———. 1978. *Eleni V*. Cmnd. 684. Fourth Report from the Select Committee on Science and Technology, Session 1977–1978. London: Her Majesty's Stationery Office.

Hutcheson, A. M., and A. Hogg. 1975. (eds.) *Scotland and Oil*. Edinburgh: Oliver and Boyd.

International Management and Engineering Group of Britain Limited. 1972. *Study of Potential Benefits to British Industry from Offshore Oil and Gas Developments*. London: Her Majesty's Stationery Office.

International Petroleum Encyclopedia. 1979. Tulsa: Petroleum Publishing Co.

Johnston, R. 1976a. (ed.) *Marine Pollution*. New York: Academic Press.

———. 1976b. "Mechanisms and Problems of Marine Pollution in Relation to Commercial Fisheries." In *Marine Pollution*, edited by R. Johnston, pp. 3–158. New York: Academic Press.

Jones, Mervyn, and Fay Godwin. 1976. *The Oil Rush*. London: Quartet Books.

Kash, D. E., et al. 1973. *Energy under the Oceans: A Technology Assessment of Outer Continental Shelf Oil and Gas Operations*. Norman: University of Oklahoma Press.

Kasperson, Roger E. 1977. "Societal Management of Technological Hazards." In *Managing Technological Hazards: Research Needs and Opportunities*, edited by Robert W. Kates, pp. 49–80. Boulder: Institute of Behavioral Sciences, University of Colorado.

Kates, Robert W. 1972. Review of *Perspectives on Resource Management*, *Annals of the Association of American Geographers*, vol. 62, pp. 519–520.

———. 1977. (ed.) *Managing Technological Hazards: Research Needs and Opportunities*. Boulder: Institute of Behavioral Sciences, University of Colorado.

Kent, P. E. 1967. "North Sea Exploration: A Case History." *Geographical Journal*, vol. 133, no. 3, pp. 289–301.

Kerr, Richard A. 1977. "Oil in the Ocean: Circumstances Control Its Impact." *Science*, vol. 198, no. 4322, pp. 1134–1136.

Keto, David B. 1978. *Law and Offshore Oil Development: The North Sea Experience*. New York: Praeger Publishers.

Klemme, H. D. 1977. "One-Fifth of Reserves Lie Offshore." *Oil and Gas Journal*, vol. 75, no. 35, pp. 108–128.

Lenihan, John, and William W. Fletcher. 1977. (eds.) *The Marine Environment*. New York and San Francisco: Academic Press.

Lewis, T. M., and I. H. McNicoll. 1978. *North Sea Oil and Scotland's Economic Prospects*. London: Croom Helm.

Lindblom, Gordon P. 1978. "Oil Spill Control Chemicals—A Current View." In *Chemical Dispersants for the Control of Oil Spills, ASTM STP 659*, edited by L. T. McCarthy, Jr., G. P. Lindblom, and H. F. Walter, pp. 127–140. Philadelphia: American Society for Testing and Materials.

Livesey and Henderson. 1973. "Sullom Voe and Swarbacks Minn Area: Master Development Plan and Report." Report prepared for the Zetland County Council.

McAuliffe, Clayton D. 1977a. "Dispersal and Alteration of Oil Discharged on a Water Surface." In *Fate and Effects of Petroleum Hydrocarbons in Marine Ecosystems and Organisms*, edited by Douglas A. Wolfe, pp. 19–35. Oxford and New York: Pergamon Press.

———. 1977b. "Evaporation and Solution of C_2 to C_{10} Hydrocarbons from Crude Oils on the Sea Surface." In *Fate and Effects of Petroleum Hydrocarbons in Marine Ecosystems and Organisms*, edited by Douglas A. Wolfe, pp. 363–372. Oxford and New York: Pergamon Press.

MacKay, D. I., and G. A. Mackay. 1975. *The Political Economy of North Sea Oil*. Boulder, Colo.: Westview Press.

McNicoll, Iain H. 1976. *The Shetland Economy*. Research Monograph No. 2. Glasgow: The Fraser of Allender Institute for Research on the Scottish Economy at the University of Strathclyde.

———. 1977a. "The Impact of Oil Supply Bases on the Economy of Shetland." *Maritime Policy Management*, vol. 4, pp. 215–226.

———. 1977b. "The Impact of Local Government Activity on a Small Rural Economy: An Input-Output Study of Shetland." *Urban Studies*, vol. 14, pp. 339–345.

———. 1977c. "The Impact of Outside Industry: A Case Study of Oil in Shetland." Mimeo.

Manners, Ian R. 1978. *Planning for North Sea Oil: The U.K. Experience*. Policy Study No. 6. Austin: Center for Energy Studies, University of Texas.

———. 1980a. *Offshore Oil: An Overview*. Policy Study No. 11. Austin: Center for Energy Studies, University of Texas.

———. 1980b. *The Coastal Energy Impact Program in Texas*. Austin: Bureau of Business Research, University of Texas.

Massachusetts Institute of Technology. 1970. *Man's Impact on the Global Environment: Assessment and Recommendations for Action*. Report of the Study of Critical Environmental Problems. Cambridge, Mass.: MIT Press.

Maxwell, Gaskin. 1977. *North Sea Oil and Scotland: The Changing Prospect.* Edinburgh: Royal Bank of Scotland.

Maycock, G. W. 1975. "Proposed Oil Refinery, Oil Storage Facilities and Marine Terminal Facility at Nigg Point, Easter Ross." Report of the public inquiry submitted to the Secretary of State for Scotland.

Meacher, Michael. 1979. "Why Are We So Secret about Our Secrets?" *Times*, November 30, p. 14.

Michael, A. D., 1977. "The Effects of Petroleum Hydrocarbons on Marine Populations and Communities." In *Fate and Effects of Petroleum Hydrocarbons in Marine Ecosystems and Organisms*, edited by Douglas A. Wolfe, pp. 129–137. Oxford and New York: Pergamon Press.

Milgram, Jerome. 1977. "The Cleanup of Oil Spills from Unprotected Waters." *Oceanus*, vol. 20, no. 4, pp. 86–94.

Milliman, John D. 1977. "Argo Merchant: A Scientific Community's Response." *Oceanus*, vol. 20, no. 4, pp. 40–45.

Ministry of Housing and Local Government. 1965. *The Future of Development Plans: A Report by the Planning Advisory Group.* London: Her Majesty's Stationery Office.

Mitchell, Bruce. 1980. "Models of Resource Management." *Progress in Human Geography*, vol. 4, no. 1, pp. 32–56.

Mitchell, James K. 1974. "Natural Hazards Research." In *Perspectives on Environment*, edited by Ian R. Manners and Marvin W. Mikesell, pp. 311–341. Washington, D.C.: Association of American Geographers.

———. 1976. "Onshore Impacts of Scottish Offshore Oil: Planning Implications for the Middle Atlantic States." *Journal of the American Institute of Planners*, vol. 42, no. 4, pp. 386–398.

Mix, Michael C., et al. 1977. "Chemical Carcinogens in the Marine Environment." In *Fate and Effects of Petroleum Hydrocarbons in Marine Ecosystems and Organisms*, edited by Douglas A. Wolfe, pp. 421–431. Oxford and New York: Pergamon Press.

Moody, J. D. 1975. "An Estimate of the World's Recoverable Crude Oil Resources." *Proceedings of the Ninth World Petroleum Congress, Tokyo*, vol. 3, pp. 11–20.

Moore, Gerald. 1976. "Legal Aspects of Marine Pollution Control." In *Marine Pollution*, edited by R. Johnston, pp. 589–698. New York: Academic Press.

Mutch, Dick. 1978a. "U.K. Setting Drilling Milestones." *Offshore*, vol. 38, no. 2, pp. 111–119.

———. 1978b. "BNOC Control Worries U.K. Operators." *Offshore*, vol. 38, no. 7, pp. 98–108.

National Academy of Sciences. 1975. *Petroleum in the Marine Environment.* Washington, D.C.

Nelson-Smith, Anthony. 1977. "Biological Consequences of Oil Spills." In *The Marine Environment*, edited by John Lenihan and William W. Fletcher, pp. 46–69. New York: Academic Press.

Norges Offentlige Utredninger (N.O.U.). 1977. *The Bravo Blow-out: The Action Command's Report*. Oslo: Universitetsforlaget.
North East Scotland Development Authority. 1975. *North East Scotland and the Offshore Oil Industry*. Aberdeen.
———. 1977. *Annual Report for 1976/77*. Aberdeen.
———. 1978. *Offshore Oil Directory*. Aberdeen.
Odell, Peter R. 1968. "The Significance of Oil." *Journal of Contemporary History*, vol. 3, no. 3, pp. 93–110.
Odell, P. R., and K. Rosing. 1974a. "Weighing Up the North Sea Wealth." *Geographical Magazine*, vol. 47, no. 3, pp. 150–155.
———. 1974b. "A Simulation Model of the Development of the North Sea Oil Province, 1969–2030." *Energy Policy*, vol. 2, no. 4, pp. 316–329.
Odum, Eugene P. 1971. *Fundamentals of Ecology*. 3d ed. Philadelphia: W. B. Saunders Co.
Organization for Economic Co-operation and Development (OECD). 1977a. *Environmental Impacts from Offshore Exploration and Production of Oil and Gas*. Paris: OECD Environment Directorate.
———. 1977b. *The Onshore Effects from Offshore Oil and Gas Development*. Paris: OECD Group on Energy and Environment.
O'Riordan, T. 1971. *Perspectives on Resource Management*. London: Pion.
———. 1976. *Environmentalism*. London: Pion.
Portmann, J. E. 1977. "International Marine Pollution Controls." *Marine Pollution Bulletin*, vol. 8, no. 6, pp. 126–132.
Pritchard, S. Z. 1978. "The North Sea as a Special Area." *Marine Policy*, vol. 2, no. 1, pp. 65–67.
Project Appraisal for Development Control. 1974. "A Critique of the Sphere Reports on Loch Carron and Loch Broom." Working Paper No. 20. Aberdeen: University of Aberdeen, Department of Geography.
Reekie, R. Fraser. 1975. *Background to Enviornmental Planning*. London: Edward Arnold.
Robinson, C. 1976. "The Economics of North Sea Oil and Gas." *Erdoel-Erdgas-Zeitschrift International Edition*, pp. 13–19.
Robinson, C., and J. R. Morgan. 1976. *Effect of North Sea Oil on the United Kingdom's Balance of Payments*. Guest Paper No. 5. London: Trade Policy Research Center.
———. 1977a. "The Economics of North Sea Oil Supplies." *Chemical Engineer*, vol. 75, pp. 388–391.
———. 1977b. "Will North Sea Oil Save the Economy?" *Petroleum Economist*, vol. 44, no. 1, pp. 7–9.
———. 1978. *North Sea Oil in the Future: Economic Analysis and Government Policy*. London: Macmillan for the Trade Policy Research Center.
Rosen, David H., and Deborah Voorhees-Rosen. 1978. "The Shetland Islands: The Effects of Social and Ecological Change on Mental Health." *Culture, Medicine and Psychiatry*, vol. 2, pp. 41–67.
Ross and Cromarty County Council. 1974. "Impact Study: Planning Application by Messrs. Fred Olsen Ltd. and Arnish Point, Stornoway."
Royal Commission on Local Government in England (Redcliffe-Maud Re-

port). 1969. *Report*. Cmnd. 4040. 3 vols. London: Her Majesty's Stationery Office.

Royal Commission on Local Government in Scotland (Wheatley Report). 1969. *Report*, Cmnd. 4150. 2 vols. Edinburgh: Her Majesty's Stationery Office.

Royal Norwegian Ministry of Finance. 1974. *Petroleum Industry in Norwegian Society*. Parliamentary Report No. 25 (1973–74).

———. 1975. *Natural Resources and Economic Development*. Parliamentary Report No. 50 (1974–75).

Royal Society for the Protection of Birds (RSPB), Conservation Planning Department. 1979. *Marine Oil Pollution and Birds*. Sandy, Bedfordshire.

Ryan, Paul R. 1977. "The Composition of Oil: A Guide for Readers." *Oceanus*, vol. 20, no. 4, p. 4.

Sanders, Howard L. 1977. "The West Falmouth Spill—*Florida*, 1969." *Oceanus*, vol. 20, no. 4, pp. 15–24.

Scottish Development Department. 1973a. *North Sea Oil Production Platform Towers: Construction Sites—A Discussion Paper*. Edinburgh: Her Majesty's Stationery Office.

———. 1973b. *North Sea Oil and Gas: Interim Coastal Planning Framework*. Edinburgh: Her Majesty's Stationery Office.

———. 1974a. *North Sea Oil and Gas: Coastal Planning Guidelines*. Edinburgh: Her Majesty's Stationery Office.

———. 1974b. *North Sea Oil and Gas Pipeline Landfalls: A Discussion Paper*. Edinburgh: Her Majesty's Stationery Office.

———. 1974c. *Examination of Sites for Gravity Platform Construction on the Clyde Estuary*. Edinburgh: Her Majesty's Stationery Office.

———. 1974d. *North Sea Oil: Oil Related Development Proposals*. Circular No. 23/1974. Edinburgh: Her Majesty's Stationery Office.

———. 1974e. *Coastal Planning Guidelines*. Circular No. 61/1974. Edinburgh: Her Majesty's Stationery Office.

———. 1975a. *North Sea Oil and Gas Developments in Scotland: A Physical Planning Resume*. Edinburgh: Her Majesty's Stationery Office.

———. 1975b. *North Sea Oil and Gas Developments in Scotland—Oil Terminals: Implications for Planning*. Edinburgh: Her Majesty's Stationery Office.

———. 1975c. *Memorandum of Guidance on the Procedure in Connection with Statutory Inquiries*. Circular No. 14/1975. Edinburgh: Her Majesty's Stationery Office.

———. 1976. *Environmental Impact Analysis: Scottish Experience, 1973–1975*. Edinburgh: Her Majesty's Stationery Office.

———. 1977a. *National Planning Guidelines for Petrochemical Developments*. Edinburgh: Her Majesty's Stationery Office.

———. 1977b. *Oil, Gas and Petrochemicals: Planning Information Notes*. Edinburgh: Her Majesty's Stationery Office.

———. 1977c. *National Planning Guidelines*. Circular No. 19/1977. Edinburgh: Her Majesty's Stationery Office.

Scottish Information Office. 1977. *Brief on North Sea Oil*. Edinburgh.

Scottish Office. 1977a. "The Impact of North Sea Oil-Related Activity on Employment in Scotland." *Scottish Economic Bulletin*, no. 11, pp. 8–14.

———. 1977b. "The Output of North Sea Oil-Related Industry." *Scottish Economic Bulletin*, no. 12, pp. 13–18.

———. 1979a. "Oil-Related Employment—Further Aspects." *Scottish Economic Bulletin*, no. 18, pp. 20–33.

———. 1979b. "Employment in Companies Wholly Related to the North Sea Oil Industry." *Scottish Economic Bulletin*, no. 19, pp. 26–28.

Sebek, Victor. 1978. "The Mixed Blessings of the Latest Marine Pollution Package." *Marine Pollution Bulletin*, vol. 9, no. 4, pp. 85–87.

Select Committee on Science and Technology, House of Commons. 1978. *Eleni V*. Report together with the Minutes of Evidence taken before the General Purposes Sub-Committee and Appendices, Session 1977–78. London: Her Majesty's Stationery Office.

Select Committee on Scottish Affairs. 1972. *Land Resource Use in Scotland*. Edinburgh: Scottish Office.

Shetland Islands Council. 1976. *Shetland Structure Plan*. 4 vols. Lerwick.

Sibthorp, M. M. 1975. (ed.) *The North Sea: Challenge and Opportunity*. London: Europa Publications.

Sills, David L. 1977. "Oil Spills: Risks and Relevance." *Science*, vol. 195, no. 4279, p. 636.

Singleton, David A. 1978. "BNOC's Oil Field Accolade." Letter to *Times*, July 5, p. 20.

Smith, Ben. 1976. "Eye on the Economy Prompts 'Pay As You Go' Policy for Britain." *Offshore*, vol. 36, no. 11, pp. 55–57.

———. 1977. "U.K. Looks to the North Sea Production to Clear Future." *Offshore*, vol. 37, no. 2, pp. 58–61.

Smith, Hance D. 1975. "Shetland." In *Scotland and Oil*, edited by A. MacGregor Hutcheson and Alexander Hogg, pp. 66–73. Edinburgh: Oliver and Boyd.

Smith, James Eric. 1968. *'Torrey Canyon' Pollution and Marine Life: A Report by the Marine Biological Association of the United Kingdom Laboratory, Plymouth*. Cambridge: Cambridge University Press.

Smith, Peter J. 1975. (ed.) *The Politics of Physical Resources*. Harmondsworth, Middlesex: Penguin Books in association with the Open University Press.

Sphere Environmental Consultants Ltd. 1973. "Impact Analysis: Oil Platform Construction at Loch Carron." Report prepared for the Scottish Development Department.

———. 1974. "Loch Carron Area: Comparative Analysis of Platform Construction Sites." Report prepared for the Scottish Development Department.

———. 1977. "Environmental Impact Analysis: Development of Beatrice Field Block 11/30." Report prepared for Mesa (UK) Limited.

Stegeman, John J. 1977. "Fate and Effects of Oil in Marine Animals." *Oceanus*, vol. 20, no. 4, pp. 59–66.

Steinhart, Carol, and John Steinhart. 1972. *Blowout*. Belmont, Calif.: Duxbury Press.

Steven, Robert R. 1979. "Heavy Oil Discoveries May Bolster North Sea." *Offshore*, vol. 39, no. 5, pp. 113–116.

———. 1981. "UK North Sea Tips 1.8 Million B/D." *Offshore*, vol. 41, no. 7, pp. 111–120.

Stewart, Robert J. 1977. "Tankers in U.S. Waters." *Oceanus*, vol. 20, no. 4, pp. 74–85.

Stewart, Robert J., and J. W. Devanney III. 1978. "Oil Spills and Offshore Drilling." *Science*, vol. 199, pp. 125–128.

Strathclyde Regional Council, Department of Physical Planning. 1977. "Environmental Impact Appraisal: Preliminary Assessment of Sites for a Petrochemical Complex."

Teal, John M. 1977. "Food Chain Transfer of Hydrocarbons." In *Fate and Effects of Petroleum Hydrocarbons in Marine Ecosystems and Organisms*, edited by Douglas A. Wolfe, pp. 71–77. Oxford and New York: Pergamon Press.

Texas House of Representatives. 1980. *Report on the Ixtoc 1 Oil Spill*. Austin: Committee on Environmental Affairs.

Travers, William B., and Percy R. Luney. 1976. "Drilling, Tankers, and Oil Spills on the Atlantic Outer Continental Shelf." *Science*, vol. 194, no. 4267, pp. 791–796.

United Nations. 1972. *Report of the U.N. Conference on the Human Environment*. New York.

U.S. Coast Guard. 1973. *Draft Environmental Impact Statement: International Convention for the Prevention of Pollution from Ships*. Washington, D.C.

U.S. Environmental Protection Agency. 1976. *Oil Spills and Spills of Hazardous Substances*. Washington, D.C.: Government Printing Office.

U.S. Fish and Wildlife Service. 1979. *Proceedings of the 1979 U.S. Fish and Wildlife Service Pollution Response Workshop, 8–10 May, St. Petersburg, Florida*. Washington, D.C.: U.S. Government Printing Office.

U.S. General Accounting Office. 1978. *Liquefied Energy Gases Safety*. Report to the Congress of the United States by the Comptroller General. Washington, D.C.

U.S. Senate, Committee on Commerce. 1974. *North Sea Oil and Gas: Impact of Development on the Coastal Zone*. Washington, D.C.: U.S. Government Printing Office.

Vandermuelen, John H. 1977. "The Chedabucto Bay Spill—Arrow, 1970." *Oceanus*, vol. 20, no. 4, pp. 31–39.

Vandermuelen, John H., et al. 1978. "Marine Environments: Recovery after Oil Spills." *Science*, vol. 202, no. 4363, p. 7.

Van Meurs, A. P. H. 1971. *Petroleum Economics and Offshore Mining Legislation*. New York: American Elsevier Publishing Co.

Vielvoye, Roger. 1978. "PRT Boost Threatens N. Sea Operators." *Oil and Gas Journal*, vol. 76, no. 31, p. 112.

———. 1979. "BNOC Aims for Bigger Share of North Sea Action." *Oil and Gas Journal*, vol. 77, no. 12, pp. 47–50.

Wassall, Harry. 1979. "Leasing Policies Need Revamping, Recharging." *Offshore*, vol. 39, no. 7, pp. 43–50.

White, Baroness. 1977. "International Action on Ship-Generated Oil Pollution." *Marine Pollution Bulletin*, vol. 8, no. 3, pp. 54–57.

White, Gilbert F. 1961. "The Choice of Use in Resource Management." *Natural Resources Journal*, vol. 1, pp. 23–40.

———. 1963. "Contributions of Geographic Analysis to River Basin Development." *Geographical Journal*, vol. 129, pp. 412–435.

———. 1969. *Strategies of American Water Management*. Ann Arbor: University of Michigan Press.

———. 1974. "Natural Hazards Research." In *Directions in Geography*, edited by Richard J. Chorley. London: Methuen.

White, Irvin L., Don E. Kash, Michael A. Chartock, Michael D. Devine, and R. Leon Leonard. 1973. *North Sea Oil and Gas: Implications for Future United States Development*. Norman: University of Oklahoma Press.

Williams, Louis G. 1977. "Oil Spills: Risks and Relevance." *Science*, vol. 195, no. 4297, p. 636.

Wolfe, Douglas A. 1977. (ed.) *Fate and Effects of Petroleum Hydrocarbons in Marine Ecosystems and Organisms*. Oxford and New York: Pergamon Press.

World Bank Group. 1979. *A Program to Accelerate Petroleum Production in Developing Countries*. Washington, D.C.: World Bank.

Wyman, T. S. 1978. "Tanker Safety." *Science*, vol. 200, no. 4347, p. 1218.

Zetland County Council. 1974. *Sullom Voe District Plan*. Lerwick: County Planning Office.

Index